Fundamentals of
HYDRAULIC ENGINEERING

Fundamentals of HYDRAULIC ENGINEERING

Alan L. Prasuhn

South Dakota State University
Brookings, South Dakota

HOLT, RINEHART AND WINSTON

New York   Chicago   San Francisco   Philadelphia
Montreal   Toronto   London   Sydney
Tokyo   Mexico City   Rio de Janeiro   Madrid

Publisher: Ted Buchholz
Acquisitions Editor: John Beck
Senior Project Manager: Suzanne Magida
Production Manager: Paul Nardi
Design Supervisior: Robert Kopelman
Illustrations: J and R Services

Prasuhn, Alan L., 1938–
    Fundamentals of hydraulic engineering.

    Includes bibliographies and index.
    1. Hydraulic engineering.   I. Title.
TC145.P695   1987   627   86-9975

ISBN 0-03-003948-7

Holt, Rinehart and Winston
The Dryden Press
Saunders College Publishing

Contents

Preface

Fundamentals of Hydraulic Engineering is intended first and foremost as a textbook for the senior-level introductory course in hydraulic engineering required in most civil engineering curricula. The length and depth of coverage is satisfactory for a two-semester sequence; however, in the more usual one-semester course, the instructor may freely select the portions that they wish to cover. While intended primarily for civil engineering students, the text is appropriate for agricultural engineering students and others interested in the subject of hydraulics. Because of the depth of coverage in many of the chapters, the book should be useful in senior elective courses or an introductory course at the masters level and provide a practical reference for the practicing engineer as well.

One of my goals in writing this text was to clearly present hydraulic engineering fundamentals based on the students' previous engineering and mathematics preparation, so that they could directly apply sound hydraulic principles in their professional practice. Many instructors will find this approach beneficial to their students because a number of current, similar textbooks tend toward theoretical concepts at the expense of realistic engineering applications.

A broad coverage of much of the field of hydraulic engineering is included in the book, with as much depth as possible within reasonable length constraints. Briefly, *Fundamentals of Hydraulic Engineering* provides a review of fluid mechanics principles, coverage of hydrologic principles and their application, a thorough development of the fundamentals of pipe and open channel flow, followed by coverage of sedimentation mechanics, modeling, hydraulic machinery, and drainage hydraulics. Extensive attention is not given to more advanced and specialized topics such as river engineering or dam design; however, many aspects of these topics are treated in the applications of hydrologic and hydraulic principles.

My goal is to provide all civil engineering students with some exposure to hydraulic engineering, an understanding of the basic principles, and *an ability to apply those principles*. It is also my hope that the book will spark an interest in some students to undertake further studies and plan for a career in what the author has found to be the fascinating and rewarding field of hydraulic engineering.

Although some of the latter chapters build on earlier chapters, in many cases chapters, sections, or subsections can be skipped without detriment to a logical development of the subject. The prerequisite for the course for which this book is intended is a one-semester, usually junior-level course in fluid mechanics. Although a review of fluid mechanics is included, it is assumed that the reader has a working understanding of the basic fluid flow principles. Some civil engineering curricula offer separate courses in hydrology and hydraulic engineering. For those wishing to use a different textbook for hydrology, chapters 3, 4, and 5 can be skipped with more than enough material to fill a one-semester course in hydraulics. Those institutions that offer only a single course, such as water resources engineering, will find that chapters 3, 4, and 5 provide a concise introduction to the practical aspects of hydrology.

Chapters 6 and 7 on pipeline and open channel hydraulics, respectively, provide a strong foundation in hydraulics. The remaining chapters on sediment transport, modeling, hydraulic machinery, and drainage can be chosen in any combination desired by the instructor. In addition, various sections or subsections can be deleted to fit the time available for the course.

The author has attempted to provide a balanced mix between hydraulic theory on the one hand and practical applications on the other. Considerable emphasis is placed the development of the principles of hydraulic engineering. As stated, the only starting assumptions are the basic principles of fluid mechanics. The theoretical development is followed by ample applications of the theory to engineering problems.

Clarity has been treated as the first priority. When felt necessary, a lucid explanation has been included, even at the expense of some conciseness. A greater than usual number of illustrative *engineering examples* are included. It has been the author's observation that students rarely feel that a textbook has enough worked-out examples.

The knowledgeable reader should find that the book is up to date. While the classic and fundamental procedures are stressed, the author includes selected recent innovations. Discussions of comprehensive computer models are included. Each of the governmental agencies and some of the larger consulting firms have their own standard procedures for computer analysis and design. It is neither practical nor appropriate to include exhaustive descriptions of these procedures. However, to place a greater emphasis on the practicalities of hydraulic engineering, a limited number of these procedures will be included, at least in a general way.

Six hydraulic engineering programs in BASIC are included in Appendix E. These programs are useful in their own right, or as a helpful programming guide for students. In addition, some of these programs may be useful as subroutines in more comprehensive programs.

The book also provides frequent opportunities for individual programming by the students. A number of chapter-ending problems refer to computer programs. Some of these computer programs can be assigned as homework for additional challenge in lieu of the more conventional problems. An instructor may also decide to use these computer problems for in-class discussion.

An extensive set of problems is provided at the end of each chapter. An ample number of problems is provided so that an instructor can freely choose between those in English and SI units and also go for three or more semesters without excessive repeating. The text makes almost equal use of both U.S. customary and SI units. I strongly feel that the engineering students of today should be equally competent in both sets of units.

I wish to express my thanks and gratitude to my editors, John J. Beck, for his skillful guidance as we prepared the manuscript, and Suzanne Magida, who turned my typed offering into a beautiful book. I would like to thank my reviewers for their thoughtful reviews and constructive comments made during the preparation of this manuscript. I appreciate the helpful critique prepared by Dr. Richard A. Denton of Berkeley, California, who reviewed an earlier version of this manuscript. I am also grateful to an anonymous reviewer from the University of Delaware who provided excellent suggestions at various stages of writing, and for the critique of the final manuscript. I also wish to specifically express my gratitude to Professor Randall J. Charbeneau, University of Texas, Austin, and Richard J. Heggen, University of New Mexico, for their constructive critiques and commentaries.

In addition, I would be especially amiss if I did not particularly acknowledge the fine civil engineering students at South Dakota State University. They bore with me through successive manuscript drafts used in the classroom. They discovered enroute many errors and omissions that required corrections and additions. Finally, and most importantly, I would like to express my sincere gratitude and thanks to my family for their patience, help, and understanding beyond the call of duty.

A. L. P.
Brookings, South Dakota

Fundamentals of
HYDRAULIC ENGINEERING

chapter
1

Introduction

Figure 1–0 To avoid a repeat of the catastropic 1953 flood tide that claimed 1800 lives in the Netherlands, the Dutch engineers have constructed the Eastern Scheldt Storm Surge Barrier. Clockwise from the top left: Construction of the piers (maximum height and weight are 40 m and 18,500 tons, respectively); a portion of the 66 piers awaiting flooding of the island and transport to the construction site; pier placement in depths up to 30 m; placement of road beam over the 40-m clear opening with only the positioning of the gates remaining. (All photos courtesy of the Rijkswaterstaat Deltadienst.)

1

1–1 OVERVIEW OF HYDRAULIC ENGINEERING

Hydraulic engineering encompasses a broad portion of the field of civil engineering. Not only is hydraulic engineering a professional discipline in its own right, but it is also a necessary tool for other civil engineers whose projects interact with water in one of the many ways described below. Thus, it forms an integral part of the civil engineering curriculum. As in almost any type of engineering, there is a significant overlap between hydraulic engineering and other fields such as agricultural engineering and mechanical engineering. Engineers in other fields must be aware when water problems may influence their work. Likewise, the hydraulic engineer must recognize when expertise of a different sort is required.

We define hydraulic engineering as the application of fluid mechanics principles to problems dealing with the collection, storage, control, transport, regulation, measurement, and use of water. Large projects such as a major dam providing for flood control, power generation, navigation, irrigation, and recreation involve many aspects of hydraulic engineering. Smaller projects, such as an individual bridge across a river, the measurement of the flow rate in a small irrigation ditch, the protection of a stream bank from erosion, or the sizing of a water supply pipe or storm drain, require different hydraulics principles. Although the above projects by no means cover the field of hydraulic engineering, taken together, they illustrate some of the range of engineering works that falls into the scope of the above definition.

Before any hydraulic engineering project can be constructed, a number of technical and semitechnical questions must be answered. For example how much water is involved? In the case of water from precipitation, the answer may involve its statistical distribution not only with time, but its distribution over the watershed area as well. This may require the estimation of the volume of water to be expected, the amount of water needed for a particular purpose, or both.

What is the quality of the water? The chemical and biological quality is more in the domain of the environmental engineer. However, the hydraulic engineer is concerned with the transport of sediment by the river, the interaction of the water with its alluvial boundary, and the occurrence of scour and disposition.

Although the emphasis in this book is placed on the relationship of hydraulic principles with the physical environment, the engineer must never forget that he or she is also working in a chemical and biological environment. For example, subsurface pipelines designed without regard to chemical or biological buildups may perform far differently and with a much lower efficiency than originally intended. The proper design of a grass-lined channel must consider the seasonal growth of vegetation if the required capacity is to be obtained.

The hydraulic design must also account for the unexpected. The design of a culvert to pass a specified discharge may be relatively straightforward. However, the culvert may be completely plugged with consequent upstream flooding if proper consideration is not given to the formation of debris jams at the entrance.

The hydraulic engineer must also collaborate with the structural engineer, since many hydraulic engineering projects involve a variety of different types of structures. Alternately, hydraulic impacts and constraints affect various kinds of structural projects. The typical questions that arise include: What kinds of hydraulic structures are required? What structural and foundation problems are associated with them? What hydraulic forces and considerations affect the structure?

There are usually a myriad of legal questions: Who owns the water? Who has the right to use it? Clearly, all of these questions are interrelated. For example, the quantitative and legal questions merge with the question of how much water may be legally used for a specified purpose or by a particular user.

Finally, what are the impacts of the project on the environment? Included in this question are fish and wildlife, scenic beauty, historical and archeological significance, and recreation. How will fish spawning be affected? Will there be an adequate supply of gravel for spawning? Is a fishway necessary? Will increased recreational use make up for the destruction of a wild river? Will historical or archeological sites be preserved or, at least adequately documented? All too often many of these factors have not received the attention they deserve. The hydraulic engineer must be prepared to answer the environmental questions to the satisfaction of the various regulatory agencies. The final design must be compatible with these constraints.

The answer to all of the above questions and considerations permits the engineer to proceed to the bottom line: Is the project technically feasible? Is the project economical? These are questions that should be answered before any project is started.

This book will approach the above problems and questions by emphasizing the fundamental principles of hydraulics and applying them to hydraulic engineering problems. Length constraints make it impossible to cover completely all topics of significance to the hydraulic engineer and still provide an adequate development of the basic principles. Consequently, a compromise has resulted, and the reader who desires more depth in a particular topic is directed to the references at the end of each chapter. After due consideration, many important topics had to be deleted. These include coastal and offshore problems, the design of intakes and outlets for thermal power plants, as well as other hydraulic structures, including dams.

Because of the introductory nature of the book, less emphasis will be placed on the planning and execution of the large project. Only after the principles are understood can they be synthesized into the larger view required for the major project.

Hydraulic engineers can expect to play an important role in the future just as they have in the past. While there may be fewer large dams built in the coming decades than in the last 50 years, the battle against flood waters is by no means over. More land will be placed under irrigation in the future. Better and more efficient uses of our limited water will be demanded. In addition, new technical areas will be opening up, such as the use of tidal power and the problem of what to do with the large dams when they become filled by sediment. The list is endless, and each item will require the skills of the hydraulic engineer.

Before proceeding through the book, the reader should become familiar with the information in the appendixes. Although most symbols are defined when first encountered, they are also included in Appendix A. Fluid properties will be found in Appendix B. In keeping with modern trends, U.S. customary and SI (Systemè International) units will be used throughout the book on an approximately 50/50 basis. Conversion factors between the two systems of units as well as additional conversion factors are found in Appendix C. In addition, geometric properties are located in Appendix D and selected computer programs are placed in Appendix E.

To provide continuity for students who have arrived at this level with different backgrounds in fluid mechanics, a brief review of fluid mechanics is included in the following chapter. No attempt is made to develop these concepts and the reader should refer to a textbook on fluid mechanics for a proper derivation and more detailed explanations.

1–2 HISTORY OF HYDRAULIC ENGINEERING

The first water-related projects involving the rudiments of hydraulic engineering date back to the first civilizations. The Sumerian civilization along the Euphrates and Tigris rivers created a network of irrigation canals dating from 4000 B.C. According to Durant, this system "was one of the greatest achievements of Sumerian civilization, and certainly its foundation."[1] Although the first canals were surely used for irrigation, their obvious use for navigation was apparent as well. Thus, canal as well as river navigation was important to the early Middle-Eastern civilizations. The degree of this importance may be gauged by the frequent association, and even identification, of the canals and rivers with their gods.[2]

Although much of the water diversion into the canals depended on manual labor, it was only natural that dams or other low control structures would also be used for this purpose, with the added advantage that some storage could be gained thereby. The first dam is certainly lost in antiquity. The oldest known dam, and the oldest such structure remaining in existence, is the Sadd el-Kafara or Dam of the Pagans built between 2950 and 2750 B.C.[3] The dam was constructed across a valley some 20 miles south of Cairo. The crest length was 348 ft and it had a maximum height of 37 ft. The lack of sediment accumulation behind the ruins of the stone-faced structure indicates that it probably was overtopped and washed out within a few years of its closure.

The legend that King Menes dammed the Nile at an even earlier date is unsubstantiated and, considering the magnitude of such an undertaking, most unlikely.[4] What was more likely accomplished, itself no mean feat, was the protection of the city of Memphis from the waters of the Nile by a levee along the river.

[1]W. Durant, *Our Oriental Heritage, The Story of Civilization: Part 1,* New York: Simon and Schuster, 1954, p. 124.
[2]R. Payne, *The Canal Builders,* New York: Macmillan, 1959, pp. 9–34.
[3]N. Smith, *A History of Dams,* London: Peter Davies, 1971, p. 1.
[4]Ibid., p. 5.

Civilizations whose very existence depended upon a single water supply, like the Nile, learned to both respect and study the river behavior. Nilometers to measure the stage of water level have been in use from as early as 3000 to 3500 B.C.[5] Not only did these gauges provide a means of recording the annual rise and fall of the Nile, but they were also used for flood control. As the flood season arrived, rowers moving in the direction of the current and more rapidly than the flood wave could bring river forecasts to the downriver regions. At the Roda gauge just north of Cairo, water level records have been kept, with some gaps, since A.D. 620.[6]

The use of groundwater from springs or dug wells dates back to antiquity, with recurring references in the Old Testament. The most notable groundwater collection system was the qanats[7] of Persia, which date back 3000 years. The qanats were a system of nearly horizontal tunnels that served the combined functions of infiltration galleries and water transmission tunnels. They provided an ideal solution in arid regions in that they collected the water over large areas and in addition kept the water cool thereby reducing evaporation. The construction procedure began when an initial well was dug in a water-bearing region, followed by a second well some distance away. The two would be connected by a tunnel and the process would continue from vertical shaft to vertical shaft. Qanats extended to lengths of 25 to 28 miles, with typical slopes of 1 to 3 ft per 100 ft, and depths as great as 400 ft.[8]

The Greek civilization brought innovations in many areas that have come down to use mainly through the writings of Archimedes and Hero of Alexandria.[9] The principles of hydrostatics and flotation postulated by Archimedes are taught in introductory fluid mechanics and Archimedean screw pumps are still used today. Hero's surviving manuscripts express an understanding of flow measurement and the role of velocity, a concept that perhaps the Romans never understood. In *Pneumatica,* Hero describes the construction of pumps for fire fighting and the use of hydrostatic and hydraulic forces to open temple doors automatically and operate other devices.

If the Greeks were the innovators and thinkers, the Romans were the builders and engineers. This was partly due to Greek abhorrence of any association with manual labor, which left all aspects of engineering entirely to slaves. The Romans, on the other hand, were organizers and developers. Engineering became a highly respected profession, including in its ranks the Emperor Hadrian.[10] Among Roman works on hydraulics are the writings of Sextus Julius Frontinus (c. A.D. 40–103) who, according to Finch[11] is "history's most famous engineer-administrator." In *De Aguis Urbis Romae,* Frontinus describes the construction of city water supplies, aq-

[5]A. K. Biswas, *History of Hydrology,* Amsterdam: North-Holland, 1970, p. 14.
[6]H. E. Hurst, R. P. Black, and Y. M. Simaika, *Long Term Storage,* London: Constable, 1964, p. 14.
[7]A variety of different spellings are found in the literature with qanat and kanat being the most common. A graduate student from Iran insists that the most phonetic spelling is ghanat.
[8]A. K. Biswas, *History of Hydrology,* Amsterdam: North-Holland, 1970, p. 29.
[9]Translations are available in English, but see first N. FitzSimons (ed.), *Engineering Classics of James Kip Finch,* Kensington, MD: Cedar Press, 1978, pp. 3–6. Also see H. Rouse and S. Ince, *History of Hydraulics,* New York: Dover, 1963, pp. 15–22.
[10]N. Upton, *An Illustrated History of Civil Engineering,* New York: Crane Russak, 1976, p. 28.
[11]N. FitzSimons (ed.), *Engineering Classics of James Kip Finch,* Kensington, MD: Cedar Press, 1978, p. 9.

ueducts, and sewers, but displays a somewhat limited understanding of hydraulics itself.

All told, 11 aqueducts served the city of Rome, discharging an estimated 84 million gallons per day,[12] or about 38 gallons per day per capita. Evidence of over 200 Roman aqueducts serving more than 40 cities has been found.[13] The outstanding example is probably the famous Pont du Gard, which supplied water to Nimes in what is now southwest France. The sewer constructed to drain the Roman Forum, the cloaca maxima, is still in service.[14]

Although great strides were not made during the Dark and Middle Ages, nevertheless, progress related to hydraulic engineering continued. More bridges were built, and wind and water power came into ever-increased usage. The Industrial Revolution in the eighteenth century and the industrialization that followed greatly increased the need for, and the scope of, engineering. Hydraulic engineering as we know it today advanced rapidly during this period. When the principles of fluid mechanics were utilized in the twentieth century, progress became even more accelerated. Our understanding of hydraulic concepts is far greater than that of our professional ancestors, which only serves to increase our admiration of their accomplishments.

The emphasis has been placed on early hydraulic engineering. Far more additional information on this fascinating subject can be obtained from the references found in the footnotes and at the end of the chapter. For a single comprehensive source, the reader is particularly directed to *History of Hydraulics* by Rouse and Ince.

References

Binnie, G. M., *Early Victorian Water Engineers,* London: Thomas Telford Ltd., 1981.

Biswas, A. K., *History of Hydrology,* Amsterdam: North-Holland, 1970.

FitzSimons, N., *Engineering Classics of James Kip Finch,* Kensington, MD: Cedar Press, 1978.

"History of Hydraulics," *Proceedings,* Seventeenth Congress International Association for Hydraulic Research, Baden-Baden: 1977, pp. 735–819.

Payne, R., *The Canal Builders,* New York: Macmillan, 1959.

Reynolds, J., *Windmills and Waterwheels,* New York: Praeger, 1970.

Robbins, F. W., *The Story of Water Supply,* London: Oxford University, 1946.

Rouse, H., *Hydraulics in the United States, 1776–1976,* Iowa City, IA: Iowa Institute of Hydraulic Research, 1976.

Rouse, H. and S. Ince, *History of Hydraulics.* New York: Dover, 1963.

Smith, N., *A History of Dams,* London: Peter Davies, 1971.

Upton, N., *An Illustrated History of Civil Engineering,* New York: Crane Russak, 1975.

Van Deman, E. B., *The Building of Roman Aqueducts,* Washington D.C.: Carnegie Institute of Washington, 1934.

[12]C. Hershel, *Frontinus and the Water Supply of the City of Rome,* New York: Longmans, Green, 1913.
[13]N. Upton, *An Illustrated History of Civil Engineering,* New York: Crane Russak, 1976, p. 34.
[14]G. M. Fair and J. C. Geyer, *Water Supply and Waste-Water Disposal,* New York: Wiley, 1954, p. 5.

chapter
2

Review of Fluid Mechanics

Figure 2–0 Thames Barrier—a flood defense for the city of London (courtesy of the Greater London Council).

2

2–1 INTRODUCTION

The purpose of this chapter is to review the more important and relevant principles and concepts covered in an introductory course in fluid mechanics. An individual who is well grounded in fluid mechanics may readily skip all or part of this material: however, at least a causal examination is recommended because much of the subsequent notation is introduced herein and other necessary information such as the fluid properties are discussed and located in the appendixes. Topics to be covered include fluid properties, hydrostatics, the equations of continuity, energy, momentum, and dimensional analysis. This is intended as a very concise review of introductory fluid mechanics, and while it should cover most of what we will need, additional fluid mechanics concepts will be required from time to time. The principles will be presented without derivation, since further details, as well as a complete derivation, can be found in any book on fluid mechanics.

A few definitions from fluid mechanics are also necessary. The most important follow: a *streamline* is a line everywhere tangent to the velocity vectors at any instant. Given that a streamline exists in a flow, the velocity at all points along the line are therefore tangent to the line. A *streamtube* is simply a bundle of streamlines and composes what we will refer to as a *control volume*. A control volume will be identified for purposes of analysis as a fixed region in space through which the fluid or a portion of the fluid passes. A flow is *steady* or *unsteady,* depending on whether the velocity at points within the flow is varying with time. If the velocity does not change with time at any point in the flow, the flow is steady. If the velocity changes with time at one or more points, the flow is unsteady. Finally, a flow with straight parallel streamlines is a *uniform* flow, and conversely, if the streamlines are curved or nonparallel, the flow is *nonuniform.*

2–2 FLUID PROPERTIES

The fluid properties are those characteristics of a specific fluid that depend on such a limited number of parameters that they are made readily accessible by graphing or tabulation. In the case of water, which is of paramount concern in hydraulic engineering, the properties will be either approximately constant or more generally a function of only the water temperature. From time to time certain properties of air will be required, so they also will be introduced in this section, but only to the extent needed. The fluid properties to be covered are density, specific weight, dynamic viscosity, kinematic viscosity, compressibility, and surface tension.

The *density* of a fluid (indicated by the Greek lower case letter rho, ρ) is defined as its mass per unit volume. The density of water varies slightly with temperature[1] as given in both U.S. customary and SI units in Table B–1, Appendix B. For most hydraulic engineering problems it is adequate to use 1.94 slugs/ft^3 or 1000 kg/m^3 as the density of water in U.S. customary or SI units, respectively. The density of air depends on both temperature and pressure according to the thermodynamic relationship known as the perfect gas law. For a standard atmospheric pressure of 14.7 psia (or 101.3 kN/m^2 abs), the density of air is tabulated as a function of temperature in Table B–2, Appendix B.

It should be noted in the foregoing, and subsequently hereafter, that absolute pressure will be denoted by either a or abs following the appropriate units. Gauge pressure (the pressure relative to the atmospheric pressure) will be indicated without appendage. Except where noted, gauge pressures will be used exclusively throughout the book.

The *specific weight* of a fluid is the weight of the fluid per unit volume. This property will be designated by the Greek lowercase letter gamma, γ. Values for water and air (at atmospheric pressure) are located in Tables B–1 and B–2, respectively. However, for water, reasonable accuracy is usually achieved by using 62.4 lb/ft^3 or 9810 N/m^3. Since the specific weight is defined on the basis of fluid weight, it is a fluid property only because the acceleration of gravity is usually assumed constant. In fact, the specific weight is related to the density by

$$\gamma = \rho g \tag{2-1}$$

where g is the gravitational acceleration generally taken as 32.2 ft/s^2 or 9.81 m/s^2.

The *viscosity* of a fluid is a measure of its resistance to deformation or flow under an applied shear stress. We will need to deal with two types of viscosity. The first, called the *absolute viscosity* or sometimes the *dynamic viscosity* will be represented by the Greek lower case letter mu, μ. The magnitude of μ is the constant of proportionality that exists between an applied shear stress and the resulting rate of deformation. This relationship may be expressed as follows:

$$\mu = \frac{\tau}{du/dy} \tag{2-2}$$

The shear stress, a shear force per unit area indicated by the Greek lowercase letter tau, τ, is shown in Fig. 2–1. The resulting rate of deformation, du/dy, is also indicated in Fig. 2–1 by the slope of the velocity distribution. A fluid with a greater absolute viscosity would require a greater applied shear stress in order to achieve an identical rate of deformation.

---

[1]Although the variation of density with temperature is very slight and can usually be ignored in engineering calculations, its significance can not be understated. Water has its greatest density at approximately 4°C. Consequently, at the freezing point its density is slightly less, a fortunate circumstance since ice therefore floats rather than accumulating on the bed of oceans and lakes. The density variation is also responsible for currents in the bodies of water.

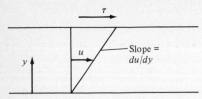

Figure 2–1 Definitional sketch for viscosity.

The absolute viscosity of a fluid is significantly influenced by its temperature. Refer to Tables B–1 and B–2 in Appendix B to obtain an accurate value. The units for the absolute viscosity are lb-s/ft^2 or N-s/m^2 as can be verified from Eq. 2–2.

The ratio of absolute viscosity to density occurs so frequently in fluid mechanics and hydraulics equations that it is convenient to define a second viscosity based on this ratio. Known as the *kinematic viscosity,* and expressed by a Greek lowercase letter nu, ν, it is simply

$$\nu = \frac{\mu}{\rho} \qquad (2\text{–}3)$$

The units are ft^2/s or m^2/s, and values for water and air (at atmospheric pressure only) are found in Tables B–1 and B–2, respectively. Computer program 1 in Appendix E provides a reasonable estimate of the kinematic viscosity of water.

Although usually treated as incompressible, there are occasions (e.g., water-hammer problems) when the compressibility of water must be taken into account. The *modulus of compressibility, K,* is then required. This measure of fluid compressibility is defined as the ratio of an incremental compressive stress, *dp,* to the relative increase in density that results,

$$K = \frac{dp}{d\rho/\rho} \qquad (2\text{–}4)$$

Since the relative change in density is dimensionless, the modulus of compressibility must have the same dimensions as the pressure term. Table B–1 in Appendix B contains values of K for water.

The final fluid property, the *surface tension* (represented by the Greek lowercase letter sigma, σ) will have even less application throughout this book, although tabulated values for water will be found in Appendix B, Table B–1. Surface tension pertains to the different attraction that exists between water molecules themselves and the molecules of other substances. The effects of surface tension are responsible for the spherical shape of water drops and bubbles, and also phenomena such as capillary action.

2–3 HYDROSTATICS

The basic principles of hydrostatics state that in a liquid at rest the pressure in an horizontal plane is everywhere constant, while in the vertical direction, the pressure decreases with increasing elevation at a rate equal to $\gamma\Delta y$, where Δy is the elevation change. Thus, in a tank of water with a free surface (i.e., an air-water interface), the pressure at a depth y_0 is given by

$$p = \gamma y_0 \qquad (2\text{–}5)$$

As discussed previously, the gauge pressure is assumed in this equation. Consequently, the pressure p is actually the pressure difference between the point in question and the overlying atmospheric pressure, now taken as zero.

The hydrostatic pressure distribution applies to actual flow conditions in certain cases. The most important example, as far as we are concerned, is in the plane perpendicular to the flow direction in uniform flow. The flow may be confined or have a free surface, but if the streamlines are straight and parallel (i.e., uniform flow), then the hydrostatic pressure distribution applies throughout this cross section.

This pressure distribution can be used to determine the force on a surface. Theoretically, the force can always be evaluated by integrating the pressure distribution over the area in question. However, in introductory fluid mechanics this theoretical approach leads to some general, readily applicable equations that alleviate the need to go through the integration process repeatedly. First, consider a plane surface such as that shown in Fig. 2–2a, which is inclined at an angle θ, and has its true projection also shown. The magnitude of the force on this surface is

$$F = \gamma \bar{h} A \qquad (2\text{–}6)$$

where $\bar{h}$ is the vertical distance from the free surface to the centroid of the area A. The line of action of this force is located a distance y_p from the surface as measured along the incline. This distance is

$$y_p = \bar{y} + \frac{\bar{I}}{\bar{y}A} \qquad (2\text{–}7)$$

where $\bar{I}$ is the centroidal moment of inertia of the surface and $\bar{y}$ like $\bar{h}$ is the distance to the centroid, but this time measured along the incline similar to y_p. These calculations will be illustrated in Example 2–1. Geometric properties such as centroids

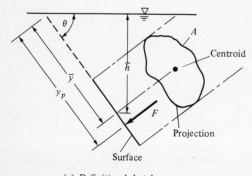

(a) Definitional sketch

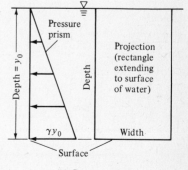

(b) Pressure prism

Figure 2–2 Hydrostatic force.

and moments of inertia are included in Appendix D for various two-dimensional and three-dimensional shapes.

This procedure can be further simplified in the situation where the plane surface is vertical and extends to the free surface itself. Then, the pressure distribution that acts on the surface (and called the pressure prism as in Fig. 2–2b) can be used directly to find the magnitude and direction of the resultant force. This also will be demonstrated in Example 2–1.

If the surface in question is curved, then the force is resolved into horizontal and vertical components. The horizontal force on the curved surface is equivalent in both magnitude and point of application to the horizontal force that would act on the

EXAMPLE 2–1

Assume that the vertical surface of Fig. 2–2b has a width of 2 m and extends 3 m below the water surface. Determine the magnitude and location of the force of the water on this surface using Eqs. 2–6 and 2–7, and then repeat the calculations using the concept of the pressure prism.

Solution

Starting with Eq. 2–6, the force is

$$F = (9810)(1.5)(3)(2) = 88,290 \text{ N}$$

Note that when the surface is vertical, dimensions h and y are interchangeable, thus Eq. 2–7 locates the force relative to the free surface, as follows:

$$y_p = 1.5 + \frac{(2)(3)^3/12}{(1.5)(3)(2)} = 2 \text{ m}$$

Observe that the triangular pressure distribution of Fig. 2–2b acts on a rectangular surface. The magnitude of the force may be obtained by integration of the pressure over the area, which in this case is simply the volume of the triangular pressure prism. Using this approach the force is

$$F = \frac{(3)(2)(3)(9810)}{2} = 88,290 \text{ N}$$

In this calculation the pressure at the bottom of the surface, 3γ, forms one of the legs of the pressure prism. Since the force is given by the volume of the prism, its location will be through its centroid, which is $\frac{2}{3}$ of the distance from the surface or 2 m. It is apparent that both procedures yield identical answers.

vertical projection of the curved surface. Thus this force can be evaluated using Eqs. 2–6 and 2–7 (or the pressure prism). The vertical component, which we will have little need to determine, is obtained on the basis of the displaced volume of fluid.

2–4 FLOW EQUATIONS

This section will cover the continuity, energy, and momentum equations as applied to the incompressible flow of water. The example that follows will illustrate the application of these principles in a single problem. The *continuity equation* is a statement of the conservation of mass. Referring to the control volume (or streamtube) of Fig. 2–3, the conservation of mass requires that the discharge into the control volume equals the discharge out. Recognizing that the discharge past a section is equal to the product of the average velocity and the cross-section area we may write

$$Q_1 = Q_2 \tag{2–8a}$$

or

$$V_1 A_1 = V_2 A_2 \tag{2–8b}$$

These equations can be generalized for control volumes in which the flow enters or leaves through more than one cross section.

The *energy equation* or *Bernoulli equation* may assume a number of different forms depending on circumstances. At this point we will only consider steady flow that is confined (as in conduit) or contains a free surface. In the former case we may write

$$\frac{p_1}{\gamma} + y_1 + \frac{V_1^2}{2g} = \frac{p_2}{\gamma} + y_2 + \frac{V_2^2}{2g} + H_L \tag{2–9}$$

between any two sections 1 and 2. As formulated, each term has dimensions of length and is called a "head." The first three terms on each side of the equation are the pressure, elevation, and velocity heads, respectively. The *pressure head* provides

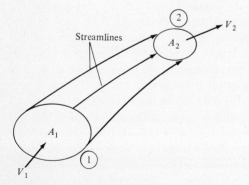

Figure 2–3 Control volume.

a means for doing work (assuming a pressure difference exists), and the *elevation head* and *velocity head* are in turn the potential energy and kinetic energy on a per-unit-weight basis (ft-lb/lb or m-N/N). The velocity is the average velocity at the cross section whereas the pressure and elevation refer to the center line. The sum of each set of three terms is the *total head H*, which is a constant in Eq. 2–9 if there are no energy losses. Thus,

$$H = \frac{p}{\gamma} + y + \frac{V^2}{2g} \tag{2–10}$$

and

$$H_1 = H_2 \tag{2–11}$$

All losses are incorporated in the *head loss* term H_L in Eq. 2–9. These losses include the effect of conduit friction as well as other losses such as at elbows and valves. The sum of the pressure and elevation heads is defined as the *piezometric head* and denoted by *h*. Thus,

$$h = \frac{p}{\gamma} + y \tag{2–12}$$

If a pump or turbine were present in the system, then the total head would be increased or decreased thereby. This change in total head would be added to Eq. 2–9 by the inclusion of a $+H_T$ or $-H_T$ to the left-hand side. The equation is applicable between two sections in steady, incompressible flow. However, the two sections must be regions of uniform (or nearly uniform flow). Further, correction factors α_1 and α_2, defined by

$$\alpha = \frac{1}{AV^3} \int_A v^3 dA$$

where *A* and *V* are as previously defined and *v* is the variable velocity across the section may be used as multipliers for the velocity head terms to take into account the effect of the actual velocity distribution. This is discussed more fully in any fluid mechanics text. Here, they will generally be assumed equal to unity and ignored.

The applicability and conditions relating to Eqs. 2-9 through 2-12 apply to free surface flows as well. For purposes of application, Eq. 2-9 is written in terms of the piezometric heads. Since sections 1 and 2 must lie in regions of uniform flow the piezometric head in each case will equal the elevation of the water surface above the datum. This will be illustrated in the example that follows.

The *momentum equation* as developed in fluid mechanics states that the net force in a given direction acting on the fluid within a control volume is equal to the net rate of flow of momentum out of the control volume in that direction plus the time rate of increase of momentum within the control volume. For steady flow, the internal momentum must remain constant and the momentum equation may be written as

$$\Sigma F_x = (\rho Q V_x)_{\text{out}} - (\rho Q V_x)_{\text{in}} \tag{2–13}$$

The x direction has been chosen arbitrarily, and the equation can be applied in any direction. The left-hand side is the algebraic sum of the x direction forces that act on the fluid. Frequently, friction and gravity (or weight component) terms may be ignored. The terms on the right-hand side represents the rate of flow of momentum out of and into the control volume in the x direction. The direction is determined in each case by the component of the average velocity.

The preceding equations and the example that follows all refer to one-dimensional flow analysis. The emphasis is on changes in the direction of the flow, and variations in the other directions are replaced by average quantities, such as the average velocity, or through the assumption of a constant piezometric head. While this is adequate to provide a satisfactory result in many cases, there are also many situations where more general equations are required. Many of these are beyond the scope of this book.

EXAMPLE 2–2

The radial gate shown in Fig. 2–4a seats on a 0.4-m-high sill in a 5-m-wide rectangular channel. The gate is positioned so that the upstream and downstream depths are 2 m and 0.4 m, respectively. Determine the horizontal force that acts on the control structure (i.e., on the gate plus the sill).

Solution

The control volume is shown in Fig. 2–4b. The forces consist of the two hydrostatic forces on either end and the total force the combined structure exerts on the water. The momentum equation alone is inadequate because of the number of unknowns. However, we may write the energy equation between sections 1 and 2. Choosing a datum at the lower channel level and ignoring losses gives

$$2 + \frac{V_1^2}{(2)(9.81)} = 0.4 + 0.4 + \frac{V_2^2}{(2)(9.81)}$$

From continuity

$$V_1 y_1 (5) = V_2 y_2 (5)$$

Combining the two equations we get

$$V_1 = 0.99 \text{ m/s}, \qquad V_2 = 4.95 \text{ m/s}$$

and

$$Q = (4.95)(0.4)(5) = 9.90 \text{ m}^3/\text{s}$$

Now writing the momentum equation

$$F_1 - F_2 - F = \rho Q(V_2 - V_1)$$

and substituting

$$\frac{(9810)(2)^2(5)}{2} - \frac{(9810)(0.4)^2(5)}{2} - F = (1000)(9.90)(4.95 - 0.99)$$

Solving, $F = 54{,}900$ N.

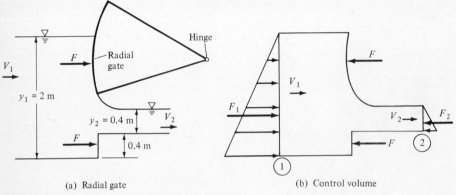

(a) Radial gate   (b) Control volume

Figure 2–4 Force on a flow control structure.

2–5 DIMENSIONAL ANALYSIS

The procedure known as *dimensional analysis* will be introduced using the Buckingham pi method that is presented in most fluid mechanics textbooks. It is included here because certain relationships (both in equation and in graphic form) that we will use are at least partially formulated using dimensional analysis. In addition both the so-called dimensionless numbers and the chapter on modeling will be introduced through the process.

The advantage of dimensional analysis is that it helps organize a problem and reduces the total number of variables that have to be considered. The Buckingham pi theorem states that if a given dimensional dependent variable depends on $n - 1$ dimensional independent variables, and if a total of m independent dimensions are required to describe the n variables, then the n dimensional variables will reduce to $n - m$ dimensionless variables. The dimensional variables may include pressure, velocity, density, etc. The independent dimensions refer to force, length, and time or mass, length, and time. The dimensionless variables are the so-called pi terms. A total of m repeating variables are chosen from the list of independent variables[2]. The

---

[2]Although it is best to exclude the dependent variable from the repeating variables, there are occasions when it is necessary to violate this rule.

repeating variables should be the more significant terms. They should be chosen so that all m dimensions are represented: however, no two can have identical dimensions. These m repeating variables are then combined with each of the $n - m$ non-repeating variables to form the dimensionless terms. The procedure will be illustrated in Example 2–3.

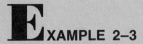

EXAMPLE 2–3

Develop an equation for the force F_p on a circular pier of diameter D placed in a wide river that has an average velocity V and depth y. Assume that the fluid properties of density, specific weight, and dynamic viscosity are also important.

Solution

Based on the foregoing statement, the force may be expressed as a function of the several independent variables

$$F_p = f(D, y, V, \rho, \gamma, \mu)$$

Upon examination, $n = 7$ and $m = 3$ (the force dimension is present in the force term, and length and time are contained in the velocity term). Thus $n - m = 4$ and we can anticipate four dimensionless pi terms. Three repeating variables are now required, and D, V, and ρ are chosen. The diameter D is selected rather than y as the more significant geometric term, V is clearly important, and ρ is the most important fluid property as explained later in the example.

The pi terms may now be formed by the following procedure:

$$\pi_1 = F_p D^a V^b \rho^c$$

Some combination of the four variables must be dimensionless, although the magnitudes of the exponents $a, b,$ and c are as yet unknown. Once a dimensionless combination is found, however, the term may be raised to any power and still remain dimensionless. Hence, there is a degree of freedom that has permitted expressing F_p to the first power. Substituting the fundamental dimensions into the equation for π_1 yields

$$F^0 L^0 T^0 = (F)^1 (L)^a \left(\frac{L}{T}\right)^b \left(\frac{FT^2}{L^4}\right)^c$$

The three fundamental dimensions are independent and if the equation is to be dimensionless as assumed on the left-hand side, then each of the three dimensions must drop out independently. Thus,

$$F^0 = (F)^1(F)^c = F^{1+c}$$

Therefore,

$$0 = 1 + c \quad \text{or} \quad c = -1$$

Similarly, the exponents on time must satisfy

$$0 = -b + 2c \quad \text{and} \quad b = 2c = -2$$

and finally, from the length dimension

$$0 = a + b - 4c \quad \text{and} \quad a = 4c - b = -2$$

Thus,

$$\pi_1 = \frac{F_p}{\rho D^2 V^2}$$

In the same way

$$\pi_2 = \frac{y}{D}, \quad \pi_3 = \frac{\gamma D}{\rho V^2} \quad \text{and} \quad \pi_4 = \frac{\mu}{VD\rho}$$

At this point, a dimensionless equation for F_p may be written

$$\frac{F_p}{\rho D^2 V^2} = \phi\left(\frac{y}{D}, \frac{\gamma D}{\rho V^2}, \frac{\mu}{VD\rho}\right)$$

which may also be solved for the actual force F_p,

$$F_p = \rho D^2 V^2 \phi\left(\frac{y}{D}, \frac{\gamma D}{\rho V^2}, \frac{\mu}{VD\rho}\right)$$

This is a satisfactory solution to the problem statement. However, to improve its usefulness some additional operations may be performed. Any combination of dimensionless terms may be used to replace an existing term provided that the total number of terms is not reduced. For example,

$$\left(\frac{F_p}{\rho D^2 V^2}\right)\left(\frac{D}{y}\right) = \frac{F_p}{\rho D y V^2}$$

can be used to replace the original dependent variable. Next consider

$$\left(\frac{\gamma D}{\rho V^2}\right)\left(\frac{y}{D}\right) = \frac{\gamma y}{\rho V^2}$$

If this is now raised to the negative one-half power, we get the more conventional term defined as the Froude number, Fr:

$$\left(\frac{\gamma y}{\rho V^2}\right)^{-1/2} = \frac{V}{\sqrt{(\gamma/\rho)y}} = \frac{V}{\sqrt{gy}} = \text{Fr}$$

Likewise, the reciprocal of the fourth term is the Reynolds number, Re:

$$\left(\frac{\mu}{VD\rho}\right)^{-1} = \frac{VD\rho}{\mu} = \frac{VD}{\nu} = \text{Re}$$

The new equation becomes

$$F_p = yD\rho V^2 \phi\left(\frac{y}{D}, \text{Fr, Re}\right)$$

Compare this equation with the drag force equation from fluid mechanics,

$$F_p = \frac{C_D A \rho V^2}{2}$$

where A is the projected area (approximately yD). The 2 may be added arbitrarily to our equation since it is only in functional form. By comparison, the drag coefficient C_D must be a function of geometry (i.e., y/D), Fr, and Re.

Experience indicates that the Froude number influences the results only to the extent that gravity affects the flow. Thus this term is insignificant unless waves are present. Further, the Reynolds number reflecting viscous effects is important at low values, but as Re increases in magnitude it ceases to be significant (this is why ρ was favored over the other fluid properties as a repeating variable). Dimensional analysis can never provide numerical values in the resulting equation. Experiments or known drag coefficients would be required to determine the actual numerical relationship. However, dimensional analysis has reduced the problem to a manageable one.

PROBLEMS

Section 2–2

2–1. Determine the density of a liquid that has a specific weight of 65.5 lb/ft^3.

2–2. What is the specific weight of a liquid that has a density of 2.0 slug/ft^3?

2–3. What is the density of a liquid that has a specific weight of 9900 N/m^3?

2–4. Determine the specific weight of a liquid that has a density of 850 kg/m^3.

2–5. If a liquid has a specific gravity of 0.88 and a dynamic viscosity of 3×10^{-4} lb-s/ft^2, what is its kinematic viscosity?

2–6. What is the kinematic viscosity of a liquid that has a dynamic viscosity of 3.1×10^{-3} N-s/m^2 and a density of 870 kg/m^3?

2–7. What is the change in density of water at 60°F that is caused by a pressure increase of 1000 psi?

2–8. What is the change in density of water at 20°C that is caused by a pressure increase of 7 MN/m^2?

Section 2–3

2–9. Determine the force on a circular disk that has a diameter of 10 ft if it is submerged with its center 20 ft below the water surface and (1) it is horizontal; (2) it is vertical; and (3) it is inclined at 45 deg.

2–10. Determine the magnitude of the force and its location on the circular gate shown. If it is hinged through its center, what moment must be applied at its axis in order to keep the gate closed?

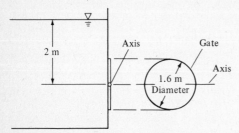

Figure P2–10

2–11. Determine the net force and the net moment on the rectangular gate shown below. The hinge is at the top and the salt water has a density of 1.99 slug/ft^3.

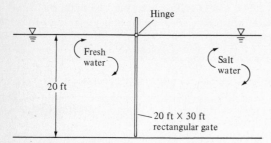

Figure P2–11

2–12. The depth of water behind a rectangular sea wall on a fresh-water lake is 4 m. Determine the force on the wall if it has a length of 10 km.

Section 2–4

2–13. The open channel shown is 5 ft wide, and the upstream depth is 6 ft. The outflow structure at the end of the channel consists of a 3-ft-diameter pipe that discharges into the atmosphere. Determine the horizontal force on the end plate of the channel. Ignore friction, but do not ignore the upstream approach velocity.

2–14. With reference to Prob. 2–13, determine the total and piezometric heads at section 1 and section 2.

2–15. Repeat Prob. 2–13 if the 3-ft-diameter outlet pipe is followed by a contraction section so that the outlet into the atmosphere is through a 2-ft-diameter pipe with centerline elevation 1.5 ft above the channel bottom. Find only the force on the end plate.

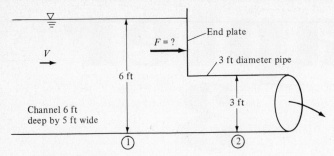

Figure P2–13

2–16. The flow of water is from left to right in the 1.5-m-by-1.5-m square duct shown. Determine the force on the circular strut that extends across the duct. The upstream and downstream pressures are 400 kN/m² and 300 kN/m², respectively, and the average velocity at the upstream section is 4 m/s. There is no flow in the side pipe (as per Prob. 2–17). Further, duct friction may be ignored.

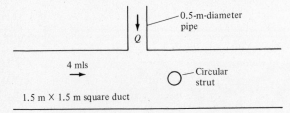

Figure P2–16

2–17. Repeat Prob. 2–16 if flow enters from the 0.5-m-diameter side pipe at the rate of 1 m³/s.

Section 2–5

2–18. Use dimensional analysis to derive an equation for the discharge per unit width q over a rectangular weir. Assume that the discharge is a function of the weir height P, the head above the crest H, the acceleration of gravity g, and the kinematic viscosity ν.

2–19. Repeat Prob. 2–18 assuming that the viscous effect is unimportant.

2–20. Use dimensional analysis to derive an equation for the power requirements associated with the towing of a barge. Assume that the power P is a function of the width w, the draft d, the length L, and the velocity V of the barge, and the density and specific weight of the water.

2–21. Repeat Prob. 2–21 if the kinematic viscosity of the water is important.

References

Prasuhn, A. L., *Fundamentals of Fluid Mechanics*, Englewood Cliffs, NJ: Prentice-Hall, 1980.

Rouse, H. (ed.), *Engineering Hydraulics*, New York: Wiley, 1950.

chapter
3

Hydrology

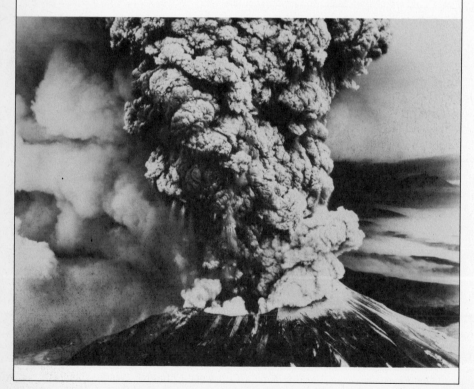

Figure 3–0 Mt. St. Helens eruption (U.S. Geological Survey).

3

3–1 INTRODUCTION AND THE HYDROLOGIC CYCLE

Hydrology is the study of the various steps through which water passes from the time it evaporates from the oceans and inland bodies of water until it returns. The various steps form the *hydrologic cycle* to be considered shortly. Depending on the approach used, hydrology may be divided into physical hydrology or statistical hydrology. Both of these topics are often graduate or undergraduate civil engineering courses in their own right. In this chapter we will concentrate on physical hydrology and attempt to describe the various processes by physical laws, or at least by physical observation. The emphasis will be placed on those portions of the hydrologic cycle that are of greatest concern to the hydraulic engineer.

Many of the processes, although dependent on physical laws, are also probabilistic in nature. Both the precipitation and the stream flow vary from day to day. While the rainfall on a given day may only be weakly dependent on the preceding or antecedent rainfall, the discharge of a river may depend quite strongly on the previous day's discharge. Consequently, both physical and statistical hydrology are necessary tools for a working understanding of the subject.

Physical hydrology has been further divided into surface hydrology, the subject of this chapter, and subsurface hydrology. Because groundwater hydraulics is essentially a subject in itself, it has been reserved for the following chapter. An introduction to statistical hydrology will follow in a subsequent chapter.

A schematic diagram illustrating the major components of the hydrologic cycle is given in Fig. 3–1. Starting with the oceans, and other bodies of water, evaporation releases water vapor that is carried over land masses by the wind. The air and water vapor may rise because of mountain ranges (orographic lifting), frontal activity (frontal lifting), or uneven heating of the air (the convective heating that leads to summer thunderstorms). As the water vapor rises, it cools, condenses, and falls back to earth as some type of precipitation, such as rain or snow. A portion of the precipitation evaporates as it falls through the air. An additional amount is intercepted by vegetation, usually to be reevaporated and returned directly to the atmosphere. This normally minor process is called *interception*. The initial portion of the water that reaches the ground forms pools in surface depressions called *depression storage*. Some of this water may evaporate while some will infiltrate into the soil, a process known as *infiltration*. If the precipitation continues until the depressions are filled, some of the water will start to flow overland as *surface runoff*. Water that infiltrates may remain near the surface as soil moisture. This water may evaporate over time or be taken into plants where it ultimately will be returned to the atmosphere by plant *transpiration*. Other subsurface water will percolate to the groundwater table where it will flow slowly until it intersects a river, lake, or ocean.

This overland flow is also subject to evaporation and infiltration, but otherwise reaches a recognizable channel within a short distance. Usually from this point the

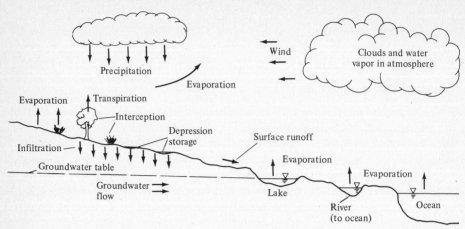

Figure 3–1 The hydrologic cycle.

water will flow in ever larger channels, from rills to ditches to streams to rivers and finally to lakes and oceans, completing the cycle. The subsequent sections in this chapter, and in fact many of the remaining chapters, will deal with various portions of the hydrologic cycle, particularly as they apply to hydraulic engineering.

3–2 PRECIPITATION

Precipitation refers to any water or ice with sufficient mass that it falls to earth. Rain and snow are of greatest importance to the engineer. The desirability of accurate forecasting of precipitation is important at all levels, from flood forecasting and agricultural planning down to the Sunday picnic. A knowledge of the amount and distribution of precipitation is of paramount importance to the hydraulic engineer if flooding is to be minimized and water efficiently utilized.

Rain is liquid precipitation that can be broken down roughly as follows:

Type of Rain	Intensity
Drizzle	< 0.04 in./hr
Light rain	0.04–0.10 in./hr
Moderate rain	0.10–0.30 in./hr
Heavy rain	> 0.30 in./hr

Snow differs significantly from rainfall because of the possible long lag that can occur between the time that the snowfall occurs and when it melts and either infiltrates or runs off. From the engineering point of view, this is frequently an advantage because of the greater opportunity it affords to forecast accurately the melt and subsequent runoff quantities.

Other forms of precipitation that are significant to such diverse groups as meteorologists, the transportation industry, and the agricultural industry include hail, sleet, mist, and fog. Hail consists of concentric layers of ice that build up to large diameters due to the buffeting and continued suspension by turbulence in a moisture-laden and freezing atmosphere. Sleet is formed by raindrops falling through freezing air. Mist and fog are airborne droplets that remain suspended in the air because they are so small.

Of the total precipitation that falls on the land surfaces of the world, approximately 35 percent comes from land area evaporation, while the remaining 65 percent comes from ocean evaporation. Roughly 25 percent of this water returns to the oceans as runoff while 75 percent returns directly to the atmosphere by evaporation and transpiration. This last 75 percent from evaporation and transpiration is divided, with 40 percent returning to the oceans to combine with the 25 percent runoff to form the original 65 percent and 35 percent remaining over the land masses as originally stated.

Measurement of Precipitation

Precipitation is measured using both nonrecording and recording precipitation gauges. The most common types are designed primarily for rainfall, but the recording type may be used, at least to some extent, for the measurement of snowfall as well. The major measurement network in the United States is supported by the U.S. National Weather Service. Simply because of cost, most of the more than 14,000 gauges in the United States are nonrecording, and most of the gauges are read by volunteer observers.

The standard nonrecording gauge is shown in Fig. 3–2. It consists of an 8-in. diameter funnel that directs the rainfall into a smaller-diameter tube that is designed so that when the calibrated measuring stick is inserted into the tube, the observer can easily read the depth of water collected by the gauge to the nearest 0.01 in. Rainfall accumulation in excess of 2 in. overflows the small-diameter tube, but is collected in the larger cylinder that holds the funnel. Thus, the observer can empty and refill the measuring tube as often as necessary to obtain the total accumulation since the last reading.

By its very nature, a nonrecording gauge gives only the total rainfall between readings, which are usually once a day. Only a rough estimate of the time, duration, or intensity of the rainfall is usually available. For those uses that require only the total rainfall and the annual distribution of rainfall, this information may be satisfactory. If more information is necessary, a recording gauge is required.

The most commonly used recording gauge, a weighing rain gauge, is shown in Fig. 3–3. This gauge consists of a galvanized bucket that sits on a scale. Precipitation is collected, again with an 8-in. diameter funnel, and directed into the bucket. As precipitation occurs it increases the weight in the bucket, and a mechanical linkage drives an ink-pen pointer across a chart. The chart, which is usually changed weekly, is wrapped around a cylindrical drum driven by a clock mechanism. The result is an ink trace of the accumulated weight of water calibrated in inches of water versus

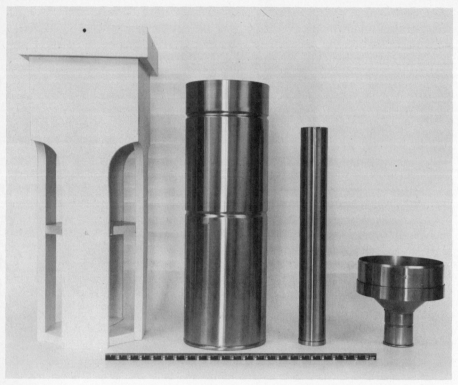

Figure 3–2 Exploded view of a nonrecording precipitation gauge showing the support, overflow can, measuring tube, and funnel with the measuring stick at the bottom (courtesy of U.S. Weather Service, NOAA).

time. The chart directly gives both the amount of rainfall during any time interval and the duration of the precipitation. In addition, the slope of the ink trace is the intensity of the rainfall.

An oil film is sometimes added to the bucket to reduce evaporation and antifreeze is used in winter to prevent freezing. Accurate rainfall measurement is not easy and some care must be taken, even in locating the gauge. It should be placed on the ground some distance from trees and buildings. The consistency of a given rain gauge is sometimes destroyed when a new building is placed too near it or as a nearby tree grows. A windbreak is desirable, however, since the catch of a gauge may drop to as low as 50 percent of the actual rainfall in a high wind. Unfortunately, most of the possible errors are accumulative.

The weighing gauge may also be used to measure the water accumulation from snowfall by using the 8-in.-diameter collecting ring with the funnel removed. This works to a degree in regions of moderate snowfall, but the many factors that contribute to the errors in rainfall measurement are generally magnified under snow conditions.

In regions of heavy snowfall, the determination of accumulated moisture content

Figure 3–3 Weighing-type recording rain gauge (automatic gauges are also in use) (courtesy of U.S. Weather Service, NOAA).

is a two-step process. Depths of snow are obtained from *snow stakes*. They are stakes that are graduated so that the depth of the snow pack is easily read, either by a ground observer or from an overhead plane. A series of these stakes are arranged into a snow course that is intended to give a broad enough coverage so that an average snow depth can be obtained. A ground observer must also collect representative samples of the snow along the snow course. By comparing the weight of the sample with its volume, the moisture content of the sample is readily calculated. The moisture content can then be applied to the snow course itself to estimate the depth of water in the snow pack.

Other devices are also available, such as the *snow pillow*. This is a liquid-filled pillow that is placed on the ground before the snowfall. As snow accumulates on it, the internal pressure increases. This is sensed by a pressure transducer that transmits the information to a receiving unit. Although it gives the moisture content directly (rather than the depth of snow), it is subject to a variety of problems such as a bridging action in the snow as the depth increases.

Average Precipitation over a Region

The 8-in.-diameter collector of a rain gauge has an area of only 0.349 ft^2. Compare that with the average density of the rain gauges, which is something like 1 in 250 mi^2. This yields an area ratio of 1 to 2 $\times$ 10^{10}. Consequently, the record of a precipitation gauge is a point sample that is, at best, expected to represent an extremely large area. We can therefore expect a relatively low accuracy when dealing with gauge data. If more than one gauge is in or near an area of interest, the amount of precipitation can be estimated more accurately. However, we need to make the best possible estimate based on the several gauges.

Because of the nonuniform distribution of the precipitation gauges, a straight arithmetic average of the gauge results is usually inadequate. A better method, known as the *Thiessen polygon method,* utilizes a weighted average based on the assumption that the precipitation in an area nearest a particular gauge is best represented by the gauge. The procedure, which is illustrated in Example 3–1, requires that the various gauging stations first be connected by straight lines. The respective areas, which are to be used as weighting factors, are then defined by constructing perpendicural bisectors of the lines. The polygonal areas that result contain the area closest to each gauge. In this method more weight is given to the more predominate gauge, while the procedure determines and excludes those gauges that have no influence on the region in question. Gauges that are outside of the region may also be included if appropriate.

The preceding procedure calculates the average precipitation over an area by a more or less purely mechanical process. If a mountain range or other physical geographic feature affects the precipitation distribution in a known fashion, a second averaging method, known as the *isohyetal method,* may be superior. In this instance, lines of constant rainfall, or isohyets, are drawn and the average precipitation is based on the relative size of the areas between each isohyet. The isohyets are constructed in the same way as elevation contour lines on a contour map. The process may again be mechanical, as in Example 3–2, or the contours can be located on the basis of topographic features or past precipitation patterns.

The preceding discussion deals with measured precipitation data. It has been treated first as a point measurement and subsequently as a spatial average. The variation of precipitation with time will be discussed further in conjunction with the prediction of runoff and the unit hydrograph method.

EXAMPLE 3–1

The annual precipitation at seven rain gauge stations in or near the drainage basin of Fig. 3–4 is as follows:

Station	Precipitation (in.)
A	33.27
B	37.21
C	35.40

(continued)

Station	Precipitation (in.)
D	39.73
E	37.80
F	37.31
G	34.70

Use the Thiessen polygon method to determine the average precipitation over the basin for the year.

Solution

With reference to Fig. 3–4, the various gauging stations are connected by straight dashed lines. The perpendicular bisectors have been constructed and shown as the heavier, nondashed straight lines. This process indicates that no portion of the basin is closer to station G than to one of the other stations, and therefore, station G precipitation will not be included. The area of the basin closest to each gauge has been measured by a planimeter[1] and tabulated (in arbritrary units) in the following table. Each of the respective values of precipitation is then multiplied by the weighting factor for that area. The sum is finally divided by the total area to get the average precipitation.

Region	Area A (units)	Precipitation P (in.)	$A \times P$
A	1.80	33.27	59.89
B	2.02	37.21	75.16
C	3.72	35.40	131.69
D	4.68	39.73	185.94
E	3.15	37.80	119.07
F	2.86	37.31	106.71
Sum	18.23		678.45

$$P = \frac{\Sigma(A \times P)}{\Sigma A} = \frac{678.45}{18.23} = 37.22 \text{ in.}$$

[1]Other techniques such as counting squares on superimposed graph paper will also suffice.

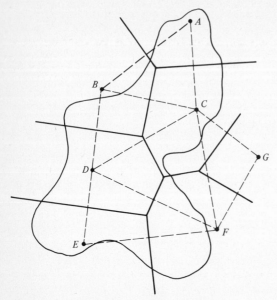

Figure 3–4 Thiessen polygon method.

EXAMPLE 3–2

Repeat Example 3–1 using the isohyetal method.

Solution

Since no additional information is available concerning the distribution of the precipitation, the lines of constant precipitation will be based on linear interpolation between the various gauging stations. The stations have been connected by straight, light-weight lines in Fig. 3–5 and labeled with the interpolated integer precipitation values. Using these values, and assuming that the isohyets will form contours around the peak, the isohyets shown in Fig. 3–5 have been completed. Each isohyet has been identified by a large Arabic numeral indicating the respective annual precipitation. Note that gauge *G* now has some influence in the estimation of the precipitation distribution, whereas it was excluded in the previous example using the Thiessen polygon method.

The regions established between the isohyets are identified by Roman numerals. For those regions that are approximately rectangular, the average precipitation is the average value of the bounding contours. The average precipitation in the other regions is simply the best estimate based on the distribution pattern. A planimeter was used to determine the area of each region. These areas are used as weighting factors for the precipitation values, and the remaining calculations are similar to those for the Thiessen polygon method.

Region	Area A (units)	Precipitation P (in.)	$A \times P$
I	1.71	39.4	67.37
II	5.76	38.5	221.76
III	5.08	37.6	191.01
IV	1.61	36.6	58.93
V	1.56	35.5	55.38
VI	1.55	34.5	53.48
VII	0.81	33.6	27.22
VIII	0.12	32.8	3.94
Sum	18.20		679.08

$$P = \frac{\Sigma(A \times P)}{\Sigma A} = \frac{679.08}{18.20} = 37.31 \text{ in.}$$

Since the isohyets are based solely on linear interpolation, the two methods give very similar results.

3–3 EVAPORATION, TRANSPIRATION, AND SNOWMELT

Precipitation that reaches the surface of the earth is subject to evaporation at almost every step of the hydrologic cycle. Water that is temporarily in depression storage, water that infiltrates but remains near the surface, and water in any stream, river, or lake is subject to evaporation. In addition, plant life returns water to the atmosphere through the process of *transpiration*. Information on this process is essential in agricultural engineering and the science of irrigation. Because of the difficulty associated with the estimation of transpiration and the various types of evaporation, they are sometimes combined in to a single term called the *evapotranspiration*.

The emphasis in this introduction to the subject of hydrology will be limited to the measuring or calculation of evaporation from bodies of water. Although the measurement of evaporation will be considered separately from the evaporation calculations, the two cannot be completely independent. If for no other reason, the evaporation equations that follow require measured data for calibration and verification.

The calculation of evaporation is of primary concern to the hydraulic engineer because the evaporated water represents a loss of available water for other purposes. A reservoir with a large surface area in a dry region can lose a significant portion of

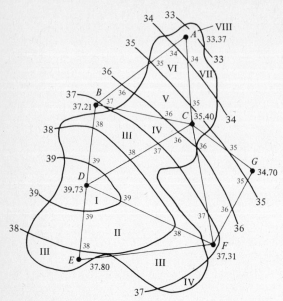

Figure 3–5 Isohyetal method.

the annual inflow to evaporation. The subsequent evaporation equations may readily be used to demonstrate this point.

Measurement of Evaporation

Evaporation from ground surfaces is usually based on the results of test plots or equations developed therefrom. *Evaporation pans* are generally used to obtain a direct measurement of evaporation from a body of water. After experimentation with buried pans and floating pans in the hope of closely duplicating natural conditions, the U.S. National Weather Service has settled on a standard, above-ground pan known as the Class A pan (see Fig. 3–6). This is a 4-ft-diameter by 10-in.-deep, unpainted, galvanized-iron pan placed just above the ground on a wood frame that allows free air circulation around and under it. Although the pan evaporation under these conditions is not identical to that of a body of water in the vicinity, it does give the most standardized results.

The elevation of the water surface is read regularly using a hook gauge and the pan is normally refilled to a depth of 8 in. whenever the depth drops below 7 in. The unit is usually installed in conjunction with a rain gauge, a wind anemometer, and a temperature station, so as to moniter the other relevant parameters. The rain gauge is necessary to correct the evaporation readings for the amount of precipitation that has been added to the pan.

At least partly because of the method of installation, the Class A pan has signif-

Figure 3–6 Evaporation pan with wind anemometer installed on wooden support (courtesy of U.S. Weather Service, NOAA).

icantly more evaporation than a body of water. A *pan coefficient* is introduced to express this difference. On an annual basis, the pan coefficient, defined as the ratio of lake evaporation to pan evaporation, has a value of approximately 0.70. The calculated annual evaporation from a lake will usually fall within about 12 percent of the actual evaporation when using this coefficient. For periods of a month or less, the use of the pan coefficient results in much larger errors. However, for planning a water project, knowledge of the annual evaporation is frequently adequate.

Calculation of Evaporation

Any textbook on hydrology will provide a wealth of material on evaporation and should be consulted if more information is desired. The process of evaporation depends on the air and water temperatures, vapor pressure, wind speed, and intensity of the turbulence in the air. Many equations, based on either the laws of physics or an empirical approach, have been formulated to take into account some or all of these parameters.

Dalton's law, first proposed in 1802, serves as a model for the following equations. In its simplest form it may be expressed as

$$E = C(e_w - e_a) \tag{3-1}$$

where E is the rate of evaporation, C is a mixing coefficent, e_w is the saturation vapor pressure at the temperature of the water surface, and e_a is the actual vapor pressure of the overlying air. The greatest difficulty, of course, is in the determination of the coefficient C, which is commonly, but probably inadequately, expressed as a function of the wind velocity. With the evaporation E in inches/day, and the vapor pressure in inches of mercury, the following are typical evaporation equations:

The Fitzgerald equation[2] (1886) is written

$$E = (0.40 + 0.199V)(e_w - e_a) \tag{3-2}$$

where V is the velocity in mph near the surface.

The Horton equation[3] (1917) is written

$$E = 0.4(2 - e^{-0.2V})(e_w - e_a) \tag{3-3}$$

where velocity is as defined for the Fitzgerald equation.

The Lake Hefner, Oklahoma equation[4] (1954) is written

$$E = 0.06V_{30}(e_w - e_a) \tag{3-4}$$

where V_{30} is the velocity in mph at 30 ft above the water surface.

In the preceding equations, the intensity of turbulent mixing is implicitly included in the numerical coefficient and wind velocity. Temperature is implicitly included in the vapor pressure terms. The Lake Mead equation[5] (1958) explicitly includes the temperature:

$$E = 0.072V_{30}(e_w - e_a) [1 - 0.03(T_a - T_w)] \tag{3-5}$$

where, V_{30} is as in the Lake Hefner equation and T_a and T_w are the average air and water surface temperatures, respectively, in °C.

Calculation of Snowmelt

The calculation of snowmelt is of major importance in areas subjected to large snow accumulations. In the western part of the United States, for example, snowmelt provides a major portion of the runoff for the region. Because of the lag between the time of the snowfall and the subsequent melt, the estimation of runoff from snow is

[2]D. Fitzgerald, "Evaporation," *Transactions,* American Society of Civil Engineers, Vol. 15, 1886.

[3]R. E. Horton, "A New Evaporation Formula Developed," *Engineering News Record,* Vol. 78, No. 4, April 26, 1917.

[4]G. E. Harbeck, and E. R. Anderson, "Water-loss Investigations: Vol. 1–Lake Hefner Studies Technical Report," U.S. Geological Survey Paper 269, Washington, D.C., 1954.

[5]M. A. Kohler, T. J. Nordenson, and W. E. Fox, "Evaporation from Pans and Lakes," U.S. Weather Bureau Research Paper 38, Washington, D.C., May, 1955.

a different problem from that of rain. The amount of melt can be calculated on the basis of available heat.

The sources of heat include direct solar radiation, warm air, rain, and condensation of moisture. The calculation is complicated by the number of additional factors involved. Initially, snow reflects most of the solar radiation, but reflection decreases as it ages and becomes dirty. Overlying still air causes very little melt because of the low heat conductivity of air. As the wind increases, so does turbulent mixing. The surface air is constantly replaced by warmer air and the melt increases.

Warm rain also provides a heat source. In U.S. customary units, 1 Btu is released when the temperature of 1 lb of water is lowered 1°F. It requires 144 Btu to melt 1 lb of water from ice (or snow) at 32°F (the heat of fusion). Hence, the melt M in inches due to rainfall is given by

$$M = \frac{P(T - 32)}{144} \tag{3–6}$$

where T is the temperature in °F and P is the precipitation in inches.

Condensation becomes important as a heat source since 1073 Btu are released when a pound of water is condensed at 32°F (the heat of condensation). By comparison with the heat of fusion, each unit of moisture available on the surface of the snow because of condensation results in the melting of about 7.5 units of water from the snowpack.

The actual computation of melt from a basin becomes very difficult and will not be pursued here. For more information the reader is directed to *Snow Hydrology* and *Handbook of Applied Hydrology*, both referenced at the end of the chapter.

3–4 INFILTRATION

The *infiltration rate f* is the actual rate at which water enters the soil at a given time. The *infiltration capacity* f_p is the maximum rate at which water can enter the soil under given soil conditions and at a given time. If the rainfall rate exceeds f_p, then $f = f_p$, whereas if the rainfall rate drops below f_p, then $f = $ the rate of rainfall. Assuming that the rate of rainfall is adequate to exceed the infiltration capacity, then the infiltration capacity can be expressed as a function of time according to

$$f_p = f_c + (f_0 - f_c)e^{-kt} \tag{3–7}$$

In this equation, f_p is the time-dependent infiltration capacity, f_c is the terminal value of f_p, f_0 is the value of f_p at the beginning of the rainfall, k is an empirical constant, and t is the elapsed time since the start of the rainfall. This relationship is sketched in Fig. 3–7.

Equation 3–7 provides a satisfactory basis for studying the behavior of the infiltration capacity during rainfall that is of sufficient magnitude that it exceeds the infiltration capacity. However, lack of adequate data precludes its use in many instances. Still assuming that the rainfall rate exceeds the infiltration capacity, the dif-

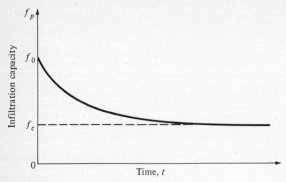

Figure 3–7 Infiltration capacity.

ference between the two rates is the rate of surface runoff. The surface runoff, which will be considered in detail later, is of primary interest to the hydraulic engineer.

The characteristics of each soil affect its infiltration capacity. In addition, external factors ranging from those that change seasonally to those that change during the actual course of the rainfall also affect the magnitude of the infiltration capacity. The soil texture reflects the difference between soil types. Vegetative cover, mechanical compaction, cultivation, boring animals, root decay, and the change of water viscosity all lead to seasonal, or at least relatively slow, changes in the infiltration capacity with time. During a given rain, the infiltration capacity is reduced by soil compaction due to raindrops and by the increasing saturation of the soil as the rain continues.

Carefully controlled test plots are frequently used to determine and study infiltration characteristics. A *ring infiltrameter* may be used to measure the infiltration capacity. This consists of a metal ring or tube with a diameter of 8 to 36 in. It is driven 18 to 24 in. into the ground as shown in the left-hand sketch in Fig. 3–8. Water is added to the ring and kept at a constant depth such as ¼ in. above the ground surface. The rate at which the water must be added to maintain the constant depth is the infiltration capacity. Because of the spreading of the streamlines as the water leaves the infiltrameter, a second concentric ring is added, as shown in the right-hand sketch. This greatly reduces the spread of the streamlines below the center ring. Water is kept at the same constant level in both rings but only the inner one is used to evaluate the infiltration capacity. Even at best, the results are only approximate because the insertion of the infiltrameter causes some disturbance of the soil. In addition, like the previously discussed rain gauge, the ring infiltrameter provides only a point measurement and the results must be treated as such.

An estimate of the average infiltration capacity over a large area is often required when relating the runoff to the precipitation. For this purpose, the Φ index is frequently used. It is defined as the rate of rainfall above which the volume of rainfall equals the volume of water that runs off during the time period under consideration. The Φ index is best obtained from rainfall-runoff records. To illustrate its determination, assume the hourly rainfall intensity distribution given in Fig. 3–9. Assume further that the total volume of runoff from the storm is equivalent to 1.20 in. over

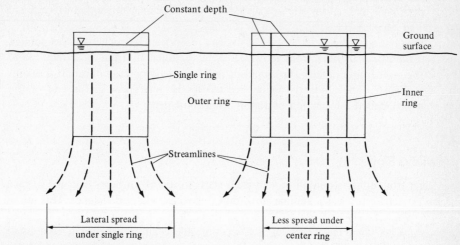

Figure 3–8 Ring infiltrameters.

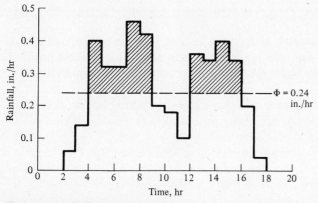

Figure 3–9 Definition of Φ index.

the basin. To calculate the Φ index, the horizontal dashed line may be continually lowered until the shaded area equals the volume of runoff. The position of the line at this point is the desired value. In this example, Φ = 0.24 in./hr since the shaded area above the line equals the 1.20 in. of runoff. A little care in the calculation of the shaded area will verify this result.

3–5 SURFACE RUNOFF

Estimating surface runoff depends first upon whether the runoff is due to precipitation as rain or snowmelt from earlier snowfalls. In the former case, we have seen that precipitation that reaches the ground as rainfall initially fills the low spots on the uneven ground surface as depression storage. The water then attempts to infiltrate into the ground, and only when the depressions are filled and the rainfall exceeds the infiltration capacity does water begin to run off.

Snowmelt was considered briefly at the end of Section 3–3, where it was pointed out that runoff calculations from snowmelt differ significantly from those required to estimate runoff from rainfall. The emphasis will be strictly on rainfall runoff throughout the chapter. The reader is again directed to the references at the end of the chapter for detailed information on runoff from snowmelt.

Overland Flow

As water from either a rainfall or snowmelt starts to run off, it flows over the surface of the ground, usually as a laminar flow until it reaches a defined channel. This initial runoff, which rarely exceeds a distance of more than a few hundred feet, is called *overland flow*. Once the water enters a recognizable channel, whether it be a small rill or a larger channel, the flow is usually turbulent. This can be demonstrated using the transition Reynolds number, $\mathrm{Re} = VR/\nu = 500$. Assuming a kinematic viscosity $\nu = 10^{-5}$ ft^2/s (in U.S. customary units) gives

$$VR = 5 \times 10^{-3} \tag{3–8}$$

This equation indicates that either the average velocity V or the hydraulic radius[6] R must be very small for laminar flow to prevail. Specifically, if the overland flow has a depth (or hydraulic radius) of 0.01 ft, the flow would remain laminar up to a velocity of about 0.5 ft/s.

It was determined in fluid mechanics that a two-dimensional laminar flow has a parabolic velocity distribution in the plane perpendicular to the flow. This leads to an average velocity of

$$V = \frac{\gamma y^2 S}{3\mu} \tag{3–9}$$

where y is the depth and S is the slope of the ground surface. Multiplying by the depth, the discharge per unit width becomes

$$q = \frac{\gamma y^3 S}{3\mu} \tag{3–10}$$

The typically uneven surface makes these equations very approximate for most applications. However, they do provide some insight into the behavior of overland flow. They have also been used in the computer modeling of portions of the hydrologic cycle.

---

[6]Here R is the hydraulic radius defined as the ratio of the cross-section area to the wetted perimeter (that portion of the boundary that resists the flow). In a wide channel the hydraulic radius becomes equal to the depth.

Hydrograph

A *hydrograph* is a plot of the discharge or flow rate as a function of time. It will be conceptualized first for a small impervious area exposed to a uniformly distributed, constant-intensity rainfall. Runoff from the area will commence at the start of the rainfall and increase as shown in Fig. 3–10, until the rate of runoff just equals the rate of precipitation. In Fig. 3–10, the runoff units have been kept the same as those of rainfall.[7] Normally, different units will be used to describe the runoff. These will be introduced shortly.

Once the runoff equals the rainfall in this simplified example, the runoff will continue at the same rate (ignoring evaporation) as long as the precipitation continues. When the precipitation ceases, the rate of runoff will start to decrease. However, runoff will continue for a period of time as the small surface area is drained and the rate of runoff drops to zero. The entire runoff plot is called a hydrograph. The region of increasing flow rate is called the rising limb of the hydrograph and the region of falling discharge is the recession limb. The hydrograph will prove to be of great importance in the analysis of stream and river flow.

The Rational Equation

The foregoing conceptual model leads to the *rational equation,* which is a method of predicting peak flow rates from small ungauged drainage basins. Assuming that the runoff rate equals the rainfall rate, the product of the drainage area A and the rainfall rate i should equal the peak discharge from an impervious area. If c is the fraction of the rainfall that runs off ($1 - c$ would be the fraction that infiltrates), then the discharge is

$$Q = ciA \tag{3–11a}$$

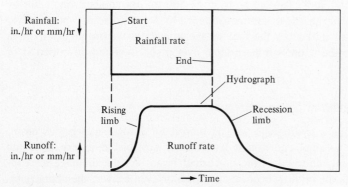

Figure 3–10 Hydrograph from small impervious area.

---

[7]Runoff units of in./hr or mm/hr imply that a volume equal to the product of the specified inches (or millimeters) times the area of drainage area runs off in one hour.

If the rainfall intensity is in in./hr and the area A is in acres, the resulting runoff will be in ac-in./hr. Discharge units will be introduced shortly, but 1 ac-in./hr = 1.008 cfs, where cfs is cubic feet per second. Thus, as a reasonable approximation, the discharge in Eq. 3–11a can be interpreted as cfs in U.S. customary units when the rainfall and area are as expressed above. In SI units the rational equation becomes

$$Q = \frac{ciA}{360} \tag{3-11b}$$

where i is in mm/hr, A is in hectares,[8] and the resulting discharge is in m^3/s. The rational equation is covered in more detail in Chapter 11 where Table 11–1 provides typical values of c for various types of ground surfaces. Equations 3–11a and 3–11b are very approximate. They are used, however, for the sizing of culverts and storm drainage structures in small drainage basins where better discharge information is usually not available.

Discharge and Volume Units

In SI units the discharge will most frequently be expressed in m^3/s and no confusion should arise in the typical calculations. In some books the cumec is used as the equivalent abbreviation for discharge in SI units. In U.S. customary units the discharge is generally expressed in cubic feet per second, which is abbreviated as cfs. Other equivalent terms occurring in the literature are the *cusec* and the *second-foot* (the latter abbreviated sf).

The obvious units of volume are the cubic meter (m^3) and cubic foot (ft^3). In addition there are two other U.S. customary units in common usage. One is the second-foot-day (sfd), which is a volume equal to a flow of 1 cfs for a period of one day. Since there are $60 \times 60 \times 24$ seconds in a day, 1 sfd equals 86,400 ft^3. The second unit is the acre-foot (ac-ft), which is a volume equivalent to an acre of area covered to a depth of 1 foot. Thus, 1 ac-ft = 43,560 ft^3. By comparison, 1 sfd = 1.98 ac-ft, a relationship that is sometimes rounded to 2 ac-ft. The advantage to sfd and ac-ft units is that it reduces the size of the numerical values. In addition, a volume of water may be expressed as a depth over a specified area in either SI or U.S. customary units.

3–6 STREAM FLOW

In the previous section a hydrograph was developed for a small impervious basin. Through use of the rational equation it was further possible to approximate the amounts of runoff and infiltration from any small basin. As the drainage area increases in size so that the outflow becomes a stream or even a river, the characteristics of the hydrograph become ever more important. However, the shape and peak of the hydrograph also become more difficult to predict. The basin is obviously not impervious and the infiltration capacity may be expected to vary throughout the area.

---

[8] A hectare is equivalent to $10^4 \, m^2$.

In addition, the rainfall itself will rarely be distributed uniformly or be of constant intensity for long periods of time. Finally, the rainfall duration will not be sufficient to allow an equilibrium to be achieved between the rainfall and runoff as was assumed in Fig. 3–10.

In spite of all the variables involved, the hydrographs still tend to adopt a characteristic shape. As a rule, the runoff collects relatively rapidly after the start of a rainfall so that the hydrograph typically has a steep rise. This is followed by a longer recession curve after the rainfall ceases. Figure 3–11 illustrates the general shape of the runoff hydrograph.

Another characteristic of the runoff hydrograph will be introduced at this point. In addition to the surface runoff, which rapidly makes its way to the stream system, there is the result of the previously discussed infiltration. As the water enters the surface of the ground some water flows just below the surface as an *interflow*. This interflow reaches the stream somewhat later than the surface runoff, but more quickly than the remainder of the groundwater flow. Because of its uncertain nature it is usually included with the surface runoff. The combined surface and interflow are together called the *direct runoff*. An additional portion of the subsurface flow continues to percolate downward until it reaches the groundwater table. There, it contributes to the more slowly moving *groundwater flow*. Although specific analysis of groundwater hydraulics will be reserved for a subsequent chapter, we need at this time to consider its contribution to the total flow hydrograph.

Starting at the beginning of the runoff from a particular storm, labeled point *A* in Fig. 3–11, the rapid rise in flow rate is due to the direct runoff quickly reaching the point in the channel where the hydrograph is measured. As the storm terminates,

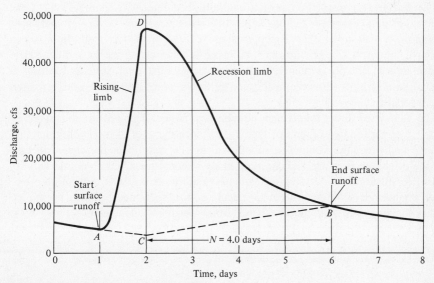

Figure 3–11 Typical river hydrograph.

the hydrograph will peak, and the recession curve follows as the basin drains. The steeper portion of the recession limb is associated with the draining of the direct runoff. After the curve begins to flatten, most of the stream flow comes from the groundwater flow into the stream through the channel banks. Usually, the surface of the groundwater table intersects the surface elevation of the stream.

In working with the hydrograph we must be able to estimate the end of the direct runoff (labeled point B) and, in addition, separate the direct runoff from the groundwater flow.[9] Different assumptions can be made with respect to the portion of the hydrograph that represents each of the two types of flow. The most common method of dividing the hydrograph involves the continuation of the preceding recession curve (the curve leading up to point A) to a point C lying directly below the peak (point D). This extension, usually drawn as a straight line, is then followed by a second straight line from C to the estimated end of the direct runoff at some point B. The end point is somewhat arbitrary and depends largely on the basin characteristics such as shape and slope. A study of several hydrographs from different storms on the drainage area may help to pinpoint the location of point B. In lieu of other alternatives, a very approximate estimation of the time N (in days) between the peak and the end of the direct runoff is given by

$$N = A^{0.2} \tag{3-12a}$$

in U.S. customary units or

$$N = 0.827 A^{0.2} \tag{3-12b}$$

in SI units. The drainage area A is in square miles or square kilometers respectively. The equation is approximate at best and is probably least accurate for smaller basins.

The foregoing construction results in the dashed line ACB in Fig. 3–11. The region under the hydrograph and above this line is the direct runoff. The groundwater contribution to the stream flow is represented by the dashed line itself. In working with the hydrograph, the groundwater discharge is often called the *base flow*. Since the base flow is usually a relatively small portion of the total discharge during a period of storm runoff, the precise location of the dividing line ACB is not a matter of great concern.

The volume of direct runoff may now be determined. This will be illustrated in Example 3–3 for the hydrograph of Fig. 3–11. Since the axes of the hydrograph are discharge and time, the volume of runoff will be the area under the curve. In this case, the discharge is in cfs and the time in days; consequently the volume will be obtained in sfd units. In SI units the volume would be expressed in m^3. In the example, a tabulated solution is used to calculate the direct runoff. Since the direct runoff is the area $ACBD$, it can also be determined using a planimeter.

---

[9]This technique becomes particularly important in the development of the unit hydrograph (see Section 3–7).

EXAMPLE 3–3

Assume that the hydrograph of Fig. 3–11 is the outflow from a drainage basin of 1020 mi^2. Calculate the direct runoff in sfd, ac-ft, and in. over the basin.

Solution

To separate the hydrograph into direct runoff and base flow, we need to estimate the end of the hydrograph (point B). Using Eq. 3–12a

$$N = (1020)^{0.2} = 4.0 \text{ days}$$

This locates point B as shown in Fig. 3–11.

The following tabulation of total discharge and base flow at 6-hr intervals gives the direct runoff in sfd units:

Time (days)	Total discharge (cfs)	Base flow (cfs)	Direct runoff (cfs)
1	5000	5000	0
1.25	9300	4800	4500
1.5	20,000	4500	15,500
1.75	35,300	4200	31,100
2	47,000	4000	43,000
2.25	46,100	4400	41,700
2.5	44,300	4800	39,500
2.75	41,400	5200	36,200
3	37,400	5600	31,800
3.25	32,700	5900	26,800
3.5	27,400	6300	21,100
3.75	22,500	6700	15,800
4	19,500	7100	12,400
4.25	17,300	7400	9900
4.5	15,700	7800	7900
4.75	14,300	8200	6100
5	13,200	8600	4600
5.25	12,200	8900	3300
5.5	11,300	9300	2000
5.75	10,600	9700	900
6	10,000	10,000	0
			354,100

The volume of direct runoff in sfd units is

$$\frac{354,100}{4} = 88,525 \text{ sfd}$$

Note that each value of direct runoff is assumed to occur over a time period of 0.25 days. The discharge of 4500 cfs that occurs after 1.25 days would be equivalent to a volume of 4500 sfd if it continued for 24 hr. Since it is assumed to represent only 6 hr, the 4500 cfs contributes only 1125 sfd to the total volume. Upon adding the entire column we get the preceding result.

This total volume is equivalent to

$$(88,525)(1.98) = 175,280 \text{ ac-ft}$$

Finally, for a drainage basin of 1020 mi^2, this volume can be converted to a depth over the basin as follows:

$$\text{depth} = \frac{(88,525)(86,400)}{(1020)(5280)^2} = 0.269 \text{ ft}$$

or $(0.269)(12) = 3.23$ in.

Measurement of Stream Flow

There are a number of ways of measuring or estimating the discharge in a river or stream. In very small streams, a low weir is sometimes placed across the channel. The hydraulic equations of Section 7-9 may then be used. In much the same way, the spillway of a dam may be used with a weir equation or the spillway may be calibrated with a model study. Various methods have been developed to measure the passage of a tracer released upstream and from the downstream concentration distribution determine the discharge. Another alternative is a direct calculation using the Manning equation (Eq. 7-5). A variety of electronic meters have also been used successfully in some instances. If a rough estimate is satisfactory, a float may be timed and the surface velocity converted to an approximate average velocity by multiplying by 0.85. However, the area of the channel cross section must still be known to estimate the discharge.

In most rivers these procedures are unsatisfactory. In addition to calculational difficulties associated with many of the foregoing methods, they do not usually provide a continuous record of discharge as a function of time. The usual time-dependent record of discharge is obtained through a two-step process using a standard stream gauging station, such as the one shown in Fig. 3-12. In the United States, these are usually operated by the U.S. Geological Survey (U.S.G.S.). The typical installation consists of a large stilling well that is connected to the stream at a point below the low water line if possible. A float within the stilling well is connected by wire with a drum in the upper chamber. As the river *stage* (or water surface elevation) changes, the rise or fall of the float rotates the drum. Independently, an ink pen

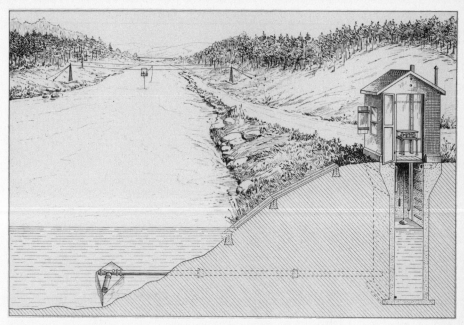

Figure 3–12 Sketch of river stage recording station with cableway-supported stream gauging in the background (U.S. Geological Survey).

is driven longitudinally along the drum by a clock mechanism. The result is a *stage hydrograph* or continuous record of the stage of the river. Alternative methods of obtaining the stage take advantage of the hydrostatic pressure distribution and simply monitor the pressure and pressure changes as the river rises and falls. This may be done directly with an inexpensive pressure transducer, or an air bubbler may be used. In the latter case, air is slowly forced into the river through a submerged outlet, thereby avoiding the high cost of the conventional gauging station. As the stage increases, a greater pressure is required to supply the air. A record of the stage can be obtained by monitoring this air pressure.

The stage may be electronically digitized and transmitted directly to an office receiver and computer. Ultimately, the stage hydrograph must be converted to a continuous record of discharge called a *discharge hydrograph* or simply a *hydrograph*. This is done through a *rating table* or *rating curve,* an example of which may be found in Fig. 3-13. Each value of the stage may be converted to the corresponding discharge through use of the rating curve.

The construction of a rating curve is a time-consuming operation, but one that must be undertaken for each gauging station. At its best, it represents a unique relationship between the stage at a particular gauging station and the discharge in the river. The potential difficulties will be addressed later. Specific data points are included in Fig. 3-13. These are obtained by a crew that has to go out to the river at the different stages and measure the discharge corresponding to that stage. This may be rather difficult, particularly at the higher discharges that often represent flood

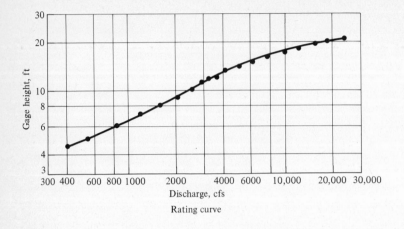

Rating curve

Gage ft	Q cfs	Gage ft	Q cfs	Gage ft	Q cfs	Gage ft	Q cfs	Gage Gage	Q cfs	Gage ft	Q cfs
4.5	413	7.2	1267	9.9	2510	12.6	3787	15.3	6600	18.0	12,300
4.6	436	7.3	1307	10.0	2550	12.7	3876	15.4	6755	18.1	12,580
4.7	460	7.4	1347	10.1	2590	12.8	3966	15.5	6910	18.2	12,870
4.8	484	7.5	1388	10.2	2629	12.9	4057	15.6	7073	18.3	13,160
4.9	509	7.6	1429	10.3	2669	13.0	4150	15.7	7238	18.4	13,450
5.0	535	7.7	1471	10.4	2710	13.1	4240	15.8	7406	18.5	13,750
5.1	562	7.8	1513	10.5	2750	13.2	4331	15.9	7577	18.6	14,050
5.2	590	7.9	1556	10.6	2790	13.3	4423	16.0	7750	18.7	14,360
5.3	618	8.0	1600	10.7	2830	13.4	4516	16.1	7944	18.8	14,670
5.4	647	8.1	1646	10.8	2870	13.5	4610	16.2	8141	18.9	14,980
5.5	677	8.2	1693	10.9	2910	13.6	4706	16.3	8342	19.0	15,300
5.6	707	8.3	1740	11.0	2950	13.7	4802	16.4	8546	19.1	15,630
5.7	738	8.4	1788	11.1	2990	13.8	4900	16.5	8754	19.2	15,960
5.8	770	8.5	1837	11.2	3030	13.9	5000	16.6	8966	19.3	16,300
5.9	802	8.6	1886	11.3	3070	14.0	5100	16.7	9181	19.4	16,650
6.0	835	8.7	1936	11.4	3110	14.1	5203	16.8	9400	19.5	17,000
6.1	868	8.8	1987	11.5	3150	14.2	5308	16.9	9623	19.6	17,350
6.2	901	8.9	2038	11.6	3196	14.3	5413	17.0	9850	19.7	17,700
6.3	935	9.0	2090	11.7	3241	14.4	5520	17.1	10,080	19.8	18,060
6.4	970	9.1	2140	11.8	3287	14.5	5629	17.2	10,310	19.9	18,430
6.5	1005	9.2	2190	11.9	3334	14.6	5738	17.3	10,540	20.0	18,800
6.6	1041	9.3	2240	12.0	3380	14.7	5849	17.4	10,780	20.1	19,070
6.7	1077	9.4	2290	12.1	3430	14.8	5962	17.5	11,030	20.2	19,830
6.8	1114	9.5	2340	12.2	3490	14.9	6075	17.6	11,270	20.3	21,140
6.9	1152	9.6	2380	12.3	3550	15.0	6190	17.7	11,520	20.4	22,530
7.0	1190	9.7	2425	12.4	3620	15.1	6310	17.8	11,780	20.5	24,000
7.1	1228	9.8	2470	12.5	3700	15.2	6450	17.9	12,040		

Figure 3–13 Rating table and curve for the Big Sioux River at Akron, Iowa (data provided by the U.S. Geological Survey).

conditions and occur during inclement weather. Even after the rating curve is established, the process continues from time to time to ensure that there is no change in the relationship.

The discharge itself is determined through the process of *stream gauging,* which is a graphic integration of the velocity distribution over the cross section. The velocity measurements are made using a *current meter* of which the most common type is the *Price meter* shown in Fig. 3-14. In shallow water the meter is positioned in the stream by a wader, but more generally it is supported by cable from above. Bridges are frequently used as measuring section, but a cableway is sometimes established from which the operator can traverse the river and make the necessary velocity mea-

Figure 3–14 Price current meter suspended above a gravel bed stream (U.S. Geological Survey).

surements. Regardless, the measuring section should be near, but not necessarily immediately at the stream gauging station.

The current-meter operation is based on holding the meter in a given location for a time period, usually in excess of 40 s to eliminate timing errors, and counting the number of revolutions of the meter wheel during the period. The counting is facilitated by the making and breaking of an electrical circuit as the cups rotate about their vertical axis. The current meter has its own calibration curve that provides the velocity as a function of the number of revolutions per minute.

To gauge a river, the cross section is divided into a number of vertical sections that approximate rectangles (Fig. 3-15). The number of sections is arbitrary, but should be such that no section has more than 10 percent of the total discharge. The discharge of each section is calculated by taking the product of the depth, width, and average velocity. The discharge of the river is finally the sum of the incremental discharges.

The velocity distribution in the vertical direction is approximately logarithmic, so the actual velocity should exactly equal the average at a point that is 62 percent of the depth as measured downward from the surface. In practice the average velocity is usually assumed to equal the average of two point velocities obtained at points located at two-tenths and eight-tenths of the depth. When the section is so shallow that two measurements are not practical, a single measurement is obtained at six-tenths of the depth, as measured vertically downward.

Considerable time is required to determine a single discharge, particularly in a large river. If the discharge is changing rapidly, some error will be introduced into the measurement. The advantage of working from a bridge, beyond increased safety, is the constant reference that it provides and the increased ease in subdividing the cross section.

It is important that the gauging station be located at a *control* section. The concept of a control will be discussed in far more detail in Chapter 7, but it is sufficient for present purposes to require that a control section is one for which there is a unique relationship between the depth and discharge. That is, for each depth (actually water surface elevation) there is one and only one discharge that can occur. One

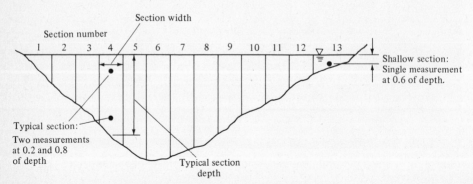

Figure 3–15 River gauging section to measure discharge.

necessary condition is an unchanging cross-sectional shape. The gauging station should be located at a section that is apparently free from scour or deposition. In addition to this requirement, the stage-discharge relation must not be altered by up-stream or downstream influences. In most rivers the downstream direction is of greater concern. A constricted downstream section, for example, can cause a back-water effect upstream that increases the depth in excess of that expected for the particular discharge. Another location to be avoided is just upstream of the conflu-ence with another river. In this case flood conditions in the other channel may lead to backwater conditions and a stage at the gauging station that is independent of the actual flow rate at the section. There are methods to correct for backwater effects, but the best procedure is to avoid suspect locations if possible.

Other problems also affect the uniqueness of the rating curve. The accumulation of debris near the gauging station may temporarily influence the stage, but may be minimized by regular inspection and maintenance of the section. The formation of an ice cover in cold climates results in periods when records must be used with extreme care. However, this is usually during a period of relatively low, nearly con-stant flow.

3-7 UNIT HYDROGRAPH

We have seen in Section 3-5 that the rational equation provides a means, however approximate, of estimating runoff as a function of rainfall in small drainage basins. In Sections 3-5 and 3-6, we also examined the characteristic form of a hydrograph. In this section we will further study these characteristics and from them develop a standardized hydrograph called the *unit hydrograph*. This unit hydrograph will pro-vide a means of satisfactorily predicting runoff from a particular basin based on a given rainstorm over the basin.

By definition, the unit hydrograph[10] is the hydrograph of 1 in. (or 1 cm in SI units) of runoff from a storm of specified duration.[11] There are three fundamental assumptions upon which the unit hydrograph method is based:

1. On a given drainage basin, all storms of the same duration have identical durations of surface runoff. Thus, the period of runoff is dependent on basin characteristics and the duration of the rainfall, but independent of the actual volume of runoff from the storm. To put it another way, two storms of iden-tical duration, but of different intensities, will have hydrographs that reflect the different intensities by the volume of direct runoff under the hydrograph. However, the time period from the start to the termination of the surface runoff will be the same for both storms. This leads to the second assumption.

[10]L. K. Sherman, "Streamflow from Rainfall by the Unit-graph Method," *Engineering News Record*. Vol. 108, pp. 501–505, 1932.

[11]Strictly speaking, the unit hydrograph must also be based on a storm of specified areal distribution. This will be discussed later in the section, but for now it can be assumed that the storm is distributed uni-formly over the drainage area.

2. On a given drainage basin, two storms of identical duration (and, for the present, assumed uniformly distributed over the basin) will produce similar direct runoff hydrographs. This is the principle of linearity. At any time t after the start of the direct runoff, the direct runoff rates of the two hydrographs will be directly proportional to the respective values of total runoff.

3. The direct runoff hydrograph from a given storm is independent of antecedent storms and the possibly concurrent runoff from them. As we will see, this assumption provides the justification for synthesizing a stream hydrograph from a sequence of rainfall events by the principle of superposition.

Development of Unit Hydrograph

The construction of the unit hydrograph will be considered first. This hydrograph will have an appearance similar to that of Fig. 3-11 and the data of Example 3-3 will be used in the following example to illustrate the procedure. It is unlikely that the hydrograph from a specified storm will have exactly 1 in. or 1 cm of direct runoff. Consequently, the foregoing assumptions must be used to generate the unit hydrograph. Once a suitable hydrograph is selected, the direct runoff must be determined by separation of the hydrograph as was done in Example 3-3. The duration of the rain that produced the runoff must also be identified. This latter point determines the *effective duration* for the resulting unit hydrograph. The unit hydrograph ordinates are immediately obtained by multiplying each direct runoff ordinate by the ratio of the unit hydrograph runoff volume (by definition 1 in. or 1 cm) to the actual volume of direct runoff in corresponding units. The procedure is illustrated in Example 3-4.

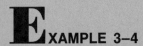

EXAMPLE 3–4

Assume that the hydrograph of Example 3-3 and Fig. 3-11 is due to a storm of 24-hr duration. Determine and tabulate the unit hydrograph for 24-hr storms on that drainage basin.

Solution

The duration is specified as 24 hr and therefore, the resulting unit hydrograph will only be usable with storms of 24-hr duration. The next step is normally to determine the direct runoff. This has previously been determined in Example 3-3 to be 3.23 in. of runoff over the basin. This is, of course, not the requisite 1 in. required for the unit hydrograph. However, assumption 2 immediately permits the adjustment of the runoff ordinates from Example 3-3 on the basis of the ratio 1/3.23. The values of direct runoff at 12-hr intervals, with the time set equal to zero at the start of the direct runoff and the resulting unit hydrograph ordinates, are tabulated in the following:

Time (days)	Direct runoff (cfs)	Unit hydrograph (cfs)
0	0	0
0.5	15,500	4800
1	43,000	13,300
1.5	39,500	12,200
2	31,800	9850
2.5	21,100	6530
3	12,400	3840
3.5	7900	2450
4	4600	1420
4.5	2000	620
5	0	0

The right-hand column is the required unit hydrograph. If the volume of runoff is determined for the unit hydrograph, it will be found to equal 1 in. of runoff over the 1020 mi^2 drainage basin area. It would be useful for the reader to verify this calculation.

To conclude, a unit hydrograph is associated with a particular duration of rain storm. It is also implicitly assumed, but not always satisfied, that the storm has a constant intensity throughout its duration and that the storm is uniformly distributed over the entire drainage area.

Theoretically, unit hydrographs could be developed for different areal storm distributions over the drainage basin, but this is usually impractical and a uniform distribution is assumed. This assumption is best satisfied for small-to-moderate-size basins with an upper limit of approximately 2000 mi^2. Large basins can be subdivided, however, and the various contributions from the different subareas combined by the routing technique to be considered in the following section.

Note that the hydrograph of Example 3-3 was tabulated at 6-hr intervals while the unit hydrograph of Example 3-4 was tabulated at 12-hr intervals. Neither of these intervals is related to the effective duration of the storm, which was assumed to be 24 hr. In fact, any convenient time interval may be chosen. The unit hydrograph, which will be called a 24-hr unit hydrograph, will still have an effective duration of 24 hr. It can only be used to synthesize runoff from storms that can reasonably be divided into 24-hr periods. In the same way, a 6-hr unit hydrograph can only be used to synthesize a hydrograph from storms that can be divided into periods of 6 hr. Given a unit hydrograph based on a particular effective duration for a specified basin, it is possible to generate unit hydrographs for the same basin that have any desired effective duration. This procedure will not be pursued herein, and reference should be made to any textbook on hydrology for further information.

Developing a Hydrograph from a Sequence of Storms

The development of a river hydrograph from a sequence of storms will be considered next. The three basic unit hydrograph assumptions will again form the basis for the procedure. The primary emphasis will be placed on the third assumption, which permits the superposition of runoff values from the different storms or different parts of the storm.

The concept will initially be introduced graphically as shown in Fig. 3-16. The unit hydrograph is drawn with dashed lines in each part of this figure. The effective rain duration is assumed to be 12 hr, and the duration of runoff associated with this rainfall is 72 hr. Each of the precipitation graphs at the top represents the rainfall rate less the infiltration rate, that is, it is the rate of direct runoff from the rainfall. The unit direct runoff from the precipitation is also shown with dashed lines. If precipitation resulting in this amount of runoff occurred, it would generate a runoff hydrograph identical to the unit hydrograph. Although actual quantities are not indicated, these would represent 1 in. or 1 cm of direct runoff. For the effective duration of 12 hr, this would imply runoff rates from the precipitaton (i.e., actual rate of precipitation minus rate of infiltration) of $\frac{1}{12}$ in./hr or $\frac{1}{12}$ cm/hr respectively.

In each of the three parts of Fig. 3-16, the actual storm runoff pattern is indicated by the continuous bar graph at the top. In part a, a 12-hr storm results in twice the runoff rate, and therefore, twice the runoff of that related to the unit 12-hr storm. In SI units, for example, this would be a total direct runoff of 2 cm or a rate of $\frac{1}{6}$ cm/hr. In this case, the duration of runoff will remain at 72 hr (from the first as-

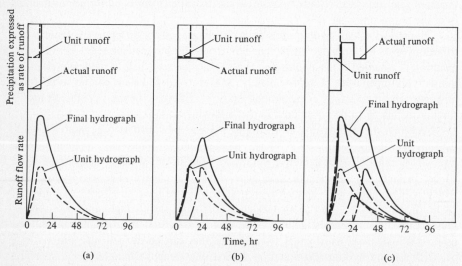

Figure 3–16 Graphic use of the unit hydrograph to develop a runoff hydrograph.

sumption) and the volume under the hydrograph will be doubled (the second assumption). In the subsequent application of the unit hydrograph method, this same effect will be achieved by doubling the value of each ordinate. The result hydrograph is shown by the continuous line in Fig. 3-16a.

In Fig. 3-16b, it is assumed that a storm provides runoff at a rate exactly that of the unit hydrograph (1 cm/12 hr in SI units), but for a period of 24 hr (as shown in the top graph). The unit hydrograph could not be applied immediately to this storm because of the different durations. However, the 24-hr storm can be thought of as two 12-hr storms, each with a runoff volume of 1 cm. The resulting total hydrograph is obtained graphically by adding a second unit hydrograph that is lagged by 12 hr to match the start of the second part of the storm. This unit hydrograph is shown by the sequence of long and short dashes. The final hydrograph, shown by the continuous line, is the sum of the other two.

The entire procedure is demonstrated in Fig. 3-16c. Here, a 36-hr storm has been divided into three 12-hr periods, which in U.S. customary units might be due to 2, 0.5, and 1 in. of runoff, respectively. The runoff hydrograph of each 12-hr storm is shown using the alternating long-and-short dashes, with the second and third portions lagged by 12 and 24 hr, respectively. Again, adding yields the total hydrograph for the storm. In addition to the direct runoff hydrograph shown, the base flow can be estimated and included to give the total river hydrograph. The usual procedure is to make the computations in tabular form as demonstrated in Example 3-5.

EXAMPLE 3–5

The unit hydrograph tabulated below has an effective duration of 12 hr.

Time (hr)	Unit hydrograph values (m³/s)
0	0
6	5.9
12	29.2
18	26.3
24	20.0
30	14.5
36	10.6
42	7.9
48	5.6
54	3.4
60	2.0
66	0.8
72	0

Estimate the total hydrograph from three days of rain. On day 1, it rained continuously at an approximate rate of 0.42 cm/hr. Day 2 may be approximated by precipitation at the rate of 0.07 cm/hr during the first 12 hr, followed by 0.38 cm/hr during the second 12-hr period. Likewise, day 3 is divided into two equal time periods with precipitation of 0.31 cm/hr and 0.19 cm/hr, respectively. The Φ index is 0.11 cm/hr and the base flow is estimated at a constant $1.7 \text{m}^3/\text{s}$.

Solution

Since the effective duration of the unit hydrograph is 12 hr, the rainfall during the first day must be divided into two periods of 12 hr each. Upon subtracting the Φ index from the rainfall rates and multiplying the result by 12 hr, the amount of runoff during each 12-hr interval is calculated as follows:

Time period	Runoff during period in cm
1	3.72
2	3.72
3	0
4	3.24
5	2.40
6	0.96

The tabular solution follows. The first column is the time in hours from the start of the rainfall. The second column is the unit hydrograph, which is repeated for convenient reference. The contribution from each 12-hr period of the storm is obtained by multiplying the depth of runoff (indicated at the top of the respective columns) by the unit hydrograph values. The columns are lagged so that the runoff hydrograph from each storm increment commences at the start of that particular portion of the storm. The different contributions at each 6-hr interval are added horizontally, along with the estimated base flow, to yield the river hydrograph.

Time (hr)	Unit hydrograph (m³/s)	Day 1 3.72 (cm)	Day 1 3.72 (cm)	Day 2 0 (cm)	Day 2 3.24 (cm)	Day 3 2.40 (cm)	Day 3 0.96 (cm)	Base flow (m³/s)	Hydrograph (m³/s)
0	0	0						1.7	1.7
6	5.9	21.9		0				↑	23.6
12	29.2	108.6	0						110.3
18	26.3	97.8	21.9						121.4
24	20.0	74.4	108.6	0					184.7
30	14.5	53.9	97.8	↑					153.4
36	10.6	39.4	74.4		0				115.5
42	7.9	29.4	53.9		19.1				104.1
48	5.6	20.8	39.4		94.6	0			156.5
54	3.4	12.6	29.4		85.2	14.2			143.1
60	2.0	7.4	20.8		64.8	70.1	0		164.8
66	0.8	3.0	12.6		47.0	63.1	5.7		133.1
72	0	0	7.4		34.3	48.1	28.0		119.4
78			3.0		25.6	34.8	25.2		90.3
84			0		18.1	25.4	19.2		64.4
90				↓	11.0	19.0	13.9		45.6
96				0	6.5	13.4	10.2		31.8
102					2.6	8.2	7.6		20.1
108					0	4.8	5.4		11.9
114						1.9	3.3		6.9
120						0	1.9		3.6
126							0.8	↓	2.5
132							0	1.7	1.7

3–8 STREAM FLOW ROUTING

The river or stream hydrograph has already been examined in some detail. We have determined how to generate a hydrograph for a particular stream using the unit hydrograph method. When a gauging station is available, the hydrograph can also be obtained by measuring the stage and converting the stage hydrograph into a discharge hydrograph using a rating curve. In either case, a hydrograph is determined at a specific section of the river. As the hydrograph moves downstream, it behaves essentially like a large wave. Ignoring the additional water that generally enters a stream from groundwater, overland flow, and tributaries, the hydrograph can be expected to attenuate as it moves downstream. That is, the base of the hydrograph (or period of direct runoff) tends to increase accompanied by a decrease in the peak flow. The process of quantitatively following the hydrograph downstream and through reservoirs is called *flow routing*. Frequently, flood conditions are a great concern and consequently the alternate terminology of *flood routing* is also used. The process of flood routing is extremely important because it provides a means of predicting the magnitude and arrival time of a peak flood discharge ahead of the actual flood. Routing also plays an important role in optimizing the design and operation of a system of reservoirs.

The passing of a flow hydrograph is an unsteady phenomenon in that the velocity as well as the depth change with time. Consequently, the most rigorous approach is to combine the unsteady flow equations from fluid mechanics, modified to describe open channel conditions, with the continuity equation. The problem is rather complex, particularly when applied to a natural water course, but can be solved with the use of the computer. A second, somewhat less-exact approach, stems from the use of the continuity equation alone. This procedure, which can be done manually, requires keeping track of the inflow, outflow, and the change in storage in a body of water. Thus, the term *storage routing* is often used to describe this technique.

Only the storage routing method will be considered in this chapter. However, the process will be applied to the passage of flood waves through both reservoirs and river channels. That portion of the development that is common to both reservoir and channel routing will be presented first. The two different applications will follow in separate subsections.

If we designate inflow by I and outflow by O, with both normally in cfs or m^3/s units, then steady flow conditions within a reservoir or a reach of river are described by $O = I$. It is assumed herein that there is no additional inflow (or outflow in the case of a diversion) between the point where the water enters and where it leaves. Continuing with this assumption, but considering unsteady flow, if $I > O$ then water is entering at a greater rate than it is leaving, and there will be an increase in the storage s within the section. If this continues for a time period Δt, then the storage (or continuity) equation may be expressed in exact form by

$$\bar{I}\Delta t - \bar{O}\Delta t = \Delta s \tag{3--13}$$

where the overbar indicates the time averaged value over the time period Δt. In addition, Δs is the resulting change in storage. This equation may be expressed in finite difference form as

$$\frac{(I_1 + I_2)}{2}\Delta t - \frac{(O_1 + O_2)}{2}\Delta t = s_2 - s_1 \tag{3--14}$$

where I and O still refer to the inflow and outflow of either a reservoir or a reach of river. The subscripts 1 and 2 identify the various quantities I, O, and s at the start and end respectively of a time period Δt. Since the finite difference form approximates Eq. 3-13 by linearizing the hydrograph, it will be most accurate for relatively short time intervals. Only one time period is considered at a time, but, as we will see in the applications that follow, the routing of an entire hydrograph is accomplished by subsequently allowing Δt to cover the complete passage of the hydrograph, time period by time period.

Equation 3-14 may be rearranged into the routing equation

$$I_1 + I_2 + \left(\frac{2s_1}{\Delta t} - O_1\right) = \left(\frac{2s_2}{\Delta t} + O_2\right) \tag{3--15}$$

Note first that every term has units of discharge, that is L^3/T. In SI units, I and O are immediately m^3/s, while $s/\Delta t$ will be consistent if the storage is in m^3 and the time is in seconds. In U.S. customary units I and O are in the conventional cfs units, but the storage is usually in sfd units, which requires that the time be expressed in days. The advantage of the arrangement in Eq. 3-15 is that the quantities on the left-hand side of the equation are known. The I values are obtained from the inflow hydrograph that is to be routed through the section, and the initial storage and outflow will either be known or conveniently estimated. After the first time period, s_1 and O_1 will be based on the results of the previous time interval. Thus, the unknown s_2 and O_2 are grouped on the right-hand side. It is apparent, however, that with two unknowns a second expression is required relating s and O. The method of obtaining this relationship differs for reservoirs and rivers and they will be considered separately from this point.

Reservoir Routing

Equation 3-15 can be used as the basic equation for routing a hydrograph through a reservoir. The inflow hydrograph will provide the I values that are obtained from the hydrograph at regular Δt time intervals. (See the routing table set up in conjunction with Example 3-6.) It will be assumed that the reservoir is of sufficient size and depth so that the water surface can be treated as horizontal. If this assumption is satisfied, the storage in the reservoir can be expressed as a function of the water-surface elevation. This relationship can be determined from a contour map without any knowledge of the inflow rate. Since the outflow from a reservoir is also related to the water-surface elevation, the two relationships can be combined to provide the necessary second relationship between the storage and the outflow.

Reservoir with Unregulated Outflow

The unregulated outflow implies that all water passes over a spillway or through an orifice for which no regulatory control is in use. The relationship between the outflow discharge and the water-surface elevation (or reservoir head) can assume many different forms. However, the two most frequent types of relationships are those given by either a weir equation ($O \sim H^{3/2}$) or an orifice equation ($O \sim H^{1/2}$). Here, H is the head above the weir crest or orifice outlet, and the constant of proportionality is either estimated or determined by a model study. A typical example of this relationship is given in Fig. 3-17. If gates are present on the spillway, they are presumed to be wide open.

Also included in Fig. 3-17 is a typical plot of reservoir area versus the water-surface elevation above the crest. This curve would be developed from a contour map by first identifying the particular contours above the reservoir that are associated with the selected water-surface elevations. The area within each contour would then be measured with a planimeter. From this information, the storage above an arbitrary datum, often the spillway crest, can be evaluated as a function of elevation. Consequently, both the storage and outflow are known at various elevations.

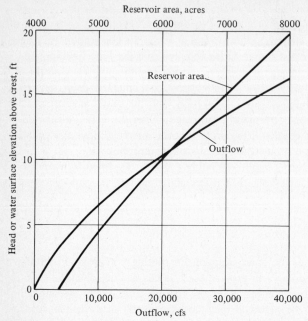

Figure 3–17 Relationship of reservoir area and outflow to reservoir head.

For routing convenience, the data is further reworked into the form of Fig. 3-18, where the quantities $(2s/\Delta t + O)$ and $(2s/\Delta t - O)$ have been plotted versus the outflow O. Note in the term $2s/\Delta t$, that Δt, in days (for the example in U.S. customary units that follows), must be identical to the routing period used in the actual routing process. In SI units, Δt, would be the number of seconds in the routing period. Both the generation of Fig. 3-18 and the routing procedure will be illustrated in Example 3-6.

The routing of the inflow hydrograph can now be carried out through the repeated use of Eq. 3-15. The initial outflow from the reservoir may be known from preceding events, or the hydrograph may be imposed on steady reservoir conditions so that the initial outflow O_1 is identical to the inflow I_1. If the reservoir was initially just filled to the crest, then O_1 would be zero. Even if the initial outflow is unknown, an estimate may be made, and after a few time steps the calculated outflows will converge on the correct value. Given or having estimated O_1, the value of $(2s_1/\Delta t - O_1)$ can be read from Fig. 3-18, whereupon I_1, I_2, and $(2s_1/\Delta t - O_1)$ may be summed according to Eq. 3-15 to yield the value of $(2s_2/\Delta t + O_2)$. Reference to the appropriate curve in Fig. 3-18 gives the value of O_2. After redefining I_2 and O_2 as I_1 and O_1 for the next step and calling the next discharge on the hydrograph I_2, one can repeat the process to calculate the next O_2. The above steps are repeated for the entire inflow hydrograph as illustrated in the routing table provided in Example 3-6.

EXAMPLE 3-6

An unregulated reservoir has a spillway crest elevation of 2015 ft above mean sea level. Reservoir outflow and surface area are given as functions of elevation in the data tabulation below. Prepare the necessary routing curves and route the inflow hydrograph through the reservoir using a 12-hr routing period. Assume that the initial outflow is 4000 cfs.

INFLOW HYDROGRAPH

Time (days)	Inflow I (cfs)
0	5000
0.5	6600
1	14,600
1.5	30,000
2	35,800
2.5	32,900
3	27,000
3.5	21,300
4	17,100
4.5	13,800
5	11,700
5.5	10,100
6	8900

Solution

The given reservoir information is tabulated in the first four columns below. These data are also plotted in Fig. 3-17. The incremental storage in ac-ft between any two elevations (see column 5 in the data tabulation table), is the product of the average area in acres at the two elevations and the difference in elevation in ft. The storage at any elevation (columns 6 and 7) is first obtained in ac-ft by accumulating the values in column 5. These values are then converted to sfd units. Columns 8 and 9 are obtained by adding or subtracting the outflow at each elevation from $2s/\Delta t$ where Δt in this case is 0.5 days. The last two columns are plotted in Fig. 3-18 and will be used as the routing curves.

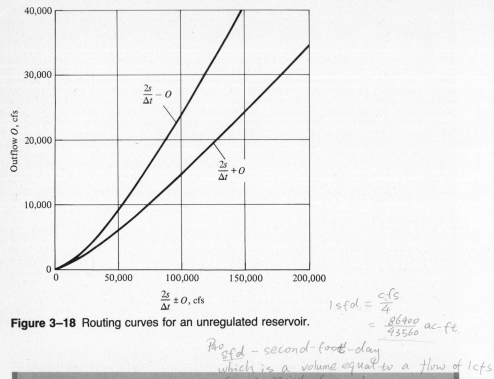

$\dfrac{2s}{\Delta t} \pm O$, cfs

Figure 3–18 Routing curves for an unregulated reservoir.

[handwritten: 1 sfd = $\dfrac{cfs}{4}$]
[handwritten: $= \dfrac{86400}{43560}$ ac-ft.]

[handwritten: P₄₀ sfd – second-foot-day]
[handwritten: which is a volume equal to a flow of 1cfs]
[handwritten: for a period of one day.]

DATA TABULATION								
Ele-vation	Head above crest	Out-flow O	Reservoir area	Δs	Storage s		$\dfrac{2s}{\Delta t}+O$	$\dfrac{2s}{\Delta t}-O$
(ft) (1)	(ft) (2)	(cfs) (3)	(acre) (4)	(ac-ft) (5)	(ac-ft) (6)	(sfd) (7)	(cfs) (8)	(cfs) (9)
2015	0	0	4360	0	0	0	0	0
2017	2	1718	4630	8990	8990	4540	19,900	16,400
2019	4	4860	4940	9570	18,560	9360	42,300	32,600
2021	6	8920	5260	10,200	28,760	14,500	66,900	49,100
2023	8	13,750	5620	10,880	39,640	19,990	93,700	66,200
2025	10	19,210	6000	11,620	51,260	25,840	122,600	84,200
2027	12	25,250	6390	12,390	63,650	32,090	153,600	103,100
2029	14	31,820	6790	13,180	76,830	38,740	186,800	123,100
2031	16	38,880	7210	14,000	90,830	45,790	222,000	144,300

The actual routing process is conducted in the routing table that follows.
Columns 1 and 2 repeat the given inflow hydrograph. Initially, we will let the
subscript 1 refer to time 0, and the subscript 2 refer to time after 0.5 days.
Next, according to Eq. 3-15 we need the quantity $(2s_1/\Delta t - O_1)$ which for O_1
= 4000 cfs can be read from Fig. 3-18 as 28,000 cfs. This value is entered in

column 3 on the first line, since subscript 1 refers to time 0. The sequence is indicated by the dashed arrow line. The three terms on the left-hand side of Eq. 3-15 may now be added to give a value of 39,600 cfs. This operation is indicated on the table by the continuous arrow lines. The result is the quantity $(2s_2/\Delta t + O_2)$, which is recorded on the second line. Recourse to Fig. 3-18 with this value gives the outflow $O_2 = 4500$ cfs as shown. The next step is from 0.5 to 1 days. The outflow O_1 now becomes 4500 cfs. The new value in column 2 can again be read from Fig. 3-18 or, as an alternative, $2O_1$ can be subtracted from the value in column 4. Either way the value in column 2 becomes 30,600 cfs. This step is again shown by the dashed arrow line. The routing continues in this manner until the entire inflow hydrograph has been considered. The resulting outflow hydrograph is given in the final column.

ROUTING TABLE $I_1 + I_2 + (\frac{2S}{\Delta t} - O_1) = (\frac{2S}{\Delta t} + O_2)$

Time	Inflow	$\dfrac{2s}{\Delta t} - \Theta$	$\dfrac{2s}{\Delta t} + \Theta$	Outflow
	I			O
(days)	(cfs)	(cfs)	(cfs)	(cfs)
0	5000	28,000		4000
0.5	6600	30,600	39,600	4500
1	14,000	38,800	51,200	6200
1.5	30,000	59,400	82,800	11,700
2	35,800	86,200	125,200	19,500
2.5	32,900	104,300	154,900	25,300
3	27,000	110,000	164,200	27,100
3.5	21,300	106,300	158,300	26,000
4	17,100	98,100	144,700	23,300
4.5	13,800	88,400	129,000	20,300
5	11,700	79,100	113,900	17,400
5.5	10,100	71,100	100,900	14,900
6	8900		90,100	12,900

Thus, the peak outflow from the reservoir of 27,100 cfs occurs 3 days after the start of the hydrograph or 1 day after the peak inflow. Finally, Fig. 3–18 indicates that the corresponding maximum reservoir elevation is 2027.6 ft.

Reservoir with Regulated Outflow

If an overflow spillway has a series of gates and one assumes that each gate is operated so that it is either wide open or closed, then the routing procedure is essentially the same as with the unregulated spillway. The only difference is that the routing curves of Fig. 3–18 are replaced by two families of curves, one set for $(2s/\Delta t + O)$ and the other for $(2s/\Delta t - O)$. The curves of Fig. 3–18 would serve when all gates are open. In addition, there would be two curves for one gate open, two for two gates open, and so on up to the number of gates. At any time in the routing process, the appropriate routing curves would be used for the specific number of open gates. The number of open gates can be changed during the reservoir routing according to a reservoir gate operating schedule.

If instead, a gate or other type of outlet work is operated so that a prescribed regulated discharge O_R is passed through the structure, Eq. 3–14 can be modified to

$$\frac{I_1 + I_2}{2} \Delta t - \frac{O_1 + O_2}{2} \Delta t - O_R \Delta t = s_2 - s_1$$

so that the routing equation becomes

$$I_1 + I_2 + \left(\frac{2s_1}{\Delta t} - O_1\right) - 2O_R = \left(\frac{2s_2}{\Delta t} + O_2\right) \tag{3-16}$$

The routing proceeds as before, except that at each time step the regulated release is subtracted from the left-hand side as indicated. This means that the unregulated release O is reduced accordingly. If all outflow is regulated, then $O_2 = O$ throughout, and the routing process would simply keep track of the changing reservoir level.

River Routing

Routing of a hydrograph down a river is complicated by two additional factors:

1. The storage in a reach of river is not merely a function of the outflow as was assumed previously for reservoir routing. This becomes immediately apparent if you consider the passage of a flood hydrograph during otherwise steady flow conditions. Initially the outflow equals the inflow, but as the flood wave enters the reach, storage increases without reference to the outflow. As the flood wave moves out of the reach, on the other hand, the decreasing storage within the reach is far more dependent on outflow rather than inflow conditions.

2. The storage can not be determined from a contour map as was done previously. The water surface can no longer be assumed to be horizontal, and even if it were, a contour map could not be read with sufficient accuracy within the confines of most river banks.

As a consequence of these complications, we must look further for the required second relationship to be used with the storage equation (Eq. 3–14). The discharge-stage rating curve, which was discussed previously, tends to plot as a straight line on log-log paper. Thus, we can write a rating equation such as

$$Q = ag^n$$

where Q is the discharge, g the stage, and a and n are constants for a given river. Assuming that a similar relation may be applied to the channel storage,

$$s = bg^m$$

where b and m are also constants. Eliminating the stage g between the two equations,

$$s = b \left(\frac{Q}{a} \right)^{m/n}$$

If x is introduced as a weighting factor that indicates the relative importance of the inflow I on the storage, then

$$s = \frac{b}{a^{m/n}} \left[xI + (1 - x)O \right]^{m/n}$$

The most common river routing procedure is the Muskingum method.[12] This approach assumes that $m/n = 1$ and $b/a = K$. On this basis, Eq. 3–17 simplifies to

$$s = K[xI + (1 - x)O] \tag{3–18}$$

This equation provides the second relationship, assuming that the constant K and the weighting factor x can be determined. This is accomplished by a rather lengthy process using available inflow and outflow hydrographs for the particular reach. At any given instant, $I - O$ is the rate of increase in storage for the reach. Thus, at the end of any time interval, the change in storage during the interval is given by

$$\Delta s = (\overline{I} - \overline{O})\Delta t \tag{3–19}$$

where Δt is the time interval between sequent values of I (and O). As before, the overbar indicates the average values during the interval (calculated using the end values). The storage s (obtained by adding the respective values of Δs) can be plotted versus the weighted discharge $[xI + (1 - x)O]$ for selected values of x. The results are similar to Fig. 3–19. A value of $x = 0$ implies that the storage is dependent solely on the outflow (the reservoir case), while $x = 0.5$ would weigh the effects of the inflow and outflow equally.

The best x is the value that results in the most nearly linear plot. Typically, this is in the range of $x = 0.2$ to 0.3. Once the x value has been identified, K is determined from the slope of the best fit line as shown. Note in Eq. 3–18, that x is dimensionless and K has dimensions of storage/discharge, or time. If the storage is in m^3 units and the discharge is in m^3/s, then time is in seconds. If the storage is in sfd units with discharge in cfs, then the time is in days. Computer program 2 in Appendix E uses a least-squares procedure to determine the best values of x and K on the basis of inflow and outflow hydrographs.

Equation 3–18, with appropriate subscripts, may be substituted into Eq. 3–14 to eliminate the storage terms s_1 and s_2. Upon rearranging and solving for O_2, we get the river routing equation

$$O_2 = c_1 I_1 + c_2 I_2 + c_3 O_1 \tag{3–20}$$

[12]"Engineering Procedure as Applied to Flood Control by Reservoirs with Reference to the Muskingum Flood Control Project," Engineer School, Fort Belvoir, U.S. Army Corps of Engineers, 1936, Appendix 8.

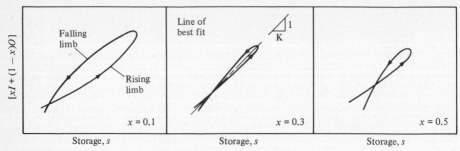

Figure 3–19 Determination of Muskingum constants.

where the three coefficients are given by

$$c_1 = \frac{Kx + 0.5\Delta t}{K - Kx + 0.5\Delta t}$$ (3–21a)

$$c_2 = -\frac{Kx - 0.5\Delta t}{K - Kx + 0.5\Delta t}$$ (3–21b)

and

$$c_3 = \frac{K - Kx - 0.5\Delta t}{K - Kx + 0.5\Delta t}$$ (3–21c)

As a check, the coefficients must also satisfy

$$c_1 + c_2 + c_3 = 1$$

As with reservoir routing, I_1, I_2, and O_1 are assumed to be known at the outset. Equation 3–20 can therefore be used to calculate O_2. The next inflow in the hydrograph is then taken as I_2 and the previous I_2 and O_2 become the new I_1 and O_1 for the new time period. The time period Δt is based on the interval between inflow values. It is imperative that K and Δt have the same units when calculating the coefficients. The routing process is illustrated in Example 3–7. Once the hydrograph is routed through a single reach, the outflow hydrograph that has just been determined will serve as the inflow hydrograph for the next reach. In this way the routing process can be carried downstream indefinitely. If there is tributary inflow, it may be added to the inflow or outflow depending on the actual location where the tributary enters the river. It is then included in the flood routing from that point downstream.

EXAMPLE 3–7

Route the inflow hydrograph which follows, through a reach of river for which it has been determined that $x = 0.30$ and $K = 0.90$ days. The initial outflow is 142 m³/s.

Inflow Hydrograph

Time (days)	Discharge (m^3/s)
0	142
0.5	187
1	396
1.5	850
2	1014
2.5	932
3	765
3.5	603
4	484
4.5	391
5	331
5.5	286
6	252

Solution

For $x = 0.3$, $K = 0.9$ days, and $\Delta t = 0.5$ days, Eqs. 3–21 yield the following: $c_1 = 0.591$, $c_2 = -0.023$, and $c_3 = 0.432$. The routing is completed in the following table. Initially, the subscript 1 refers to time 0 days, and subscript 2 refers to time 0.5 days. Thus, the first calculated outflow from the reach is

$$O_2 = c_1 I_1 + c_2 I_2 + c_3 O_1$$
$$= (0.591)(142) + (-0.023)(187) + (0.432)(142)$$
$$= 141 \text{ m}^3/s$$

The slight decrease in the initial outflow is due to the approximation associated with the Muskingum method and can be ignored.

ROUTING TABLE

Time (days)	I (m^3/s)	$c_1 I_1$ ($c_1 = 0.591$)	$c_2 I_2$ ($c_2 = -0.023$)	$c_3 O_1$ ($c_3 = 0.432$)	O (m^3/s)
0	142	—	—	—	142
0.5	187	84	−4	61	141
1	396	111	−9	61	163
1.5	850	234	−20	70	284
2	1014	502	−23	123	602
2.5	932	599	−21	260	838
3	765	551	−18	362	895
3.5	603	452	−14	387	825
4	484	356	−11	356	701

ROUTING TABLE (continued)				
Time (days)	I (m^3/s)	c_1I_1 ($c_1 = 0.591$)	c_2I_2 ($c_2 = -0.023$)	c_3O_1 ($c_3 = 0.432$)	O (m^3/s)
4.5	391	286	-9	303	580
5	331	231	-8	251	474
5.5	286	196	-7	205	394
6	252	169	-6	170	333

The outflow hydrograph peaks at 895 m^3/s after 3 days.

3–9 THE MASS CURVE

The routing of a hydrograph through a reservoir or river has just been covered. These procedures are useful to determine how a reservoir or natural channel affects the hydrograph. We also want to make use of the hydrograph to determine how large a reservoir is required, or to put it another way, how much storage is necessary to provide a given dependable flow from the reservoir. We have observed that even the unregulated reservoir attenuates the hydrograph, stretching it out as it reduces the peak. Two of the major downstream benefits resulting from the construction of a large dam are (1) the truncation of the hydrograph to reduce flooding and (2) the augmentation of the low flows. Improved navigation is just one of the myriad of benefits that follows as a result.

The minimum discharge of a river, the flow rate that is equaled or exceeded 100 percent of the time, is called the firm flow. Only in an intermittent stream does the firm flow drop to zero. The storage behind the dam of excess water from the larger discharges permits the use of this water during periods of normally low flow, and consequently, a considerable increase in the firm flow may be achieved. Thus, we are interested in the size of reservoir for a given stream that is required to provide a particular firm flow. Alternatively, we must also be able to determine the firm flow that can be obtained from a reservoir of a given size. A number of different techniques, all falling within the general heading of *operation studies,* are available to assist in answering these questions. If a number of years of river discharge data are available, the mass curve is one such reliable procedure.

The *mass curve* is a cumulative plot of the river discharges. Thus, mathematically it is the integration of the hydrograph with respect to time. It is usually plotted for the period of record, and the longer the period the more useful and reliable the results are likely to be. To explain the use of the mass curve, only three years have been included in Fig. 3–20. The accumulated inflow becomes a volume of water, with m^3, ac-ft, and sfd as commonly used units. The slope of the mass curve at any point in time, that is, its derivative, represents the discharge. A region of greater slope refers to a period of greater discharge. When applied to a reservoir study, the discharges are frequently reduced by the estimated amounts of evaporation, seepage, and other reservoir losses before plotting.

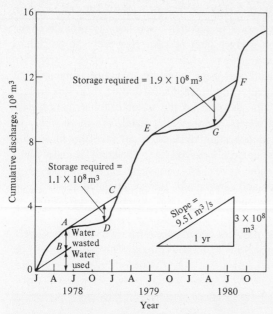

Figure 3–20 Mass curve for a reservoir study.

Let us assume that we wish to produce a constant yield or demand of 9.51 m³/s from the river data plotted in Fig. 3–20. On an annual basis, this corresponds to 3.0 × 10⁸ m³/year. This demand is plotted as the demand triangle in the lower right-hand corner. This slope must be equaled or exceeded at all points on the mass curve to satisfy the demand. The original curve clearly does not meet this requirement, and a reservoir is required. If a reservoir were initially full at the origin of the graph and water utilization occurred at the demand rate shown (and plotted from the origin), then at time *A* the reservoir would remain full, the vertical distance *AB* would be the volume of water wasted (or not used) and *B* down to the abscissa would be the volume used during the period. Up to point *A*, the mass curve has a slope that is greater than the demand curve and water is not required from storage.

At point *A* the slopes of the two curves are identical, and at that instant the river flow exactly equals the demand rate. Just beyond point *A*, however, the demand exceeds the supply and the continued demand (curve *AC*) can only be met by withdrawing water from reservoir storage. At point *D*, the supply again equals the demand (as indicated by the identical slopes), and in the time period from *D* to *C*, the reservoir is refilling. The storage required to maintain the demand during the period *A* to *D* is the vertical distance from point *D* to the demand curve *AC*. This amounts to 1.1 × 10⁸ m³. At *C*, the reservoir is again full and remains so to point *E*, where demand once again exceeds the supply. Usually this type of cycle is repeated on an annual basis. To meet the required demand during the period *E* to *F*, 1.9 × 10⁸ m³ of storage is necessary. To maintain the demand through the limited period of record of Fig. 3–20, the minimum usable reservoir capacity must equal 1.9 × 10⁸ m³. The storage requirement needed at point *D* would not be adequate during the second period of low flow. A much longer period of record is necessary to provide confidence in the long-term ability to meet a given demand rate. However, the procedure is identical to the preceding one.

Thus far, only a constant demand has been considered. Other demands that change with the time of year are equally important. One example of a nonconstant demand would arise when much of the demand on the water is for irrigation purposes. Industrial or domestic requirements, on the other hand, are more likely to create nearly constant demands. A nonconstant demand could take the form of a series of different slope straight lines, or even a curved line. The analysis under these conditions would be essentially the same as that for a constant demand. However, it is imperative that the nonconstant demand curve match the accumulated discharge on a month-by-month basis, with the January demand, for example, exactly lined up with January on the abscissa.

A second type of demand-related problem occurs when the reservoir size is specified and the maximum possible firm yield is to be determined. This situation will be covered in Example 3–8.

EXAMPLE 3–8

Based on the period of record given in Fig. 3–20, determine the maximum yield that can be sustained from a reservoir that has a usable capacity of 2.0×10^8 m^3. For convenience, the mass curve has been reproduced as Fig. 3–21.

Solution

A reservoir capacity of 2.0×10^8 m^3 can be identified by the vertical line included on the right-hand side of Fig. 3–21. This is shown for reference purposes. The solution is obtained by simultaneously adjusting the location of this vertical line (or capacity) and the demand lines (such as *AC* and *EF* in Fig. 3–20) until a demand line of minimum slope results. This is shown in Fig. 3–21 by lines *AB* and *CD*. These two lines are the maximum demands that could be satisfied by the specified reservoir during each period of low flow.

The solution is obtained by selecting the minimum slope from these demand lines. In this case, line *CD* must be chosen. The slope of 3.22×10^8 m^3/year gives a yield of 10.21 m^3/s. Note that if line *AB* had been chosen, a greater slope and yield would be indicated. This yield would have been satisfactory during the first dry period, but the yield could not have been maintained during the second period.

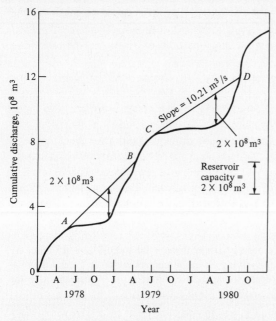

Figure 3–21 Maximum yield from a given reservoir.

P

PROBLEMS

Section 3–2

3–1. The depths of the snow pack as observed at six snow stakes in a snow course are 8.5 ft, 5.9 ft, 7.1 ft, 9.2 ft, 10.3 ft, and 7.7 ft. A series of samples indicates that the snow has an average specific weight of 15 lb/ft^3. What is the average depth of water represented by the snow pack?

3–2. The depths of the snow pack as observed at five snow stakes in a snow course are 2.5 m, 3.3 m, 2.8 m, 4.1 m, and 3.7 m. If the snow has an average density of 120 kg/m^3, what is the average depth of water represented by the snow pack?

3–3. Repeat Examples 3–1 and 3–2 assuming that gauging stations C and D are not present. Compare the results with the arithmetic average of the remaining gauges.

3–4. Repeat Examples 3–1 and 3–2 assuming that gauging stations B and F are not present. Compare the results with the arithmetic average of the remaining gauges.

3–5. Determine the precipitation over a basin provided by your instructor using (1) an arithmetic average, (2) the Thiessen polygon method, and (3) the isohyetal method.

3–6. Repeat Examples 3–1 and 3–2 if for a given storm the seven gauges record the following amounts of precipitation:

Gauge	Precipitation (mm)
A	42.8
B	44.3
C	62.1
D	35.5
E	11.0
F	17.4
G	28.9

Section 3–3

3–7. Determine the evaporation rate from a reservoir if the wind speed V_{30} = 15 mph, the surface temperature is 15°C, the overlying air has a temperature of 25°C, and the relative humidity is 30 percent. Repeat for a relative humidity of 90 percent. Use the Fitzgerald, Horton, Lake Hefner, and Lake Mead equations.

3–8. Repeat Prob. 3–7 if the wind speed V_{30} = 40 mph.

3–9. Convert Eq. 3–4 to SI units with vapor pressure in mm of mercury, the velocity at 30 ft in m/s, and the evaporation rate in mm/day.

3–10. Repeat Prob. 3–9 with respect to Eq. 3–5.

3–11. Determine the evaporation rate from a lake if the wind velocity V_{30} = 25 mph, the surface temperature is 50°F, the temperature of the overlying air is 85°F, and the relative humidity is 40 percent. Use the Fitzgerald, Horton, Lake Hefner, and Lake Mead equations.

3–12. Repeat Prob. 3–11 if the relative humidity is 10 percent.

3–13. Annual Class A pan evaporation for a year is tabulated below:

Month	Evaporation (in.)	Month	Evaporation (in.)
Jan	5	Jul	10
Feb	6	Aug	11
Mar	7	Sep	9
Apr	7	Oct	7
May	8	Nov	6
Jun	9	Dec	6

The annual precipitation in the region is 35 in., of which 30 percent reaches the river system. If a reservoir with a surface area of 30 mi^2 is proposed, calculate the net change in average stream flow (in cfs) that will result from the construction of the reservoir.

3–14. Repeat Prob. 3–13 if the annual precipitation in the region was 60 in.

3–15. Repeat Prob. 3–13 if the annual precipitation in the region was only 15 in.

3–16. If the pan evaporation data in Prob. 3–13 apply in a region where the lake evaporation is 59.5 in., what is the value of the pan coefficient?

Section 3–4

3–17. Determine the Φ index associated with the storm given in Fig. 3–9 if the total surface runoff from the basin is equal to a depth of 2 in.

3–18. Repeat Prob. 3–17 if the runoff equals 0.8 in.

3–19. A storm of 12-hr duration had a rainfall distribution as follows:

Hour	Rainfall (mm/hr)	Hour	Rainfall (mm/hr)
1	1	7	9
2	4	8	5
3	8	9	8
4	7	10	6
5	9	11	5
6	5	12	2

What is the Φ index if the total surface runoff from the basin equals 3 cm? Hint: plot a graph similar to that of Fig. 3–9.

3–20. Repeat Prob. 3–19 if the total runoff equals 5.5 cm.

Section 3–5

3–21. Water at 70°F flows over a wide plane surface that has a 2 percent slope. Determine the depth and average velocity corresponding to Re $= Vy/\nu = 500$.

3–22. Water at 10°C flows over a wide plane surface, such as a parking lot, that has a 2 percent slope. Determine the average velocity and discharge per unit width if the depth is 1 mm. Is the flow laminar or turbulent?

3–23. The average daily discharges in a river with a watershed area of 310 mi^2 are 4610 cfs, 7190 cfs, and 6100 cfs for a three-day period. Determine the volume of water discharged in sfd, ac-ft, and in. over the basin.

3–24. Convert the results of Prob. 3–23 into m^3 and mm over the basin.

3–25. The average daily discharges in a river with a watershed area of 2000 km^2 are 31 m^3/s, 59 m^3/s, and 15 m^3/s for a three-day period. Determine the volume of water discharged in m^3 and mm over the basin.

3–26. Convert the results of Prob. 3–25 into sfd, ac-ft, and in. over the basin.

3–27. A discharge of 15,000 cfs occurs during a time interval of 1 hr. Determine the volume of water in sfd and ac-ft units.

3–28. A discharge of 23,000 cfs occurs during a time interval of 2.5 hr. Determine the volume of water in sfd and ac-ft units.

Section 3–6

3–29. The hydrograph tabulated below is for a river that drains an area of 810 mi^2. Plot the hydrograph, separate the direct runoff from the base flow, and determine the volume of direct runoff in sfd, ac-ft and in. over the basin. Use Eq. 3–12 to determine the duration of the direct runoff. Tabulate the direct runoff hydrograph at 12-hr intervals.

Day	Hour	Discharge (cfs)	Day	Hour	Discharge (cfs)
0	0	850	4	0	3270
	12	820		12	2500
1	0	790	5	0	1960
	6	935		12	1510
	12	2620	6	0	1240
	18	6350		12	1020
2	0	6970	7	0	850
	6	6800		12	680
	12	6370	8	0	580
3	0	5250			
	12	4100			

3–30. The hydrograph tabulated below is for a river that drains an area of 4780 km^2. Plot the hydrograph, separate the direct runoff from the base flow, and determine the volume of direct runoff in m^3 and mm over the basin. Tabulate the direct runoff hydrograph at 12-hr intervals.

Day	Hour	Discharge (m^3/s)	Day	Hour	Discharge (m^3/s)
0	0	85	4	0	612
	12	82		12	488
1	0	79	5	0	400
	12	75		12	331
	18	103	6	0	273
2	0	305		12	230
	6	470	7	0	197
	12	642		12	168
	18	766	8	0	143
3	0	797		12	119
	6	770	9	0	96
	12	734			
	18	680			

3–31. The river cross section given in Fig. 3–15 was gauged using a Price current meter. The meter equation is given by $V = 2.1N + 0.16$, where N is the angular velocity of the meter in rev/s and V is the water velocity in ft/s. The meter was positioned in the center of each of the subsections and velocity measurements made at one or two depths. In each case, the number of revolutions during a 40-s interval was counted. The subsection depth and number of revolutions per 40 s are tabulated below. Each section has a width of 8 ft. What is the discharge?

Section	Depth (ft)	Number of revolutions	Section	Depth (ft)	Number of revolutions
1	0.9	21	8	13.2	144
2	2.4	43			111
		20	9	19.1	157
3	4.1	47			123
		25	10	15.1	153
4	3.2	48			119
		31	11	9.0	137
5	4.9	75			110
		47	12	4.7	99
6	9.3	119			55
		76	13	2.3	65
7	11.0	129			
		93			

3–32. Convert the current meter equation in Prob. 3–31 into SI units so that V is in m/s when N is in rev/s.

3–33. Repeat Prob. 3–31 if the current meter equation is $V = 2.3N + 0.14$.

3–34. Repeat Prob. 3–31 if a period of 50 s was used to obtain the number of revolutions given at each position.

3–35. Calculate the discharge in a river in which the cross section has been subdivided into 10 sections. The measured width, depth, and velocity (or velocities) at each section are as follows:

Section	Width (m)	Depth (m)	Velocity (m/s)	
1	1.5	0.4	0.8	
2	1.3	0.7	1.1	0.5
3	1.3	0.9	1.2	0.7
4	1.3	1.1	1.4	0.8
5	1.2	1.4	1.7	1.0
6	1.2	2.2	1.9	1.4
7	1.3	2.1	1.8	1.2
8	1.3	1.4	1.3	0.8
9	1.4	0.9	1.1	0.6
10	1.6	0.2	0.3	

Section 3–7

3–36. Develop a unit hydrograph (1 in. of runoff) from the hydrograph given in Prob. 3–29 and tabulate the discharge values at 12-hr intervals.

3–37. Develop a unit hydrograph (1 cm of runoff) from the hydrograph given in Prob. 3–30 and tabulate the discharge values at 12-hr intervals.

3–38. Determine the peak discharge in a unit hydrograph if the hydrograph is approximately triangular with a peak of 7000 cfs that occurs 24 hr after the start of runoff. The base flow is a constant 1000 cfs and the direct runoff ceases 96 hr after the peak. The drainage basin has an area of 800 mi^2.

3–39. Repeat Prob. 3–38 if the drainage basin has an area of 1200 mi^2.

3–40. Repeat Prob. 3–38 if the drainage basin has an area of 400 mi^2.

3–41. Determine the peak discharge in a unit hydrograph if the hydrograph is approximately triangular with a peak of 800 m^3/s that occurs 36 hr after the start of runoff. The base flow is a constant 100 m^3/s and the direct runoff ceases 96 hr after the peak. The drainage basin has an area of 5000 km^2.

3–42. Repeat Prob. 3–41 if the drainage basin has an area of 8000 km^2.

3–43. Assume that the unit hydrograph developed in Prob. 3–36 has an effective duration of 12 hr and use it to calculate the river hydrograph for the two-day storm given below.

Time period (hr)	Rainfall (in.)
0–12	0.9
12–24	1.8
24–36	0.4
36–48	1.9

Assume a Φ index of 0.05 in./hr and a base flow of 800 cfs. Tabulate the entire hydrograph and indicate the time and magnitude of the peak flow. Plot the hydrograph.

3–44. Assume that the unit hydrograph developed in Prob. 3–36 has an effective duration of 12 hr and use it to calculate the river hydrograph for the 60-hr storm given below.

Time period (hr)	Rainfall (in.)
0–12	1.7
12–24	0.3
24–36	1.6
36–48	2.1
48–60	0.8

Assume a Φ index of 0.04 in./hr and a base flow of 1200 cfs. Tabulate the entire hydrograph and indicate the time and magnitude of the peak flow. Plot the hydrograph.

3–45. Assume that the unit hydrograph developed in Prob. 3–36 has an effective duration of 24 hr and use it to calculate the river hydrograph for the five-day storm given below.

Time period (day)	Rainfall (in.)
1	1.9
2	1.4
3	1.1
4	2.2
5	0.8

Assume a Φ index of 0.02 in./hr and a base flow of 1500 cfs. Tabulate the entire hydrograph and indicate the time and magnitude of the peak flow. Plot the hydrograph.

3–46. Assume that the unit hydrograph developed in Prob. 3–37 has an effective duration of 12 hr and use it to calculate the river hydrograph for the two-day storm given below.

Time period (hr)	Rainfall (cm)
0–12	2.4
12–24	5.5
24–36	1.0
36–48	4.8

Assume a Φ index of 1.1 mm/hr and a base flow of 100 m^3/s. Tabulate the entire hydrograph and indicate the time and magnitude of the peak flow. Plot the hydrograph.

3–47. Assume that the unit hydrograph developed in Prob. 3–37 has an effective duration of 12 hr and use it to calculate the river hydrograph for the 60-hr storm given below.

Time period (hr)	Rainfall (cm)
0–12	3.1
12–24	4.1
24–36	2.8
36–48	5.5
48–60	2.2

Assume a Φ index of 1.2 mm/hr and a base flow of 80 m^3/s. Tabulate the entire hydrograph and indicate the time and magnitude of the peak flow. Plot the hydrograph.

3–48. Assume that the unit hydrograph developed in Prob. 3–37 has an effective duration of 24 hr and use it to calculate the river hydrograph for the five-day storm given below.

Time period (day)	Rainfall (cm)
1	4.2
2	1.8
3	5.0
4	2.9
5	4.7

Assume a Φ index of 0.9 mm/hr and a base flow of 120 m³/s. Tabulate the entire hydrograph and indicate the time and magnitude of the peak flow. Plot the hydrograph.

Section 3–8

3–49. The following information is available for an unregulated reservoir. The spillway crest is at an elevation of 1000 ft and the discharge is related to the head above the crest by the equation

$$Q = 120 \, h^{3/2} \text{ (cfs)}$$

At the 1000-ft contour, the reservoir has a surface area of 4100 ac. Above this elevation, the reservoir area A increases according to the equation

$$A = 4100 + 12 \, h^2 \text{ (acres)}$$

Prepare the reservoir routing curves for a routing interval of 12 hr and route the hydrograph of Prob. 3–29 through the reservoir. Initially, the reservoir has an outflow of 800 cfs. Determine the time of the peak outflow and the corresponding discharge and stage. Plot inflow and outflow hydrographs.

3–50. Repeat Prob. 3–49, and route the hydrograph of Prob. 3–29 through the reservoir. Assume that the reservoir is just full at the start.

3–51. A reservoir has area and outflow characteristics as tabulated below.

Elevation (ft)	Reservoir area (acres)	Outflow (cfs)
2000	2100	0
2002	2140	200
2004	2290	563
2006	2540	1050
2010	3210	2220
2014	4400	3700
2018	5910	5330
2022	6540	7230
2026	7410	9370

Prepare routing curves for a routing interval of 12 hr and route the hydrograph of Prob. 3–29 through the reservoir. Initially assume that the reservoir is just full. Determine the time of the peak outflow and the corresponding discharge and stage. Plot inflow and outflow hydrographs.

3–52. Repeat Prob. 3–51 assuming that the initial reservoir outflow is 2000 cfs.

3–53. The following information is available for an unregulated reservoir. The spillway crest is at an elevation of 100 m and the discharge is related to the head above the crest by the equation

$$Q = 35.5 \, h^{1.45} \, (m^3/s)$$

At the 100-m contour, the reservoir has a surface area of 4000 ha. Above this elevation, the reservoir area A increases according to the equation

$$A = 4000 + 375 \, h^{1.2} \, (hectares)$$

Prepare the reservoir routing curves for a routing interval of 12 hr and route the hydrograph of Prob. 3–30 through the reservoir. Initially, the reservoir has an outflow of 85 m^3/s. Determine the time of the peak outflow and the corresponding discharge and stage. Plot inflow and outflow hydrographs.

3–54. Repeat Prob. 3–53 if initially the reservoir is just full.

3–55. Repeat Prob. 3–53, but route the hydrograph of Example 3–7 through the reservoir. Assume that initially the reservoir is just full.

3–56. Repeat Prob. 3–55, but with an initial outflow of 200 m^3/s.

3–57. A reservoir has area and outflow characteristics as tabulated below.

Elevation (m)	Reservoir area (ha)	Outflow (m^3/s)
1600	3600	0
1602	4400	90
1604	5610	244
1606	7230	446
1608	9180	672
1610	12,110	990

Prepare routing curves for a routing interval of 12 hr and route the hydrograph of Prob. 3–30 through the reservoir. Initially assume that the reservoir is just full. Determine the time of the peak outflow and the corresponding discharge and stage. Plot inflow and outflow hydrographs.

3–58. Repeat Prob. 3–57 if the initial outflow is 100 m^3/s.

3–59. Repeat Prob. 3–50 if there is an additional constant regulated release of 750 cfs.

3–60. Repeat Prob. 3–57 if there is an additional constant regulated release of 80 m^3/s.

In Problems 3–61 through 3–69, assume that the initial river outflow equals the initial inflow.

3–61. A reach of river is determined to have routing parameters $x = 0.25$ and $K = 0.85$ days. Route the hydrograph of Prob. 3–29 through the reach. Use a routing interval of 12 hr. Plot inflow and outflow hydrographs.

3–62. Repeat Prob. 3–61 if $x = 0.15$ and $K = 0.7$ days.

3–63. Repeat Prob. 3–61 if $x = 0.20$ and $K = 1.3$ days.

3–64. A reach of river is found to have routing parameters $x = 0.22$ and $K = 0.90$ days. Route the hydrograph of Example 3–3 and Fig. 3–12 through the reach. (Note that the first two discharges must be obtained from Fig. 3–12.) Use a routing interval of 12 hr. Plot inflow and outflow hydrographs.

3–65. Repeat Prob. 3–64 if $x = 0.17$ and $K = 1.15$ days.

3–66. A reach of river is found to have routing parameters $x = 0.30$ and $K = 100,000$ s. Route the hydrograph of Prob. 3–30 through the reach. Use a routing period of 12 hr. Plot inflow and outflow hydrographs.

3–67. Repeat Prob. 3–66 using a routing period of 24 hr.

3–68. Repeat Prob. 3–66, but route the hydrograph of Example 3–7 through the reach.

3–69. Repeat Prob. 3–68 using a routing period of 24 hr.

3–70. Use the inflow and outflow hydrographs tabulated below to determine the routing parameters x and K.

Day	Hour	Inflow (cfs)	Outflow (cfs)
1	0	790	790
	12	2620	800
2	0	6970	1580
	12	6370	3830
3	0	5250	6000
	12	4100	5560
4	0	3270	4710
	12	2500	3780
5	0	1960	3220
	12	1510	2650
6	0	1240	2280
	12	1020	2060
7	0	850	1930
	12	680	1680
8	0	580	1510

Section 3–9

3–71. Using the mass curve of Fig. 3–20, determine the maximum flow rate that can be sustained from a reservoir with a capacity of 2.5×10^8 m^3.

3–72. Using the mass curve of Fig. 3–20, determine the reservoir capacity required to sustain a flow rate of 13 m³/s.

3–73. The monthly runoff from a small, 3.05 mi² basin has been converted into in. over the basin and tabulated below for an eight-year period.

Date	Runoff (in.)	Date	Runoff (in.)	Date	Runoff (in.)	Date	Runoff (in.)
1976		1978		1980		1982	
Jan	0.8	Jan	0.8	Jan	0.2	Jan	0.1
Feb	1.4	Feb	3.1	Feb	2.9	Feb	3.9
Mar	1.5	Mar	1.2	Mar	1.4	Mar	2.6
Apr	0.5	Apr	1.4	Apr	0.9	Apr	1.4
May	0.3	May	0.9	May	1.1	May	1.8
Jun	0.6	Jun	0.8	Jun	0.3	Jun	0.3
Jul	0.05	Jul	0.9	Jul	0.2	Jul	0.4
Aug	0	Aug	0.1	Aug	0	Aug	0.4
Sep	0.1	Sep	0	Sep	0	Sep	0.1
Oct	0.05	Oct	0.2	Oct	0	Oct	0.8
Nov	0.05	Nov	0.4	Nov	0	Nov	0.6
Dec	0.9	Dec	0.2	Dec	0	Dec	0.9
1977		1979		1981		1983	
Jan	0.8	Jan	0.8	Jan	0	Jan	2.0
Feb	2.1	Feb	0.8	Feb	0.2	Feb	0.4
Mar	2.3	Mar	2.8	Mar	0.1	Mar	1.7
Apr	0.6	Apr	0.9	Apr	0.1	Apr	2.0
May	0.9	May	1.0	May	0.1	May	1.6
Jun	1.0	Jun	0.9	Jun	0	Jun	2.2
Jul	2.8	Jul	0.1	Jul	1.0	Jul	1.2
Aug	0.3	Aug	0.1	Aug	0.9	Aug	0.1
Sep	0.1	Sep	0	Sep	0.2	Sep	0.1
Oct	0.1	Oct	0	Oct	0	Oct	0.6
Nov	0.2	Nov	0.9	Nov	0.1	Nov	1.1
Dec	0.1	Dec	0.5	Dec	0	Dec	0.2

Based on the eight-year period, plot the mass diagram and answer the following questions:

1. What is the maximum possible yield in cfs?
2. What storage capacity in ac-ft would be necessary to maintain this yield?
3. What period of storage in months would be required?
4. Is this yield feasible?
5. Comment on the results of an analysis based on the first four years.

3–74. Plot the mass diagram for the data in Prob. 3–73 and answer the following questions.

1. If the largest feasible reservoir has a storage capacity of 100 ac-ft, what yield (in gpm) could be maintained?

2. If the largest feasible capacity is 200 ac-ft, what yield could be maintained?

3. If the reservoir with a capacity of 100 ac-ft is built and the reservoir is full at the end of January 1978, what is its condition at the end of July 1978?

4. If the demand rate in part 1 is maintained from January 1978 to July 1978, how much water (in ac-ft) is wasted?

References

Chow, V. T. (ed.), *Handbook of Applied Hydrology*, New York: McGraw-Hill, 1964.

Eagleson, P. S., *Dynamic Hydrology*, New York: McGraw-Hill, 1970.

Hjelmfelt, A. T. and J. J. Cassidy, *Hydrology for Engineers and Planners*, Ames, IA: Iowa State University Press, 1975.

Linsley, R. K., M. A. Kohler, and J. L. H. Paulhus, *Hydrology for Engineers*, 3d ed., New York: McGraw-Hill, 1982.

Rodda, J. C., R. A. Downing, and F. M. Law, *Systematic Hydrology*, London: Newnes-Butterworths, 1976.

Snow Hydrology, Northern Pacific Division, U.S. Army Corps of Engineers, Portland, OR, 1956.

Viessman, W., J. W. Knapp, G. L. Lewis, and T. E. Harbaugh, *Introduction to Hydrology*, 2d ed., New York: Intext, 1977.

chapter 4

Groundwater Hydraulics

Figure 4–0 Center-pivot irrigation system (courtesy of Valmont Industries, Valley, NE).

4

4-1 INTRODUCTION

At this point we return to the hydrologic cycle to consider the portion that deals with the flow of water through soil. Although many aspects of groundwater and groundwater flow will be discussed, the emphasis will be placed on engineering applications. Groundwater is used to meet domestic, industrial, or agricultural needs in most portions of the United States. In fact, the economies of many regions depend on an adequate supply of groundwater. Fortunately, groundwater is a renewable resource in that there is usually a continual recharge. However, in many instances the use or consumption is significantly greater than the recharge, and the situation is similar to the actual mining of a resource. The groundwater table drops progressively lower and lower and the users must drill ever deeper to reach the water. As with surface water, the engineer must be concerned not only with the hydraulic conditions, but with the impact of his or her activities as well.

Initially, the various types and forms of groundwater and the accompanying terminology appropriate to this area of hydraulic engineering will be introduced. The hydraulics of groundwater flow will be divided into two parts: steady flow and unsteady flow. In both cases, typical pump tests, well yield, and the effect of wells on the groundwater will receive the greatest attention.

4-2 GROUNDWATER CHARACTERISTICS

The groundwater is generally divided into the different zones shown in Fig. 4–1. The major distinction is between portions of the ground that are saturated with water and those that are not. The dividing boundary is the *groundwater table*. We will restrict the term groundwater to mean that water in the *saturated zone* below the groundwater table. Just above the water table is an irregular region that may be nearly saturated called the capillary fringe. As the name implies, this water is lifted above the water table by surface tension and capillary action.

The remainder of the region between the groundwater table and the surface of the ground is the *zone of aeration* or unsaturated zone. The water in this zone is either gravity water percolating downward toward the water table or *soil moisture* held in the soil by capillary and hygroscopic action. This region may be temporarily saturated during a period of heavy infiltration. The zone of aeration is sometimes further divided into the *root zone,* an irregular region near the surface based on the depth of vegetation, and the *intermediate zone,* a region extending from the root zone down to the capillary fringe and water table. This distinction is very important to agricultural interests but need not concern us further.

The groundwater table, sometimes called the *phreatic surface,* is essentially at atmospheric pressure. Below the water table the pressure is hydrostatic. If a well extends below the water table, water will rise up to the line of zero pressure as shown

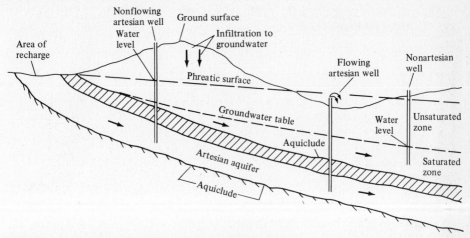

Figure 4–1 Definitional sketch for groundwater aquifers.

in Fig. 4–1. If the groundwater can flow through the ground below the water table, the region is called an *aquifer*. When the aquifer is bounded above by the groundwater table, the aquifer is an unconfined aquifer. It is eventually bounded below by an impervious layer or *aquiclude*.

Again referring to Fig. 4–1, water may enter an aquifer that is bounded both above and below by aquicludes. In this case the aquifer is an *artesian* or confined aquifer. Considerable pressure may build up in the confined aquifer and the water will rise above the aquifer limits if the aquifer is penetrated by a well. Under these conditions the well is known as an artesian well. Under a sufficiently high pressure the well may become a flowing artesian well.

The static head or level that the water reaches in a nonflowing artesian well is again called the phreatic surface and may be treated in much the same way as the actual groundwater table. Note that in either case the phreatic surface is identical to the piezometric head introduced in Chapter 2.

A number of soil characteristics affect the rate at which water flows through the aquifer or can be removed from the aquifer. These include the *porosity,* which is the volume of the voids or open spaces in a unit volume of the aquifer material. In a given sample, it is the ratio of the volume of the voids within the sample to the total volume of the sample. The porosity is a somewhat misleading parameter since the size and interconnection of the voids have more effect on the flow rate than does the porosity. Clay, which contains a large number of very small voids, has the largest porosity but usually yields the least water of the various soil types.

A second soil characteristic is the *specific yield,* defined as the volume of water that drains freely per unit volume of aquifer. This is directly significant to engineers because only that portion of the water that drains out of the aquifer under gravity will flow into a well. For a given soil sample, the specific yield is the ratio of the volume of water that drains readily from the sample to the total sample volume. The specific yield of a soil sample will always be less than its porosity. The additional

**Table 4–1 APPROXIMATE AND TYPICAL VALUES
OF AQUIFER PARAMETERS**

Aquifer material	Porosity (%)	Specific yield (%)	Permeability K (gal/day/ft^2)
Clay	45–55	3	10^{-4}–10^{-3}
Silt	40–50	5	10^{-3}–1
Sand	30–40	20–30	1–10^3
Gravel	25–40	15–30	10^3–10^6
Gravel and sand	20–35	16–30	10^2–10^4
Sandstone	10–20	8	10^{-3}–10
Shale	1–10	2	10^{-4}–1
Limestone	1–10	2	10^{-4}–1

water is held in the sample by capillary or hygroscopic forces and cannot be used. The ratio of the volume of water retained to the total sample volume is called the *specific retention.*

A third characteristic is the *permeability* or hydraulic conductivity of the soil. This term will be discussed in more detail in the following subsection and applied in the next two sections. Briefly, it is a measure of the ease with which water flows through the soil. For any given aquifer, the greatest accuracy will be achieved if the foregoing soil parameters are determined by direct measurement or calculation. As a guide, approximate values of these parameters are included in Table 4–1.[1] To repeat, reliable values can only be obtained by testing samples of the actual aquifer.

Darcy's Law

With the possible exception of flow through gravel deposits, groundwater flow is laminar. Recall from fluid mechanics that in both the Poiseuille equation for laminar flow through a circular conduit and its two-dimensional counterpart, the average velocity is directly proportional to the first power of the slope of the piezometric head line (or hydraulic grade line). It should come as no surprise that the same linear relationship applies to groundwater flow. This was first applied to the analysis of groundwater flow by Henri Darcy.[2] The usual expression, now known as Darcy's law, may be written

$$V = KS \tag{4–1}$$

where K is the coefficient of permeability, S is the slope of the groundwater table or phreatic surface, and V is the apparent velocity. Rather than attempting to define an actual velocity as the water winds its way through the openings in the aquifer mate-

---

[1]The values in Table 4–1 were collected from several sources, including Viessman, Linsley, Fair, Bouwer, and Chow. See the appropriate references at the end of the chapter.
[2]H. Darcy, *Les Fontaines publiques de la Ville De Dijon,* Paris: 1856.

rial, it is more convenient to deal with an apparent velocity. This velocity is defined as the ratio of the actual discharge through a portion of the cross section of the aquifer to the area of that cross section.

The coefficient of permeability has dimensions of velocity and, in fact, is the velocity corresponding to a gradient of unity. A velocity may be interpreted as a discharge per unit area, which explains the units of K included in Table 4–1. The coefficient of permeability may also be expressed in the more typical velocity units of m/s or ft/s. The velocity in laminar flow varies inversely with viscosity. Consequently, the effect of the groundwater temperature must be considered. The temperature is frequently in the vicinity of 60°F, and this temperature is often used to standardize the permeability. The coefficient of permeability at temperatures other than 60°F is related to that at 60°F by

$$K_T = K_{60}\left(\frac{v_{60}}{v_T}\right) \tag{4–2}$$

where the subscript T refers to a temperature other than 60°F. The coefficient of permeability K_{60} is referred to as the *standard coefficient of permeability*.

A final term to be introduced at this point is the *coefficient of transmissibility, T*. The transmissibility is defined as the discharge through a unit width of aquifer and may be expressed as

$$T = KY \tag{4–3}$$

where Y is the depth of the water flowing in the aquifer. The typical units are discharge per unit width (for example gpd/ft or m^3/day/m).

Measurement of Permeability

The coefficient of permeability may be determined by either laboratory or field measurements. The latter are obtained by running well pumping tests and applying the well equations introduced in the following sections. A *permeameter* is used for the laboratory measurements. Although the laboratory measurements are themselves more precise than the field tests, it is very difficult to obtain an undisturbed representative sample.

There are several types of permeameters operating with either horizontal or vertical flow and under either a constant or falling head. An upward flowing, constant-head permeameter is shown in Fig. 4–2. The aquifer sample placed in the permeameter has a cross section area A and flow length through the sample l. Water is supplied to the upper reservoir, where the head is maintained at a constant level by the overflow. Flow passes through the sample and the discharge is determined by measuring the volume of water leaving the permeameter during a given time interval. The slope in this case is the ratio h/l. Thus, the coefficient of permeability may be calculated according to

$$K = \frac{Q/A}{h/l} \tag{4–4}$$

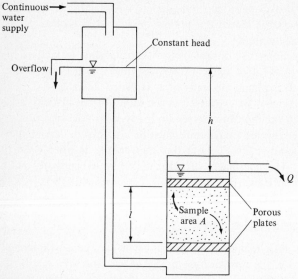

Figure 4–2 Constant-head permeameter.

If the test is run at a temperature other than 60°F, the results may be standardized using Eq. 4–2.

EXAMPLE 4–1

Determine the standard coefficient of permeability of a soil sample if the permeameter has a diameter of 6 in., a flow length of 12 in., and a constant head of 40 in. A volume of 0.05 ft³ flowed through the sample during a 15-min. period. The water temperature was 70°F.

Solution

The water discharge is

$$Q = \frac{0.05}{(15)(60)} = 5.56 \times 10^{-5} \text{cfs}$$

From Eq. 4–4 the coefficient of permeability is

$$K = \frac{Ql}{Ah} = \frac{(5.56 \times 10^{-5})(1)}{(\pi/4)(6/12)^2(40/12)} = 8.49 \times 10^{-5} \text{ft/s}$$

Correcting for temperature with Eq. 4–2, the standard coefficient is

$$K_{60} = K_{70}\frac{v_{70}}{v_{60}} = \frac{(8.49 \times 10^{-5})(1.059 \times 10^{-5})}{1.217 \times 10^{-5}} = 7.39 \times 10^{-5} \text{ft/s}$$

Converting to the more usual units.

$$K_{60} = (7.39 \times 10^{-5} \text{ ft/s})(7.48 \text{ gal/ft}^3)(86,400 \text{ s/day}) = 47.7 \text{ gpd/ft}^2$$

Therefore, the standard coefficient of permeability is 47.7 gpd/ft^2. If so desired, the corresponding coefficient of transmissibility could be obtained by multiplying this result by Y, the depth of water in the aquifer.

4–3 STEADY GROUNDWATER FLOW

Darcy's law will provide the basis for the analysis of steady groundwater flow. The emphasis will be placed on the hydraulic aspects associated with the removal of groundwater by wells. Two-dimensional flow will be considered first, followed by the three-dimensional, but frequently symmetrical, flow into a well. The hydraulic analysis will include both unconfined and confined aquifers. Although the flow of water through unsaturated soil is certainly of some importance, only flow through saturated soil will be considered.

Two-Dimensional Steady Flow

Under undisturbed conditions, groundwater flow is often nearly two-dimensional. We will be concerned here with the two-dimensional flow into an infiltration gallery such as the one shown in Fig. 4–3. These may be constructed along a hillside or parallel to, but above, a stream to intercept the groundwater flow. The original classic analysis by Dupuit[3] follows from five basic assumptions:

1. The flow is steady. Actually, in some cases it may take years before approximate equilibrium conditions are reached in an infiltration gallery or well.

2. The soil is homogeneous and isotropic. This means that the soil is the same at every point, and that there is no preferred flow direction due to soil characteristics themselves.

3. The aquifer is incompressible. This simply means that compaction of the soil does not occur as the water is withdrawn.

4. The tangent (or slope) of the water table is equal to its sine. This is the same assumption that will be made in the equilibrium analysis of open channel flow.

[3]J. Dupuit, *Etudes theoriques et pratiques sur le mouvement des eaux,* Paris: 2nd ed., 1863.

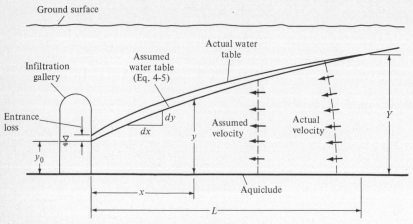

Figure 4–3 Two-dimensional groundwater flow.

5. The flow is horizontal at all points. The actual flow pattern will curve slightly, as shown in Fig. 4–3. The water in the vicinity of the water table must have the same slope as the water table itself. This assumption will be most closely satisfied when the infiltration gallery penetrates the entire depth of the aquifer and when the thickness of the aquifer is large relative to the drawdown.

The above assumptions are essential for all the derivations throughout this section. Since the assumptions are not completely satisfied, there is a small discrepancy between the calculated and actual groundwater profile. In addition, there is a small head loss, ignored in the following analysis, as the water enters the gallery.

With the x and y directions defined as positive to the right and upward as shown in Fig. 4–3, Darcy's law becomes

$$V = K \, dy/dx$$

and the discharge (per unit length of gallery) that passes through the vertical section at x is

$$q = Ky \, dy/dx$$

Separating variables and integrating,

$$qx = \tfrac{1}{2}Ky^2 + C$$

which gives rise to the parabolic water table profile shown in Fig. 4–3. Assuming a known aquifer thickness Y, at a distance L from the gallery,[4] and a depth y_0 in the gallery itself,

$$C = -\tfrac{1}{2}Ky_0^2$$

---

[4]Under ideal conditions this might be based on a reservoir that supplies water to the aquifer. Then Y would be the vertical distance up to the reservoir surface.

and

$$q = \frac{K(Y^2 - y_0^2)}{2L} \qquad (4\text{–}5)$$

At intermediate values of x

$$q = \frac{K(y^2 - y_0^2)}{2x}$$

There is considerable approximation in the analysis and only limited application. However, it serves as an introduction to the type of analysis used throughout the remainder of this section.

EXAMPLE 4–2

An infiltration gallery has a length of 1600 ft. The aquifer, which has a thickness of 25 ft, is fed by a lake that is located 1000 ft from the gallery. The aquifer has a coefficient of permeability $K = 750$ ft/day and the depth in the gallery is 1 ft. What discharge is collected by the gallery?

Solution

The coefficient of permeability is equivalent to

$$K = \frac{750}{(3600)(24)} = 8.68 \times 10^{-3} \text{ft/s}$$

From Eq. 4–5,

$$q = \frac{(8.68 \times 10^{-3})[(25)^2 - (1)^2]}{(2)(1000)} = 2.708 \times 10^{-3} \text{ cfs/ft}$$

or

$$Q = (2.708 \times 10^{-3})(1600) = 4.33 \text{ cfs}$$

over the entire length of the gallery.

Steady Flow to a Well

Figure 4–4 is a sketch of unconfined steady flow to a well. The original solution developed by Thiem[5] is similar to that of the two-dimensional case considered pre-

[5]G. Thiem, *Hydrologische Methoden,* Leipzig: J. M. Gebhardt, 1906.

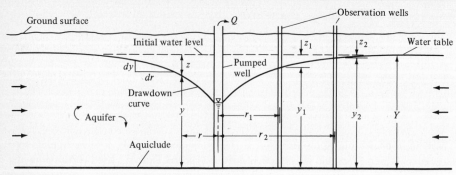

Figure 4–4 Unconfined steady flow to a well.

viously, and the same assumptions are required. In addition, and consistent with the assumption of horizontal flow, the initial water table is assumed to be horizontal.[6] The original undisturbed depth of water in the aquifer is Y and the well is pumped at a constant rate Q until equilibrium is reached. At any radial distance r from the well the *drawdown* is z and the water depth is y. Finally, if two observation wells are located at radial distances of r_1 and r_2 from the pumped well, the observed drawdowns will be z_1 and z_2 respectively.

The analysis proceeds from Darcy's law, now written as

$$V = K \, dy/dr$$

The discharge through a circumferential ring of radius r and height y is

$$Q = (2\pi r)yK \, dy/dr$$

For steady flow this is the discharge through each circumferential ring. Thus, the equation may be integrated as follows:

$$\int_{r_1}^{r_2} \frac{dr}{r} = \frac{2\pi K}{Q} \int_{y_1}^{y_2} y \, dy$$

or

$$\ln \left(\frac{r_2}{r_1} \right) = \frac{\pi K}{Q}(y_2{}^2 - y_1{}^2)$$

Solving for the discharge,

$$Q = \frac{\pi K(y_2{}^2 - y_1{}^2)}{\ln (r_2/r_1)} = \frac{\pi K \, (y_2{}^2 - y_1{}^2)}{2.303 \log_{10} (r_2/r_1)} \tag{4–6}$$

[6]It has been suggested that the resulting equations may be used, even if the aquifer is originally sloped, provided that the aquifer depths are properly defined. See L. K. Wenzel, "The Thiem Method For Determining the Permeability of Water-bearing Materials," *U.S. Geological Survey Water Supply Paper 679-A*, 1936.

If the drawdown is relatively small,

$$y_2^2 - y_1^2 = (y_2 + y_1)(y_2 - y_1) \approx 2Y(y_2 - y_1)$$

and

$$Q = \frac{2.729KY (y_2 - y_1)}{\log_{10} (r_2/r_1)} = \frac{2.729KY (z_1 - z_2)}{\log_{10} (r_2/r_1)} \tag{4-7}$$

After introducing the coefficient of transmissibility (Eq. 4–3), the above equations may also be written

$$Q = \frac{2.729T (y_2 - y_1)}{\log_{10} (r_2/r_1)} = \frac{2.729T (z_1 - z_2)}{\log_{10} (r_2/r_1)} \tag{4-8}$$

The equations result in a drawdown curve that asymptotically approaches the static phreatic surface. For calculational convenience, a zero drawdown is frequently assumed to occur at some radial distance r_0 called the *radius of influence*.

These equations provide some insight into well behavior. As mentioned before, complete equilibrium is rarely achieved. However, the assumption of steady conditions may frequently provide a satisfactory approximation. An additional error is also introduced if the pumped well does not completely penetrate the aquifer. This factor is usually small relative to the errors associated with the other assumptions. The equations may also be rearranged to solve for K or T based on the results of a pumping test. When these equations are applied to the pumping test results of a single isolated well, the drawdown will often be reasonably small and the equations satisfactorily accurate. To the extent that the aquifer is not homogeneous and isotropic, a calculated coefficient of permeability or transmissibility will represent average aquifer conditions.

Figure 4–5 is a sketch of confined steady flow to a well. The well is again assumed to penetrate the aquifer completely as shown. The well is pumped at a constant rate Q until equilibrium is reached. The drawdown in this case is simply a depression in the phreatic surface. As stated previously, the water will rise in the observation wells to the level of this surface. The required assumptions are as before;

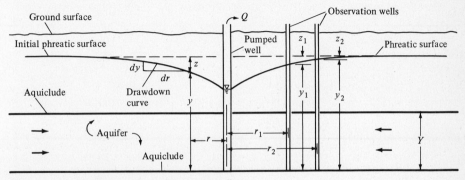

Figure 4–5 Confined steady flow to a well.

however, in this case the flow will remain horizontal with depth Y unless the drawdown curve intersects the aquifer itself. The derivation, which is left as an exercise, results in the foregoing Eqs. 4–7 and 4–8.

EXAMPLE 4–3

A confined aquifer (see Fig. 4–5) has a thickness of 50 ft. It is penetrated by a 1-ft-diameter well that is pumped at a constant rate of 1.4 cfs. Observation wells are located 100 and 300 ft from the pumped well. Under steady conditions, measured drawdowns of the phreatic surface equal 8 ft and 4.5 ft, respectively. Determine the coefficients of permeability and transmissibility for the aquifer and the drawdown of the phreatic surface at the well.

Solution

Rearrange Eq. 4–7 so as to solve for the coefficient of permeability

$$K = \frac{Q \log_{10} (r_2/r_1)}{2.729Y (z_1 - z_2)} = \frac{(1.4) \log_{10} (300/100)}{(2.729)(50)(8 - 4.5)} = 1.40 \times 10^{-3} \text{ft/s}$$

Alternately,

$$K = (1.40 \times 10^{-3})(86,400) = 121 \text{ ft/day}$$

or

$$K = (121)(7.48 \text{ gal/ft}^3) = 904 \text{ gpd/ft}^2$$

The coefficient of transmissibility is

$$T = KY = (904)(50) = 45,200 \text{ gpd/ft}$$

The drawdown at the well may be determined by again substituting into Eq. 4–7,

$$1.4 = \frac{(2.729)(1.40 \times 10^{-3})(50)(z_1 - 4.5)}{\log_{10} (300/0.5)}$$

Upon solving, $z_1 = 24.9$ ft.

The coefficient of permeability will automatically be standardized if the water temperature is 60°F. Otherwise the calculated K may be converted to the standard coefficient using Eq. 4–2.

Method of Images

In more complicated problems involving multiple wells, areas of recharge, imperme-
able barriers, and a sloping groundwater table, the flow net is an invaluable tool. Its
use will not be discussed herein, but recourse is recommended to the references at
the end of the chapter. The subject of complex well and aquifer interaction will not
be covered exhaustively; however, one technique known as the *method of images*
will be introduced.

The fact that flow-net analysis is applicable to groundwater flows means that
superposition principles apply. This can be used to advantage in certain instances
without the necessity of constructing the actual flow-net. The pumping of a well in
the vicinity of an approximately vertical impervious barrier is considered in Fig. 4–
6. The presence of the barrier means that the water cannot be drawn equally from all
directions. Rather than the symmetrical drawdown curve associated with an infinite
aquifer, there will be increased drawdown, particularly between the well and the
barrier.

This situation can be analyzed by placing an hypothetical *image well* on the line
that is perpendicular to the barrier and passes through the actual well. If the actual
well is a distance x from the barrier, the image well is placed an equal distance x
behind it, as shown in Fig. 4–6. The image well is considered to have the same
discharge Q as the actual well. The solution is obtained by first calculating the drawdown curve, assuming an infinite aquifer. This curve is sketched for both wells. The
actual drawdown curve is obtained by graphically or mathematically adding the two
curves together to get the resultant drawdown. Note that the consequence of this
procedure is to get a horizontal curve at the intersection with the barrier that is
consistant with zero flow through the barrier. An increased drawdown results elsewhere with the increase disappearing with distance from the barrier.

If a second barrier were located to the left of the well in Fig. 4–6, a second

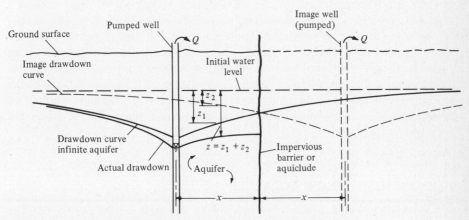

Figure 4–6 Increased drawdown due to an impervious barrier.

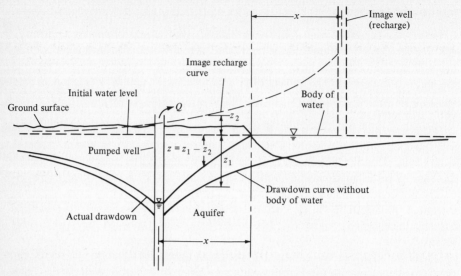

Figure 4–7 Reduced drawdown due to recharge from a body of water.

image well would also be required beyond that barrier. If the drawdown curves from the image wells should intersect the opposing barriers, additional image wells matching the first set of image wells would be required to account for the effect of the barriers on the image wells themselves. As the positioning of the barriers becomes more complicated, some care is required to locate properly the necessary image wells.

A second type of topographic interference with the drawdown curve is shown in Fig. 4–7. In this case a river or other body of water in the vicinity of the well provides direct recharge to the aquifer. The surface of the body of water remains a fixed point through which the drawdown curve must pass. To duplicate actual conditions, a negative or recharge image well is located a distance x behind this fixed point. As before, the drawdown curve is drawn for the pumped well based on an infinite aquifer. An identical curve is drawn with dashed lines representing the recharge. The resulting drawdown is obtained by subtracting the image curve from the well curve associated with an infinite aquifer. As indicated in the figure, the body of water forces the drawdown curve to match the water surface while reducing the drawdown at all points.

4–4 UNSTEADY FLOW TO WELLS

In this section selected aspects of unsteady flow to a well will be considered. The aquifer will still be assumed homogeneous and isotropic. Further, the initial phreatic surface and the flow itself will continue to be assumed horizontal. The original solution developed by Theis[7] was for artesian conditions. However, the procedure is applicable for unconfined flow, provided the drawdown is relatively small. The approximation for unconfined flow enters in because the coefficient of transmissibility $(T = KY)$ is treated as constant, even as the water table drops.

An unsteady continuity equation can be expressed for the cylindrical ring of Fig. 4–8. This ring represents a portion of the aquifer, the inner radius of which is located a distance r from the centerline of the well. The small segment of phreatic surface shown applies to either a confined or unconfined aquifer. The drawdown z relative to the original horizontal phreatic surface is now a function of both the radial distance from the well and time, that is, $z = z(r,t)$. Starting with Darcy's law, the discharge passing the inner ring may be written

$$Q_1 = T(2\pi r)S_1 = T(2\pi r)\frac{\partial y}{\partial r} = -T(2\pi r)\frac{\partial z}{\partial r}$$

where the slope must now be expressed by the partial derivative due to the time dependency of the drawdown. The slope S_2 at the outer ring may be related to the slope S_1 at the inner ring by

$$S_2 = S_1 + \frac{\partial S}{\partial r}dr$$

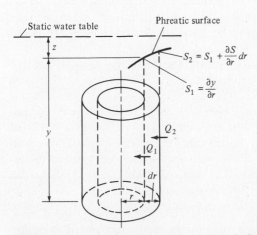

Figure 4–8 Definitional sketch for unsteady flow.

[7]C. V. Theis, "The Relation between the Lowering of the Piezometric Surface and the Rate and Duration of Discharge of a Well Using Ground-Water Storage," *Trans. American Geophysical Union* Vol. 16, 1935.

Consequently, the discharge through the outer ring is

$$Q_2 = -T(2\pi)(r + dr)\left(\frac{\partial z}{\partial r} + \frac{\partial^2 z}{\partial r^2} dr\right)$$

The difference between the two discharges must equal the rate of change of the volume of water within the ring, or

$$Q_2 - Q_1 = \frac{d\mathcal{V}}{dt}$$

The rate of change in water volume is given by

$$\frac{d\mathcal{V}}{dt} = 2\pi r \, dr \, \frac{\partial y}{\partial t} S_c = -2\pi r \, dr \, \frac{\partial z}{\partial t} S_c$$

A *storage coefficient* S_c, defined as the volume of water per unit volume, which will drain from the aquifer when the phreatic surface is lowered a unit distance, has been introduced to reflect the actual rate of change in the volume of water. In the unconfined aquifer, the storage coefficient equals the specific yield. Combining the above equations,

$$-T(2\pi)(r + dr)\left(\frac{\partial z}{\partial r} + \frac{\partial^2 z}{\partial r^2} dr\right) + T(2\pi r)\frac{\partial z}{\partial r} = -2\pi r \, dr \, \frac{\partial z}{\partial t} S_c$$

Dividing by the quantity $2\pi r \, dr$, canceling equal terms, and dropping a lower-order term leads to

$$\frac{\partial^2 z}{\partial r^2} + \frac{1}{r}\frac{\partial z}{\partial r} = \frac{S_c}{T}\frac{\partial z}{\partial t} \tag{4-9}$$

This equation governs the unsteady flow to a well. For a constant pumping rate Q, the Theis solution to this equation is

$$z = \frac{Q}{4\pi T}\int_u^\infty \frac{e^{-u}}{u}\,du \tag{4-10}$$

where u is defined by

$$u = \frac{r^2 S_c}{4Tt} \tag{4-11}$$

Before offering further explanation, note that S_c is dimensionless and Q and T must have consistent units. Equations 4–10 and 4–11 can be used directly in SI units, for example, if Q is in m^3/day and T is in m^3/day/m. Likewise, in English units Q in ft^3/day is consistent with T in ft^3/day/ft. In either case, t would be in days. If Q is in gpm and T is in gpd/ft, the two equations must be modified to

$$z = \frac{114.6Q}{T}\int_u^\infty \frac{e^{-u}}{u}\,du \tag{4-12}$$

and

$$u = \frac{1.87r^2S_c}{Tt} \tag{4-13}$$

The integral in Eqs. 4–10 and 4–12 is called the *well function of u* and written $W(u)$.

$$W(u) = \int_u^\infty \frac{e^{-u}}{u}\, du \tag{4-14}$$

It may be evaluated by the convergent series,

$$W(u) = 0.5772 - \ln u + u - \frac{u^2}{2 \times 2!} + \frac{u^3}{3 \times 3!} - \cdots \tag{4-15}$$

or values may be read from Table 4–2.[8]

Equations 4–10 and 4–11 or 4–12 and 4–13 are generally calibrated for a particular aquifer by running a nonequilibrium pumping test to determine the constants T and S_c. Once the constants are determined, the estimated drawdown at any point in the aquifer can be calculated as a function of time. To determine T and S_c, note for either set of the above equations that for constant Q, T, and S_c, the equations may be written as

$$z = C_1W(u)$$

and

$$\frac{r^2}{t} = C_2u$$

where C_1 and C_2 are both constants. Thus, a curve of r^2/t versus z must be similar to the relationship between $W(u)$ and u. The latter expression, called a *type curve* can be plotted on log-log graph paper. The procedure used to obtain T and S_c is to also plot a log-log graph, based on a pumping test, of z versus r^2/t. If one of the graphs is on transparent paper, it can be slid over the other, while maintaining a strict horizontal and vertical alignment, until similar portions of the two curves overlap (Fig. 4–9). A pinprick through both graphs within the region of similarity identifies a match point between the two functions. The respective match point values of z, r^2/t, $W(u)$, and u can be read from the two graphs and T and S_c calculated. At this point, the well equations may be used to study the impact of the well on the aquifer and predict future drawdown as required.

---

[8]L. K. Wenzel, "Methods for Determining the Permeability of Water-bearing Materials," *U.S. Geological Survey Water Supply Paper 887*, 1942.

Table 4-2 VALUES OF W(u)

u	1.0	2.0	3.0	4.0	5.0	6.0	7.0	8.0	9.0
$\times 1$	0.219	0.049	0.013	0.0038	0.0011	0.00036	0.00012	0.00004	0.00001
$\times 10^{-1}$	1.82	1.22	0.91	0.70	0.56	0.45	0.37	0.31	0.26
$\times 10^{-2}$	4.04	3.35	2.96	2.68	2.47	2.30	2.15	2.03	1.92
$\times 10^{-3}$	6.33	5.64	5.23	4.95	4.73	4.54	4.39	4.26	4.14
$\times 10^{-4}$	8.63	7.94	7.53	7.25	7.02	6.84	6.69	6.55	6.44
$\times 10^{-5}$	10.94	10.24	9.84	9.55	9.33	9.14	8.99	8.86	8.74
$\times 10^{-6}$	13.24	12.55	12.14	11.85	11.63	11.45	11.29	11.16	11.04
$\times 10^{-7}$	15.54	14.85	14.44	14.15	13.93	13.75	13.60	13.46	13.34
$\times 10^{-8}$	17.84	17.15	16.74	16.46	16.23	16.05	15.90	15.76	15.65
$\times 10^{-9}$	20.15	19.45	19.05	18.76	18.54	18.35	18.20	18.07	17.95
$\times 10^{-10}$	22.45	21.76	21.35	21.06	20.84	20.66	20.50	20.37	20.25
$\times 10^{-11}$	24.75	24.06	23.65	23.36	23.14	22.96	22.81	22.67	22.55
$\times 10^{-12}$	27.05	26.36	25.96	25.67	25.44	25.26	25.11	24.97	24.86
$\times 10^{-13}$	29.36	28.66	28.26	27.97	27.75	27.56	27.41	27.28	27.16
$\times 10^{-14}$	31.66	30.97	30.56	30.27	30.05	29.87	29.71	29.58	29.46
$\times 10^{-15}$	33.96	33.27	32.86	32.58	32.35	32.17	32.02	31.88	31.76

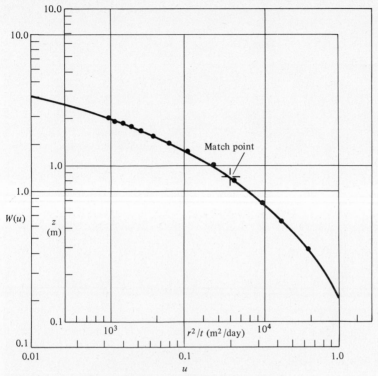

Figure 4–9 Theis drawdown and well function curves.

EXAMPLE 4–4

A 10-hour well-pumping test at a 0.5-m-diameter well is conducted at the rate of 0.06 m³/s. The resulting drawdown at an observation well located 20 m from the pumped well is tabulated below. Determine the coefficient of transmissibility and the storage coefficient of the aquifer from this test. What is the drawdown at the pumped well after one-half year? After one year?

Time t (hr)	Drawdown (m)	r^2/t (m²/day)	Time t (hr)	Drawdown (m)	r^2/t (m²/day)
0.25	0.10	38,400	4	1.39	2400
0.5	0.29	19,200	5	1.55	1920
0.75	0.44	12,800	6	1.66	1600
1	0.58	9600	7	1.78	1371

(continued)

Time t (hr)	Drawdown (m)	r^2/t (m²/day)	Time t (hr)	Drawdown (m)	r^2/t (m²/day)
1.5	0.80	6400	8	1.84	1200
2	1.00	4800	9	1.92	1067
3	1.21	3200	10	2.00	960

Solution

The parameter r^2/t, with t in days, has been calculated and tabulated in the above table. The two log graphs, z versus r^2/t and $W(u)$ versus u (from Table 4–2) have been plotted and overlaid in Fig. 4–9. The match point values are as follows:

$$W(u) = 1.22$$
$$u = 0.20$$
$$z = 0.84 \text{ m}$$
$$r^2/t = 6000 \text{ m}^2/\text{day}$$

Substituting into Eq. 4–10,

$$T = \frac{QW(u)}{4\pi z} = \frac{(0.06)(1.22)}{4\pi(0.84)} = 0.00693 \text{ m}^3/\text{s/m}$$

or $T = 599$ m³/day/m. From Eq. 4–11,

$$S_c = \frac{4Tu}{r^2/t} = \frac{(4)(599)(0.20)}{6000} = 0.080$$

At the pumped well after one half year, or 183 days

$$u = \frac{(0.25)^2(0.080)}{(4)(599)(183)} = 1.14 \times 10^{-8}$$

Hence, $W(u) = 17.74$ from Table 4–2, and the drawdown at the well is

$$z = \frac{(0.060)(17.74)}{4\pi(0.00693)} = 12.22 \text{ m}$$

After one year $u = 5.72 \times 10^{-9}$, $W(u) = 18.40$, and the corresponding drawdown is

$$z = \frac{(0.060)(18.40)}{4\pi(0.00693)} = 12.68 \text{ m}$$

This indicates that equilibrium has essentially been reached.

A somewhat simpler solution, often called the modified Theis method, has been achieved[9] by noting that for large t (and therefore small u), only the first two terms of Eq. 4–15 are significant. Thus,

$$z = \frac{Q}{4\pi T}\left(0.5772 - \ln\frac{r^2 S_c}{4Tt}\right) \tag{4-16}$$

The drawdown after time t_1 and t_2 becomes

$$z_1 = \frac{Q}{4\pi T}\left(0.5772 - \ln\frac{r^2 S_c}{4Tt_1}\right)$$

and

$$z_2 = \frac{Q}{4\pi T}\left(0.5772 - \ln\frac{r^2 S_c}{4Tt_2}\right)$$

Subtracting one from the other yields the change in drawdown during the interval,

$$z_2 - z_1 = \frac{Q}{4\pi T}\ln\frac{t_2}{t_1} = \frac{2.303Q}{4\pi T}\log_{10}\frac{t_2}{t_1} \tag{4-17}$$

This may be used to advantage by plotting the drawdown versus the $\log_{10} t$ on semi-log graph paper as in Fig. 4–10. A straight line plot verifies the original assumption. At lower values of t the deviation is apparent. Equation 4–17 may be applied by determining the change in drawdown during one log cycle of the straight portion (or the extension thereof). In one log cycle $\log_{10}(t_2/t_1) = 1$, and Eq. 4–17 reduces to

$$\Delta z = \frac{2.303Q}{4\pi T} \tag{4-18}$$

which may be solved for T.

Equation 4–16 is the equation of the straight line in Fig. 4–10. At $z = 0$ it reduces to

$$\frac{r^2 S_c}{4Tt_0} = e^{-0.5772} = 0.561$$

or

$$S_c = \frac{2.25Tt_0}{r^2} \tag{4-19}$$

Here t_0 is the time of zero drawdown, as given by the straight line extrapolation. Consequently t_0 may be read directly from Fig. 4–10, and Eq. 4–17 may be used to calculate the storage coefficient S_c. The modified Theis method is illustrated in the following example.

[9]C. E. Jacob, "Drawdown Test to Determine Effective Radius of Artesian Well," *Trans. ASCE* Vol. 112, 1947.

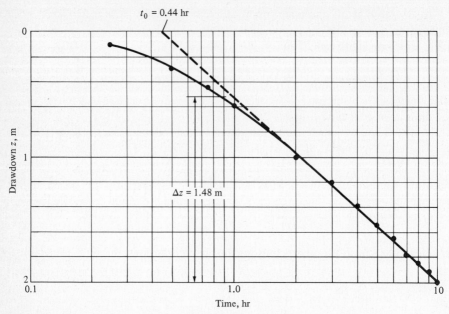

Figure 4–10 Drawdown using modified Theis method.

Equations 4–17, 4–18, and 4–19 may be modified to accommodate Q in gpm and T in gpd/ft similar to Eqs. 4–12 and 4–13. They can, of course, be adjusted to other sets of units as well.

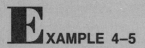XAMPLE 4–5

Repeat Example 4–4 using the modified Theis method.

Solution

Using the data of Example 4–4, the drawdown z has been plotted versus the time in hours in Fig. 4–10. From the straight line and the cycle from 1 to 10 hours, the change in drawdown z is 1.48 m. From Eq. 4–18

$$T = \frac{2.303Q}{4\pi\Delta z} = \frac{(2.303)(0.06)}{4\pi(1.48)} = 0.0074 \text{ m}^3/\text{s/m}$$

or $T = 642 \text{ m}^3/\text{day/m}$. As indicated on Fig. 4–10, $t_0 = 0.44$ hr, and

$$S_c = \frac{(2.25)(642)(0.44/24)}{(20)^2} = 0.066$$

The well radius is 0.25 m, thus the actual drawdown at the main well after one-half year (based on Eq. 4–16) is

$$z = \frac{0.06}{4\pi(0.0074)}\left[-0.5772 - \ln \frac{(0.25)^2(0.066)}{(4)(0.0074)(183)(86,400)} \right] = 11.6 \text{ m}$$

After one year

$$z = \frac{0.06}{4\pi(0.0074)}\left[-0.5772 - \ln \frac{(0.25)^2(0.066)}{(4)(0.0074)(365)(86,400)} \right] = 12.0 \text{ m}$$

There is a degree of approximation in the modified Theis procedure, but reasonably accurate results are obtained.

PROBLEMS

Section 4–2

4–1. An oil with specific gravity of 0.87 is used to determine the porosity of a soil sample. The dry and oil-saturated weights are 3.416 lb and 3.732 lb, respectively. When submerged in the oil the saturated sample displaces 28.66 in.^{3} of oil. Determine the porosity of the soil.

4–2. Repeat Prob. 4–1 if the saturated sample displaces 0.675 lb when submerged in the oil.

4–3. An oil with specific gravity of 0.855 is used to determine the porosity of a soil sample. The dry and oil-saturated weights are 2.527 N and 2.811 N respectively. When submerged in the oil the saturated sample displaces 137,100 mm^3 of oil. Determine the porosity of the soil.

4–4. Repeat Prob. 4–3 if the saturated sample displaces 0.840 N when submerged in the oil.

4–5. Determine the specific yield and specific retention of the soil in Prob. 4–1 if 40 percent of the water will drain freely from the saturated soil.

4–6. Determine the specific yield and specific retention of the soil in Prob. 4–3 if 50 percent of the water will drain freely from the saturated soil.

4–7. A soil has a standard coefficient of permeability of 4500 gpd/ft^2. What is the permeability at 40°F in corresponding units and in ft/day?

4–8. A soil has a coefficient of permeability of 0.027 gpd/ft^2 at 72°F. What is the standard coefficient of permeability in corresponding units and in ft/s?

4–9. A soil has a standard coefficient of permeability of 0.55 m/day. What is the permeability of the soil at 5°C in corresponding units and in l/day/m^2?

4–10. A soil has a coefficient of permeability of 5.0 m/day at 25°C. What is the standard coefficient of permeability of the soil in corresponding units and in l/day/m²?

4–11. If the aquifer of Prob. 4–7 has a thickness of 40 ft, what is the coefficient of transmissibility in gpd/ft?

4–12. If the aquifer of Prob. 4–9 has a thickness of 5 m, what is the coefficient of transmissibility in m³/day/m and l/day/m?

4–13. Repeat Example 4–1 if 6 lb of water at 50°F passed through the sample during a 25-min. period.

4–14. Determine the volume of water at 50°F that would pass through the permeameter and soil sample in Example 4–1 during a 15-min. period.

4–15. Repeat Prob. 4–14 for a temperature of 80°F.

4–16. Determine the coefficient of permeability in l/day/m² of a soil tested in the permeameter of Fig. 4–2 if the permeameter has a diameter of 5 cm, a flow length of 15 cm, amd a constant head of 1 m. A volume of 1.2 l flowed through the test section during a period of 1 hr.

4–17. Determine the standard coefficient of permeability of the soil in Prob. 4–16 if the permeability test was conducted at 23°C.

Section 4–3

4–18. Determine the discharge that could be collected from the infiltration gallery of Example 4–2 if the water level was maintained at a 2-ft depth within the gallery.

4–19. Plot the water surface profile in Example 4–2.

4–20. Plot the water surface profile in Prob. 4–18.

4–21. An infiltration gallery similar to Fig. 4–3 is fed from a parallel stream located 1000 m away. Determine the discharge available in the 100-m-long gallery if the depth within the gallery is 1 m and the water level of the stream is 10 m above the invert or bottom of the gallery. The soil has a coefficient of permeability of 27.1 m/day.

4–22. Repeat Prob. 4–21 if the stream was only 150 m from the infiltration gallery.

4–23. Determine the discharge from the infiltration gallery in Prob. 4–21 if the coefficient of permeability was 17 m/day.

4–24. Determine the coefficient of permeability in m/day of the soil in Prob. 4–21 if the discharge from the gallery is 0.0022 m³/s and the depths and lengths are as given.

4–25. An unconfined aquifer has a static thickness of 150 ft. A 12-in.-diameter well is pumped at a constant rate of 500 gpm. The drawdown in observation wells located 70 and 140 ft from the main well is 8.0 and 5.0 ft, respectively. Determine the coefficients of permeability (gpd/ft²) and transmissibility (gpd/ft) of the aquifer. What is the drawdown at the pumped well?

4–26. Repeat Prob. 4–25 if the aquifer thickness is 250 ft.

4–27. Estimate the error in K introduced in Prob. 4–25 by using Eq. 4–7 rather than Eq. 4–6.

4–28. Estimate the error in K introduced in Prob. 4–26 by using Eq. 4–7 rather than Eq. 4–6.

4–29. An unconfined aquifer has a static thickness of 75 m. A 20-cm-diameter well is pumped at a constant rate of 10 l/s. The drawdown in observation wells located 10 and 50 m from the main well is 4.1 and 1.3 m, respectively. Determine the coefficients of permeability (l/day/m^2) and transmissibility (l/day/m) of the aquifer. What is the drawdown in the pumped well?

4–30. Repeat Prob. 4–29 if the drawdown in the observation wells is 3.1 and 1.3 m, respectively.

4–31. Estimate the error in K introduced in Prob. 4–29 by using Eq. 4–7 rather than Eq. 4–6.

4–32. An unconfined aquifer has a static thickness of 200 ft. A 12-in.-diameter well is pumped at a constant rate of 750 gpm. The coefficient of permeability is 410 gpd/ft^2 and the radius of influence of the well at that discharge is 3500 ft. Determine and plot the drawdown curve for the well.

4–33. Repeat Prob. 4–32 if the well is pumped at the rate of 2 cfs.

4–34. Repeat Prob. 4–32 if the well is pumped at the rate of 1 cfs.

4–35. An unconfined aquifer has a static thickness of 200 m. A 30-cm-diameter well is pumped at a constant rate of 0.8 m^3/min. The coefficient of permeability is 0.31 m/day and the radius of influence of the well at that discharge is 1000 m. Determine and plot the drawdown curve for the well.

4–36. Repeat Prob. 4–35 if the coefficient of permeability is 1 m/day.

4–37. Repeat Prob. 4–35 if the coefficient of permeability is 0.6 m/day.

4–38. Using the sketch in Fig. 4–5, derive the confined steady-flow-well equation (Eq. 4–7).

In Probs. 4–39 through 4–43, assume that the stated drawdowns are measured from the static phreatic surface. Also assume that the drawdown curve does not intersect the confined aquifer.

4–39. Repeat Prob. 4–25 if the aquifer is confined with a thickness of 150 ft.

4–40. Repeat Prob. 4–25 if the aquifer is confined with a thickness of 50 ft.

4–41. Repeat Prob. 4–29 if the aquifer is confined with a thickness of 20 m.

4–42. Repeat Prob. 4–32 if the aquifer is confined with a thickness of 50 ft.

4–43. Repeat Prob. 4–35 if the aquifer is confined with a thickness of 70 m.

4–44. An 8-in.-diameter well penetrates an unconfined aquifer in which the static water table is 150 ft thick. The coefficient of permeability of the soil is 400 gpd/ft^2 and the well is located 100 ft from an impervious vertical barrier. The well is pumped at 0.8 cfs. Determine and plot the equilibrium drawdown curve on a line through the well and perpendicular to the barrier. Assume that the radius of influence of the well would be 2000 ft if the barrier were not present.

4–45. Repeat Prob. 4–44 if the pumping rate is 1.2 cfs.

4–46. Repeat Prob. 4–44 if the impervious barrier is replaced by a stream whose surface is located at the static level of the aquifer.

4–47. Repeat Prob. 4–46 if the pumping rate is 1.2 cfs.

4–48. Repeat Prob. 4–32 for the case where two wells located 1000 ft apart are both pumped at the rate of 750 gpm. Plot the profile along the line of the two wells.

4–49. Repeat Prob. 4–48 if both wells are pumped at the rate of 1 cfs.

4–50. Repeat Prob. 4–35 if there are three wells in line. The wells are 100 m apart and each is pumped at the rate of 0.8 m³/min. Plot the profile along the line of the three wells.

4–51. Repeat Prob. 4–50 if the three wells are 250 m apart.

4–52. Given the coefficient of permeability, the well diameter, pumping rate, and radius of influence, write a computer program to calculate the equilibrium drawdown profile of an unconfined aquifer.

4–53. Extend the computer program of Prob. 4–52 to calculate the drawdown at selected coordinate points if the aquifer contains multiple wells at specified locations each with its own steady pumping rate.

Section 4–4

The two data sets below represent the drawdown at an observation well as a function of time for Problems 4–54 through 4–57 that follow.

DATA SET 1		DATA SET 2	
Time (min.)	Drawdown (m)	Time (hr)	Drawdown (ft)
10	6.7	1	1.5
20	10.7	2	3.9
30	13.3	4	7.2
45	15.8	7	10.5
60	17.7	10	12.8
90	20.3	15	15.3
120	22.4	20	17.2
150	24.0	30	19.9
180	25.3	40	21.8
		50	23.4
		60	24.9
		70	25.9

4–54. Pump test data set 1 is obtained from an observation well located 7.5 m from a 30-cm-diameter well pumped at the rate of 450 l/min. Determine the coefficient of transmissibility (m³/day/m) and the storage coefficient using (1) the Theis method and (2) the modified Theis method. In both cases determine the drawdown in both the pumped well and the observation well after (1) 6 months, (2) 1 year, and (3) 10 years.

4–55. Repeat Prob. 4–54 if the pumping rate is 650 l/min.

4–56. Pump test data set 2 is obtained from an observation well located 35 ft from a 10-in.-diameter well pumped at the rate of 400 gpm. Determine the coefficient of transmissibility (gpd/ft^2) and the storage coefficient using (1) the Theis method and (2) the modified Theis method. In both cases determine the drawdown in both the pumped well and the observation well after (1) 3 months, (2) 6 months, and (3) 1 year.

4–57. Repeat Prob. 4–56 if the pumping rate is 1.5 cfs.

References

Bouwer, H., *Ground Water Hydrology,* New York: McGraw-Hill, 1978.

Chow, V. T. (ed.), *Handbook of Applied Hydrology,* New York: McGraw-Hill, 1964.

Davis, S. N. and R. J. M. DeWeist, *Hydrogeology,* New York: Wiley, 1966.

Fair, G. M. and J. C. Geyer, *Water Supply and Waste-Water Disposal,* New York: Wiley, 1954.

Linsley, R. K., M. A. Kohler, and J. L. H. Paulhus, *Hydrology for Engineers,* New York: McGraw-Hill, 1982.

Remson, I., G. Hornberger, and F. Molz, *Numerical Methods in Subsurface Hydrology,* New York: Wiley, 1971.

Rodda, J. C., R. A. Downing, and F. M. Law, *Systematic Hydrology,* London: Newnes-Butterworths, 1976.

Rouse, H. (ed.), *Engineering Hydraulics,* New York: Wiley, 1950.

Scheidegger, A. E., *The Physics of Flow through Porous Media,* New York: Macmillan, 1960.

Todd, D. K., *Ground Water Hydrology,* New York, Wiley, 1959.

Viessman, W., T. E. Harbaugh, and J. W. Knapp, *Introduction to Hydrology,* New York: Intext, 1972.

Walton, W. C., *Groundwater Resource Evaluation,* New York: McGraw-Hill, 1970.

chapter
5

Statistical Analysis of Hydrological Data

Figure 5–0 Developing storm clouds (courtesy of the U.S. Weather Service, National Oceanic and Atmospheric Administration).

5

5-1 INTRODUCTION

Although the preceding chapter dealt primarily with deterministic hydrology, much of the hydrologic data are statistical in nature. Consequently, statistical methods frequently need to be used, often with the goal of fitting a standard probability distribution to the data. This is the case with both precipitation and runoff data. We have already considered the generation of a unit hydrograph, but is the selected storm the correct storm for design purposes? What is the possibility of the occurrence of a larger storm that our culvert or bridge is not capable of handling? The routing of a flood hydrograph through a reservoir or down a river was treated in a deterministic fashion, but is a dam or downstream community at risk because of the possibility of an unanticipated extreme flood peak? Since any engineered project must also reflect a sound economic decision, we need to answer these types of questions. In particular, we must be able to evaluate the *probability* of occurrence of hydrologic events.

Statistical methods will not be derived in the chapter, and only those techniques necessary for present purposes will be introduced. The terms will be defined and the procedures will all be illustrated by example. Following a definition of terms and a discussion of probability concepts, they will be applied to selected hydrologic data.

The emphasis is placed on frequency analysis of discharge data. This concept has been developed at some depth, but at the expense of other types of statistical analysis. Not covered is the analysis of precipitation data, which has the added complexity that the precipitation intensity-frequency relationship must be developed for a range of precipitation durations. Some results of this type of analysis are included in Chapter 11. The reader should also be aware that statistical analysis is also useful for many types of data analysis, error analysis, sampling planning, and risk analysis.

5-2 PROBABILITY, RETURN PERIOD, AND PROBABILITY DISTRIBUTIONS

The concepts of probability and statistics are closely associated, particularly when dealing with a set of data. Before considering the more complete problem of predicting the probability of occurrence of a hydrologic event on the basis of a set of hydrologic data, we will develop the necessary basic principles of probability. The probability of the occurrence of a particular event is simply the chance that the event will occur. If there is a finite number of events, a not-necessarily-equal probability may be assigned to each. If the possible outcomes cover a continuous range of values, the probability can only be expressed by a mathematical function. We will impose additional constraints such as the occurrence of a particular event within a certain time period. However, to introduce the subject we will first look at an elementary example. What is the probability of rolling a particular number, say 1, with a single roll of a six-faced die? Assuming a fair die, there are six equally probable results, hence the probability of rolling a 1 is $\frac{1}{6}$ or approximately 0.167. The probability of rolling any number with a fair die is, of course, the same. What, then, is the probability of rolling a 1 or a 2? This result represents two of the six choices, and the probability of this outcome is $\frac{2}{6}$ or 0.333. To put it another way, the probability of a 1 or a 2 is the sum of the probability of a 1 plus the probability of a 2. The probability of rolling a number from 1 to 6 inclusive is the sum of each individual probability or 1. That one of the six outcomes will occur is a certainty.

Using this single die, we can examine other questions that are somewhat similar to our hydrologic interests. For example, what is the probability of not rolling a 1? If the probability of occurrence is $\frac{1}{6}$, then the probability of not rolling a 1 is $1 - \frac{1}{6} = \frac{5}{6}$. As another example, the probability of rolling a number greater than 2 is $1 - \frac{1}{6} - \frac{1}{6} = \frac{4}{6}$ or 0.667. If the possible outcomes have different probabilities, the procedure is identical. Assume that the die is so weighted that the probability of a 2 is twice the probability of any of the other equally probable outcomes. Then the probability of rolling a number greater than 2 reduces to $1 - \frac{1}{7} - \frac{2}{7} = 0.571$.

If two or more dice are rolled, then a particular outcome may be achieved by different combinations. For example, a total of 4 on a single roll of two dice results from a $3 + 1$, $1 + 3$, or $2 + 2$. In this type of problem both the number of combinations and the probability of each must be considered. We will have no need to make use of this type of analysis, and it will not be pursued further. On the other hand, we do have to examine a situation similar to multiple rolls of our single die. To illustrate this problem, determine the probability of rolling a 1 on each of 2 rolls. Each roll is independent and the chance of rolling a 1 on each is $\frac{1}{6}$. Having achieved the first 1 (of which there was only 1 chance in 6), there is still only a 1 in 6 chance of rolling the second 1. Thus, the probability is $(\frac{1}{6})(\frac{1}{6}) = \frac{1}{36} = 0.028$. In other words, there are 36 outcomes of which only one yields two 1's.

In design the most pressing need is to determine the probability of an event being equaled or exceeded. In terms of the die, this may be illustrated by the probability that the result of a given roll equals or exceeds, say, 5. From the equally probable die results, we know that the probability of a roll equaling or exceeding a value of 1 is 1, equaling or exceeding a value of 2 is 0.833, and so on. Thus, the probability that a value of 5 is equaled or exceeded is 0.333, while the probability that it is not equaled or exceeded is 0.667.

The probability of equaling or exceeding a particular value is a cumulative probability. Most of our interest will center on the probability that an event will be equaled or exceeded within a given time frame. We will define probability p as the likelihood that an event of a specified magnitude is equaled or exceeded in a given year. Immediately, the probability that the event is not equaled or exceeded in that year is $1 - p$. The event may relate to precipitation, discharge, temperature, wind speed, or any other hydrologic parameter. We will concentrate on discharge as the variable of most interest in hydraulic design. In terms of discharge data, if a discharge of 15,000 cfs has a probability $p = 0.02$, then the probability is 2 percent that this discharge will be equaled or exceeded in any one year. The reciprocal of p is the *return period* or *recurrence interval* t_p. This is equivalent to the average time interval between discharges that are equal to or greater than a specified discharge. The expression

$$t_p = \frac{1}{p} \qquad\qquad (5\text{–}1)$$

states that if an event has a probability of exceedance of p in any one year, then the average interval in years between events that equal or exceed the specified magnitude will be t_p. If the above discharge of 15,000 cfs has a 2 percent chance of being equaled or exceeded in a given year, then discharges of that magnitude or greater can be expected to occur, on the average, every 50 years.

The return period expresses the average interval between events, but it does not give the design engineer specific information concerning the likelihood of occurrence during the design life of the project. In other words, what is the probability that the previously considered discharge of 15,000 cfs with a return period of 50 years will occur during the life of a dam? Will it occur at all, or for that matter, can it occur more than once? To answer these questions we must relate probability and recurrence interval to a period of interest. Note that $1 - p$ is the probability that the discharge is not equaled or exceeded in a given year, and that the probability is again $1 - p$ that it will not occur in the second year. Thus, the probability that the discharge is not equaled or exceeded in two years is $(1 - p)(1 - p)$, and in a period of N years this probability becomes $(1 - p)^N$. The probability J that the event will occur at least once during N years is

$$J = 1 - (1 - p)^N \qquad\qquad (5\text{–}2)$$

This equation may be used to generate the results listed in Table 5–1. It is also possible to calculate the probability that an event of a specific magnitude will be equaled or exceeded, say, three times during N years. This type of calculation is usually of less interest and will not be pursued further.

Table 5–1 may be used to determine the probability that the 15,000-cfs discharge will occur in a given interval of time, such as the design life of a dam. It was previously assumed that this discharge has a recurrence interval of 50 years. If the dam has a design life of 50 years, the probability of a discharge equal to or greater than 15,000 cfs during that period is 63.6 percent. That is, it is not a sure thing that a 50-year flood will occur in any 50-year period. However, the possibility also exists that during a given 50-year period, a flood of this magnitude could occur on two or more occasions. If the design life of the dam were 100 or 200 years, then the corresponding probabilities of occurrence would be 86.7 percent and 98.2 percent, respectively. In other words, if the engineer wants to keep the risk of occurrence of a particular event very low throughout the 100-year life of a project, an extremely long recurrence interval must be chosen. For example, a design discharge limited to a 9.5 percent chance of being exceeded must have a return period of 1000 years. Note that this type of analysis provides no information concerning when an event will occur, only a best estimation of the likelihood of it occurring.

Table 5-1 PROBABILITY THAT AN EVENT WITH RETURN PERIOD t_p WILL BE EQUALED OR EXCEEDED DURING A TIME INTERVAL OF LENGTH N

t_p, year	Time N (years)							
	1	5	10	25	50	100	200	500
	Probability J							
1	1.000	1.000	1.000	1.000	1.000	1.000	1.000	1.000
2	0.500	0.969	0.999	1.000	1.000	1.000	1.000	1.000
5	0.200	0.672	0.893	0.996	1.000	1.000	1.000	1.000
10	0.100	0.410	0.651	0.928	0.995	1.000	1.000	1.000
20	0.050	0.226	0.401	0.723	0.923	0.994	1.000	1.000
30	0.033	0.156	0.288	0.572	0.816	0.966	0.999	1.000
40	0.025	0.119	0.224	0.469	0.718	0.920	0.994	1.000
50	0.020	0.096	0.183	0.397	0.636	0.867	0.982	1.000
100	0.010	0.049	0.096	0.222	0.395	0.634	0.866	0.993
150	0.007	0.033	0.065	0.154	0.284	0.488	0.738	0.965
200	0.005	0.025	0.049	0.118	0.222	0.394	0.633	0.918
250	0.004	0.020	0.039	0.095	0.182	0.330	0.551	0.865
300	0.003	0.017	0.033	0.080	0.154	0.284	0.487	0.812
350	0.003	0.014	0.028	0.069	0.133	0.249	0.436	0.761
400	0.002	0.012	0.025	0.061	0.118	0.221	0.394	0.714
450	0.002	0.011	0.022	0.054	0.105	0.199	0.359	0.671
500	0.002	0.010	0.020	0.049	0.095	0.181	0.330	0.632
1000	0.001	0.005	0.010	0.025	0.049	0.095	0.181	0.394

Probability Distributions

A *probability density function* (PDF) is a continuous mathematical expression that determines the probability of occurrence of a particular event. If such a prediction is to be based on a set of hydrologic data, then the distribution that best fits the set of data may be expected to give the best estimate, usually an extrapolation, of the probability of an event occurring. When there is a finite number of discrete outcomes or events and a probability can be assigned to each, the previous analyses using p and J are appropriate. The actual rainfall or discharge distribution over a period of years becomes a continuous function because any value is possible, at least within a broad range. In the next section we will consider the fitting of different distribution functions to actual data. For the present we will consider the distribution functions themselves.

Uniform Distribution

To introduce the concept of distributions we will begin with a simple uniform distribution. This is of little interest in hydrology, but is similar in nature to the probabilities associated with the die considered previously. The uniform probability density function is defined by

$$f(x) = \frac{1}{b - a} \quad\quad\quad\quad (5\text{–}3)$$

where b and a are the upper and lower limits of x, respectively. The function is graphed in Fig. 5–1a. The uniform PDF is appropriate when there is an equally likely outcome of any numerical value between a and b. The integral of $f(x)$ between limits of x_1 and x_2 gives the probability of an outcome within that range. If the integral is evaluated from the lowest possible value (in this case a, but more generally $-\infty$) up to x, the result is the *cumulative distribution function* (CDF). The general CDF will be written

$$F(x) = \int_{-\infty}^{x} f(x)\, dx \quad\quad\quad\quad (5\text{–}4)$$

The interpretation of Eq. 5–4 is as follows: For any specified value of x, $f(x)$ may be determined and integrated. The resulting $F(x)$ is the probability that an event less than or equal to x will occur.

The CDF for the uniform distribution of Eq. 5–3 is

$$F(x) = \int_{a}^{x} \frac{dx}{b - a} = \frac{x - a}{b - a}$$

which is plotted in Fig. 5–1b. If the integral in Eq. 5–4 is evaluated as $x \to \infty$, the result must equal unity. To demonstrate, if the CDF is evaluated for the uniform PDF of Fig. 5–1 and Eq. 5–3 (where the upper limit may be replaced by $x = b$), the result is $(b - a)/(b - a) = 1$. The probability of a specific numerical value x_1, on the other hand, is

$$\int_{x_1}^{x_1} f(x)\, dx = 0$$

The probability p that an outcome will be equal to or greater than a particular value of x is

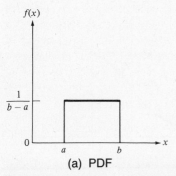

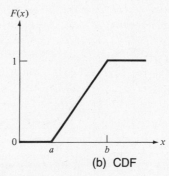

(a) PDF   (b) CDF

Figure 5–1 Uniform probability distribution.

$$p = 1 - F(x) = 1 - \int_{-\infty}^{x} f(x) \, dx \tag{5-5}$$

This equation relates the PDF to the earlier analysis of probability concepts. To illustrate, assume an equally likely result in the range $0 < x < 6$, and no values outside of this range. Then the uniform PDF in Eq. 5–3 becomes $f(x) = \frac{1}{6}$. The probability of an outcome $x > 2$ is

$$p = 1 - \int_{0}^{2} \frac{dx}{6} = 0.667$$

which is analogous to the probability that a roll of a die would produce a result greater than 2. The subsequent distribution functions are all handled in the same fashion.

Normal Distribution

The most familiar distribution is the bell-shaped normal distribution. It is sketched in Fig. 5–2 along with its S-shaped CDF. The PDF is given by

$$f(x) = \frac{1}{\sigma\sqrt{2\pi}} \exp\left[-\frac{(x - \bar{x})^2}{2\sigma^2} \right] \tag{5-6}$$

It is defined by two distribution parameters, the mean and standard deviation. These terms will be discussed further with respect to data sets in the following section. The *mean* (or average) $\bar{x}$ is evaluated by

$$\bar{x} = \frac{1}{N} \sum_{i=1}^{N} x_i \tag{5-7}$$

where x_i is the magnitude of the ith event and N is the total number of events. The *standard deviation* σ, which is a measure of the dispersion or spread of the data set, is given by

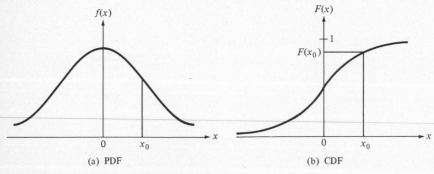

(a) PDF   (b) CDF

Figure 5–2 Normal probability distribution.

$$\sigma = \left[\frac{\displaystyle\sum_{i=1}^{N} (x_i - \bar{x})^2}{N - 1} \right]^{1/2} \tag{5–8}$$

The normal distribution describes many random processes, but it generally does not provide a satisfactory fit to flood discharges and other hydrologic data. The distribution function extends from negative to positive infinity and thereby assigns a probability to negative flow rates. A particular event x can be related to the probability of exceedance p by

$$x = \bar{x} + K\sigma \tag{5–9}$$

where K is the *frequency factor* for which values are tabulated versus p in Table 5–2. Note that $K = (x - \bar{x})/\sigma$, and therefore, according to Eq. 5–6, $f(x)$ is a function of K. In addition, if actual hydrologic data is to be expressed with respect to p, the probability of exceedance in a year, the data set will often be based on the peak value of each year. This *annual series* will be discussed in some detail Section 5–3. The use of the normal distribution will be demonstrated in Example 5–1.

Table 5–2 FREQUENCY FACTOR FOR THE NORMAL DISTRIBUTION

Exceedance probability p	Frequency factor K	Exceedance probability p	Frequency factor K
0.0001	3.719	0.550	−0.126
0.0005	3.291	0.600	−0.253
0.001	3.090	0.650	−0.385
0.005	2.576	0.700	−0.524
0.010	2.326	0.750	−0.674
0.020	2.054	0.800	−0.842
0.030	1.881	0.850	−1.036
0.040	1.751	0.900	−1.282
0.050	1.645	0.925	−1.440
0.075	1.440	0.950	−1.645
0.100	1.282	0.960	−1.751
0.150	1.036	0.970	−1.881
0.200	0.842	0.980	−2.054
0.250	0.674	0.990	−2.326
0.300	0.524	0.995	−2.576
0.350	0.385	0.999	−3.090
0.400	0.253	0.9995	−3.291
0.450	0.126	0.9999	−3.719
0.500	0.000		

EXAMPLE 5–1

Determine the probability that a discharge of 20,000 cfs will be equaled or exceeded in any one year, if the mean of an annual series of river discharges is 10,000 cfs and the standard deviation is (1) 3000 cfs and (2) 6000 cfs. What is the return period in each case?

Solution

In part (1) K is given by

$$K = \frac{20,000 - 10,000}{3000} = 3.33$$

This is in the range of Table 5–2 where interpolation is very inaccurate, but p must be somewhat below 0.0005. Reference to a more extensive table yields a value of $p = 0.00043$. For this value the return period is

$$t_p = \frac{1}{p} = \frac{1}{0.00043} = 2300 \text{ years}$$

Hence, a discharge of 20,000 cfs is an extremely rare event if the standard deviation is only 3000 cfs. In part (2) the value of K is

$$K = \frac{20,000 - 10,000}{6000} = 1.67$$

Interpolating in Table 5–2 yields a probability $p = 0.048$, which corresponds to a return period of 20.8 years. Although this example is mathematically correct, the results will be reasonable only if the actual data that establish the mean and standard deviation are normally distributed.

Log-Normal Distribution

Although the normal distribution is not well suited to hydrologic data, a related distribution called the log-normal distribution works reasonably well. This distribution assumes that the logarithms of the discharge (or whatever hydrologic parameter is being analyzed) are themselves normally distributed. The preceding equations describing the normal distribution may be used if the following substitution is made,

$$y_i = \log x_i \tag{5–10}$$

Upon taking the logarithm of each value the mean may be obtained by using Eq. 5–

7, with x replaced by y. The result is the mean of the logarithms themselves, and the operation can best be expressed directly as

$$\overline{\log x} = \frac{1}{N} \sum_{i=1}^{N} \log x_i \qquad (5\text{–}11)$$

As an alternative, the mean can also be found by taking the logarithm of the geometric mean of the set of values. This expression is

$$\overline{\log x} = \log (x_1 x_2 x_3 \ldots x_N)^{1/N}$$

Likewise, the standard deviation may be found from Eq. 5–8 where x_i and $\bar{x}$ are both based on the logarithms of the actual data. Specifically,

$$\sigma_{\log x} = \left[\frac{\sum_{i=1}^{N} (\log x_i - \overline{\log x})^2}{N - 1} \right]^{1/2} \qquad (5\text{–}12)$$

Equation 5–9 may again be used to relate the probability of exceedance to the occurrence of particular values provided that log values are used. This expression may be rewritten as

$$\log x = \overline{\log x} + K\sigma_{\log x} \qquad (5\text{–}13)$$

Since the log-normal distribution is similar to the normal distribution once the x has been replaced by $\log x$, the frequency factors may again be determined from Table 5–2. The log-normal distribution is also a special case of the distribution that follows.

Log Pearson Type III Distribution

The problem with most hydrologic data is that an equal data spread does not occur above and below the mean. The lower side is limited to the range from the mean to zero (although in many cases the minimum may be well above zero), while there is theoretically no limitation on the upper range. This contributes to what is called a *skewed* distribution. The coefficient of skew a is defined mathematically by

$$a = \frac{N \sum_{i=1}^{N} (x_i - \bar{x})^3}{(N - 1)(N - 2)\sigma^3} \qquad (5\text{–}14)$$

The use of log values in the log-normal distribution tends to reduce the skewness because those values of the frequency factor K between 0 and 1 lead to negative logarithms. Thus, data that range from 0 to $+\infty$ are transformed to a log distribution that ranges from $-\infty$ to $+\infty$. Even then, some skew usually remains. To determine the skew when using log values, Eq. 5–14 is replaced by

$$a_{\log x} = \frac{N \sum_{i=1}^{N} (\log x_i - \overline{\log x})^3}{(N - 1)(N - 2)\sigma^3_{\log x}} \qquad (5\text{–}15)$$

Table 5–3 FREQUENCY FACTOR *K* FOR THE LOG PEARSON TYPE III DISTRIBUTION

Skew	Recurrence interval (years)								
	1.010	1.111	2	5	10	25	50	100	200
	Probability of exceedance (%)								
a	99	90	50	20	10	4	2	1	0.5
				Positive Skew					
3.0	−0.667	−0.660	−0.396	0.420	1.180	2.278	3.152	4.051	4.970
2.5	−0.799	−0.771	−0.360	0.518	1.250	2.262	3.048	3.845	4.652
2.0	−0.990	−0.895	−0.307	0.609	1.302	2.219	2.912	3.605	4.298
1.8	−1.078	−0.945	−0.282	0.643	1.318	2.193	2.848	3.499	4.147
1.6	−1.197	−0.994	−0.254	0.675	1.329	2.163	2.780	3.388	3.990
1.4	−1.318	−1.041	−0.225	0.705	1.337	2.128	2.706	3.271	3.828
1.2	−1.449	−1.086	−0.195	0.732	1.340	2.087	2.626	3.149	3.661
1.0	−1.588	−1.128	−0.164	0.758	1.340	2.043	2.542	3.022	3.489
0.9	−1.660	−1.147	−0.148	0.769	1.339	2.018	2.498	2.957	3.401
0.8	−1.733	−1.166	−0.132	0.780	1.336	1.993	2.453	2.891	3.312
0.7	−1.806	−1.183	−0.116	0.790	1.333	1.967	2.407	2.824	3.223
0.6	−1.880	−1.200	−0.099	0.800	1.328	1.939	2.359	2.755	3.132
0.5	−1.955	−1.216	−0.083	0.808	1.323	1.910	2.311	2.686	3.041
0.4	−2.029	−1.231	−0.066	0.816	1.317	1.880	2.261	2.615	2.949
0.3	−2.104	−1.245	−0.050	0.824	1.309	1.849	2.211	2.544	2.856
0.2	−2.178	−1.258	−0.033	0.830	1.301	1.818	2.159	2.472	2.763
0.1	−2.252	−1.270	−0.017	0.836	1.292	1.785	2.107	2.400	2.670
0.0	−2.326	−1.282	0.000	0.842	1.282	1.751	2.054	2.326	2.576
				Negative Skew					
−0.1	−2.400	−1.292	0.017	0.846	1.270	1.716	2.000	2.252	2.482
−0.2	−2.472	−1.301	0.033	0.850	1.258	1.680	1.945	2.178	2.388
−0.3	−2.544	−1.309	0.050	0.853	1.245	1.643	1.890	2.104	2.294
−0.4	−2.615	−1.317	0.066	0.855	1.231	1.606	1.834	2.029	2.201
−0.5	−2.686	−1.323	0.083	0.856	1.216	1.567	1.777	1.955	2.108
−0.6	−2.755	−1.328	0.099	0.857	1.200	1.528	1.720	1.880	2.106
−0.7	−2.824	−1.333	0.116	0.857	1.183	1.488	1.663	1.806	1.926
−0.8	−2.891	−1.336	0.132	0.856	1.166	1.448	1.606	1.733	1.837
−0.9	−2.957	−1.339	0.148	0.854	1.147	1.407	1.549	1.660	1.749
−1.0	−3.022	−1.340	0.164	0.852	1.128	1.366	1.492	1.588	1.664
−1.2	−3.149	−1.340	0.195	0.844	1.086	1.282	1.379	1.449	1.501
−1.4	−3.271	−1.337	0.225	0.832	1.041	1.198	1.207	1.318	1.351
−1.6	−3.388	−1.329	0.254	0.817	0.994	1.116	1.166	1.197	1.216
−1.8	−3.499	−1.318	0.282	0.799	0.945	1.035	1.069	1.087	1.097
−2.0	−3.605	−1.302	0.307	0.777	0.895	0.959	0.980	0.990	0.995
−2.5	−3.845	−1.250	0.360	0.711	0.771	0.739	0.798	0.799	0.800
−3.0	−4.051	−1.180	0.396	0.636	0.660	0.666	0.666	0.667	0.667

The normal and log-normal distributions assume zero skew. If some skew exists in the data, these distributions cannot be expected to match the data set exactly.

The log Pearson Type III distribution was developed to improve the fit.[1] Instead of using just the two parameters of mean and standard deviation, this distribution is also based on the magnitude of the skew. Equations 5–11 and 5–12 are used to obtain the mean and standard deviation, and Eq. 5–15 provides the coefficient of skewness. As with the log-normal distribution, Eq. 5–13 is used to define the frequency factor. In this case, however, the relationship between the probability p and the frequency factor K also includes the skew a. The numerical values are provided in Table 5–3.

Note in this table that the line corresponding to zero skew ($a = 0$) provides the log-normal distribution described previously. Example 5–2, which follows, illustrates the use of the table and equations. The significance of this as well as the other distributions will be considered further in conjunction with the statistical analysis of an actual data set in the next section.

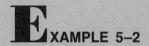XAMPLE 5–2

It has been determined that the mean of the logarithms of the annual series of river discharges is 2.700 (which corresponds to a geometric mean for the peak flows of 501 m^3/s). In addition, the standard deviation of the same log values is 0.65. Determine the discharge with a 100-year return period if the coefficient of skew is (1) -0.4, (2) 0, and (3) $+0.4$.

Solution

Equation 5–9 becomes

$$\log x = 2.7 + 0.65\,K$$

With recourse to Table 5–3, the values of K corresponding to a recurrence interval of 100 years and coefficients of skew of -0.4, 0, and $+0.4$ are 2.029, 2.326, and 2.615, respectively. Thus, the three discharges are found as follows: (1) Log $Q = 4.019$ or $Q = 10{,}400$ m^3/s; (2) Log $Q = 4.212$ or $Q = 16{,}300$ m^3/s; (3) Log $Q = 4.400$ or $Q = 25{,}100$ m^3/s.

[1] "A Uniform Technique for Determining Flood Flow Frequencies," *Bulletin 15* U.S. Water Resources Council, December 1967.

> The foregoing example clearly indicates that a large amount of skew has a pronounced effect on the magnitude of rare events. Note that the distribution specified in part (2) is log-normal since the skew is zero. If either the probability of exceedance or the recurrence interval is required for a given discharge, interpolation is required. The accuracy of this process can be improved by plotting the values of K versus log t_p for the given value of skew coefficient a. A reasonably accurate value of K can then be read from this plot.

Extreme Value Distribution

The final distribution is based on the theory of extreme values. Since an annual series is composed of the peak (or extreme) values for each year, this method of analysis should be appropriate. This distribution is sometimes called the Gumbel distribution after the first person to fit extreme value theory to hydrologic data.[2] As with the previous distributions, the extreme value distribution may be constructed in the form of Eq. 5–9 and a frequency factor tabulated. However, the distribution equation is easily applied and the frequency factor will not be introduced.

The PDF for the extreme value distribution is

$$f(x) = \frac{\pi}{\sqrt{6}\sigma} \exp\left[-b - \exp(-b)\right] \tag{5–16a}$$

where b, called the reduced variate, is

$$b = \frac{\pi}{\sqrt{6}\sigma} (x - \bar{x} + 0.45\sigma) \tag{5–16b}$$

and $\bar{x}$ and σ are the arithmetic mean and standard deviation. The CDF of this distribution is

$$F(x) = \exp\left[-\exp(-b)\right] \tag{5–17}$$

and the usual probability of exceedance p equals $1 - F(x)$, or

$$p = 1 - \exp\left[-\exp(-b)\right] \tag{5–18}$$

The extreme value distribution may be used to relate either the probability of exceedance or the recurrence interval to the magnitude of a design parameter such as the discharge. The distribution is frequently quite similar to the log-normal distribution.

---

[2]E. J. Gumbel, "Statistical Theory of Extreme Values for Some Practical Applications," Applied Mathematical Series 33, U.S. National Bureau of Standards, 1954.

EXAMPLE 5–3

Repeat Example 5–1 using the extreme value distribution.

Solution

In part (1) the mean is 10,000 cfs and the standard deviation is 3000 cfs. Thus the reduced variate for the 20,000 cfs discharge becomes

$$b = \frac{\pi}{\sqrt{6}\ (3000)}[20,000 - 10,000 + (0.45)(3000)] = 4.852$$

and p is

$$p = 1 - \exp[-\exp(-4.852)] = 0.0078$$

This corresponds to a return period $t_p = 129$ years. In part (2) the reduced variate is

$$b = \frac{\pi}{\sqrt{6}\ (6000)}[20,000 - 10,000 + (0.45)(6000)] = 2.715$$

and

$$p = 1 - \exp[-\exp(-2.715)] = 0.0641$$

In this case the return period becomes $t_p = 15.6$ years. A comparison of the results in Examples 5–1 and 5–3 indicates a large disagreement. The various distributions display markedly different characteristics, even with the same mean and standard deviation. The fault in this case would most likely lie in the inability of the normal distribution to fit the hydrologic data.

5–3 STATISTICAL ANALYSIS

In this section statistical methods will be used to analyze hydrologic data. The results of this analysis will be combined with the probability distributions discussed in the previous section with the goal of establishing design values. The results must often be based on limited data, although the accuracy will drop accordingly. We will concentrate on the annual series defined in Section 5–2. This is a series that uses the maximum value from each year. The drawback to the annual series is that some years may have several large peaks, while the peak value in other years may be of a much lower magnitude. Regardless, the large second and third peaks from a given year must be discarded and the lower peak value included if it is the largest for the year. This problem is avoided through the use of the *partial duration series,* which is based on all values above a selected cutoff value. More effort is required to obtain the data and the results are usually similar, particularly for the rarer events. Consequently, the annual series is recommended when the peak values are of interest. The partial duration series, on the other hand, is best for the analysis of the more frequent occurrences, such as recurrence intervals of less than one year.

A typical set of peak river discharges is given in Table 5–4. This data, provided by the U.S. Geological Survey, is for the Big Sioux River at Akron, Iowa. Included in this table are the maximum rates of flow for each year of record and the additional peaks that have a discharge in excess of 9000 cfs. Thus the data are suitable for either an annual series or a partial duration series analysis.

The data in Table 5–4 will be used as an illustrative example throughout this section. Table 5–5 contains a portion of the statistical summary required to analyze the data. Repeated reference will be made to this information. The analysis can be carried out by simple computer programs and the reader is advised to make use of either an existing program or to write his or her own. Much of the statistical analysis of the data can be completed without arraying the data, but if we wish to compare it with a probability distribution, the data must be arrayed so that a best estimate of p and t_p can be obtained. The annual series has been selected from Table 5–4 and arranged in descending order in Table 5–5. Column 1 is the rank m of each discharge with rank 1 associated with the largest value, and column 2 contains the respective ranked discharges.

Table 5–4 PEAK DISCHARGE DATA FOR THE BIG SIOUX RIVER AT AKRON, IOWA.

Year	Peak Q (cfs)	Year	Peak Q (cfs)	Year	Peak Q (cfs)	Year	Peak Q (cfs)
1929	20,800	1944	15,900	1954	21,700		13,100
1930	3740		11,600		15,600	1970	8580
1931	1390		11,100	1955	4940	1971	7310
1932	16,900		9840	1956	1840	1972	10,500
1933	14,200	1945	12,300	1957	19,400		10,200
1934	10,600		9820	1958	1120	1973	12,500
1935	3000	1946	8970	1959	8430	1974	3000
1936	18,000	1947	10,500	1960	49,500	1975	2920
1937	5760	1948	10,800	1961	9050	1976	3250
1938	12,700	1949	11,400	1962	54,300	1977	5270
	11,200		11,400		9010	1978	18,200
	9800		9170	1963	1650	1979	30,500
1939	6300	1950	5450	1964	2540		13,100
1940	11,700	1951	28,800	1965	21,000		12,600
1941	5820	1952	33,000	1966	16,500	1980	8730
1942	21,400		9650	1967	5300	1981	3180
	16,600		16,500	1968	635		
1943	12,000	1953	21,800	1969	80,800		

Table 5–5 STATISTICAL ANALYSIS OF THE ANNUAL SERIES FOR THE BIG SIOUX RIVER AT AKRON, IOWA.

Rank (m) (1)	Flow rate (cfs) (2)	t_p (yr) (3)	p (%≥) (4)	$\log x$ (5)	$(\log x - \overline{\log x})$ (6)	$(\log x - \overline{\log x})^2$ (7)	$(\log x - \overline{\log x})^3$ (8)
1	80,800	54.00	1.85	4.907	0.9582	0.9182	0.8799
2	54,300	27.00	3.70	4.735	0.7856	0.6172	0.4849
3	49,500	18.00	5.56	4.695	0.7454	0.5557	0.4142
4	33,000	13.50	7.41	4.519	0.5693	0.3241	0.1845
5	30,500	10.80	9.26	4.484	0.5351	0.2864	0.1532
6	28,800	9.00	11.11	4.459	0.5102	0.2603	0.1328
7	21,800	7.71	12.96	4.338	0.3893	0.1515	0.0590
8	21,700	6.75	14.81	4.336	0.3873	0.1500	0.0581
9	21,400	6.00	16.67	4.330	0.3812	0.1453	0.0554
10	21,000	5.40	18.52	4.322	0.3730	0.1392	0.0519
11	20,800	4.91	20.37	4.318	0.3689	0.1361	0.0502
12	19,400	4.50	22.22	4.288	0.3386	0.1147	0.0388
13	18,200	4.15	24.07	4.260	0.3109	0.0967	0.0300
14	18,000	3.86	25.93	4.255	0.3061	0.0937	0.0287
15	16,900	3.60	27.78	4.228	0.2787	0.0777	0.0216
16	16,500	3.38	29.63	4.217	0.2683	0.0720	0.0193
17	15,900	3.18	31.48	4.201	0.2522	0.0636	0.0160
18	14,200	3.00	33.33	4.152	0.2031	0.0413	0.0084
19	12,700	2.84	35.19	4.104	0.1546	0.0239	0.0037
20	12,500	2.70	37.04	4.097	0.1477	0.0218	0.0032
21	12,300	2.57	38.89	4.090	0.1407	0.0198	0.0028
22	12,000	2.45	40.74	4.079	0.1300	0.0169	0.0022
23	11,700	2.35	42.59	4.068	0.1190	0.0142	0.0017
24	11.400	2.25	44.44	4.057	0.1077	0.0116	0.0013
25	10,800	2.16	46.30	4.033	0.0842	0.0071	0.0006
26	10,600	2.08	48.15	4.025	0.0761	0.0058	0.0004
27	10,500	2.00	50.00	4.021	0.0720	0.0052	0.0004
28	10,500	1.93	51.85	4.021	0.0720	0.0052	0.0004
29	9050	1.86	53.70	3.957	0.0075	0.0001	0.0000
30	8970	1.80	55.56	3.953	0.0036	0.0000	0.0000
31	8730	1.74	57.41	3.941	−0.0082	0.0001	−0.0000
32	8580	1.69	59.26	3.933	−0.0157	0.0002	−0.0000
33	8430	1.64	61.11	3.926	−0.0233	0.0005	−0.0001
34	7310	1.59	62.96	3.864	−0.0853	0.0073	−0.0006
35	6300	1.54	64.81	3.799	−0.1498	0.0225	−0.0034
36	5820	1.50	66.67	3.765	−0.1843	0.0339	−0.0063
37	5760	1.46	68.52	3.760	−0.1888	0.0356	−0.0067
38	5450	1.42	70.37	3.736	−0.2128	0.0453	−0.0096
39	5300	1.38	72.22	3.724	−0.2249	0.0506	−0.0114
40	5270	1.35	74.07	3.722	−0.2274	0.0517	−0.0118
41	4940	1.32	75.93	3.694	−0.2554	0.0653	−0.0167

Table 5–5 STATISTICAL ANALYSIS OF THE ANNUAL SERIES FOR THE BIG SIOUX RIVER AT AKRON, IOWA. (continued)

Rank (m)	Flow rate (cfs)	t_p (yr)	p (%≥)	$\log x$	$(\log x - \overline{\log x})$	$(\log x - \overline{\log x})^2$	$(\log x - \overline{\log x})^3$
(1)	(2)	(3)	(4)	(5)	(6)	(7)	(8)
42	3740	1.29	77.78	3.573	−0.3763	0.1416	−0.0533
43	3250	1.26	79.63	3.512	−0.4373	0.1912	−0.0836
44	3180	1.23	81.48	3.502	−0.4467	0.1996	−0.0892
45	3000	1.20	83.33	3.477	−0.4721	0.2228	−0.1052
46	3000	1.17	85.19	3.477	−0.4721	0.2228	−0.1052
47	2920	1.15	87.04	3.465	−0.4838	0.2341	−0.1132
48	2540	1.12	88.89	3.405	−0.5443	0.2963	−0.1613
49	1840	1.10	90.74	3.265	−0.6844	0.4683	−0.3205
50	1650	1.08	92.59	3.217	−0.7317	0.5354	−0.3917
51	1390	1.06	94.44	3.143	−0.8062	0.6499	−0.5239
52	1120	1.04	96.30	3.049	−0.9000	0.8099	−0.7289
53	635	1.02	98.15	2.803	−1.1464	1.3142	−1.5066

Summations for Table 5–5

$$\Sigma x = 735{,}875 \text{ (col. 2)}$$

$$\Sigma(x - \bar{x}) = 1.144 \times 10^{-4} \text{ (col. not printed)}$$

$$\Sigma(x - \bar{x})^2 = 1.0940 \times 10^{10} \text{ (col. not printed)}$$

$$\Sigma\log x = 209.306 \text{ (col. 5)}$$

$$\Sigma(\log x - \overline{\log x}) = 1.565 \times 10^{-7} \text{ (col. 6)}$$

$$\Sigma(\log x - \overline{\log x})^2 = 9.9743 \text{ (col. 7)}$$

$$\Sigma(\log x - \overline{\log x})^3 = -1.5454 \text{ (col. 8)}$$

Statistical Results for Table 5–5

$$\text{Mean of } x = 13{,}844 \text{ cfs}$$

$$\text{Standard deviation of } x = 14{,}505 \text{ cfs}$$

$$\text{Mean of } \log x = 3.949$$

$$\text{(Geometric mean of } x = 8896 \text{ cfs)}$$

$$\text{Standard deviation of } \log x = 0.4380$$

$$\text{Skew of } \log x = -0.368$$

The best estimate of the recurrence interval for N years of record is obtained by

$$t_p = \frac{N + 1}{m} \tag{5-19}$$

As before, the probability of exceedance p is the reciprocal of the return period, or

$$p = \frac{m}{N + 1} \tag{5-20}$$

Thus, a rank of 1 in a period of record of 10 years leads to a probability $p = \frac{1}{11} = 0.0909$. The use of $N + 1$ rather than N in the above equations provides the best statistical estimate for limited data sets. As N gets larger, the distinction between the use of $N + 1$ and N disappears. Return periods based on Eq. 5–19 are tabulated in column 3 of Table 5–5, and the probability that the discharge in column two will be equaled or exceeded in a given year follows in column 4. It should be noted that Eqs. 5–19 and 5–20 apply to the partial duration series as well. The rank m and the years of record N continue to have the same meaning, although N is no longer equal to the number of values in the sample.

The necessary calculations and statistical results are given at the end of Table 5–5. The arithmetic mean and standard deviation are calculated using Eqs. 5–7 and 5–8. The mean requires the adding of column 2. The standard deviation requires the calculation of $(x_i - \bar{x})^2$ values, the result of which is given although the necessary column is not included. The only other column that is not printed is that of the $x_i - \bar{x}$ values. The summation of the $x_i - \bar{x}$ values as well as the summation of $\log x_i - \overline{\log x}$ (column 6) should equal zero. The small deviation from zero that is observed for both sets of values reflects the rounding off errors. The mean, standard deviation, and coefficient of skew for the tabulation of $\log x$ are obtained from Eqs. 5–11, 5–12, and 5–15. These calculations require the summations of columns 5, 7, and 8, respectively.

A word needs to be said concerning computational methods. The use of the computer is recommended for generating a table such as Table 5–5. The ranking process, otherwise a tedious operation, can be performed using a bubble sort or other sorting routine. The remainder of the table can be evaluated by either of two processes. A two-pass algorithm was used in which the first pass through the data summed the x and $\log x$ values to calculate the two means. The means are required in the calculations necessary for Eqs. 5–8, 5–12, and 5–15. The second pass through the data performed the remaining summations. An alternative procedure involves multiplying out the equations for standard deviation and coefficient of skew so that only a single pass is required. This advantage is not a major one with modern computers and in some cases there is a loss in significant figures. This alternative is discussed in the references at the end of the chapter and will not be pursued further.

It is now possible to fit the distribution functions discussed earlier to this data set. Graph paper can be devised by adjusting the probability scale so that the distribution functions plot as a straight line. This type of graph paper is available commercially for the normal and log-normal distributions. The Big Sioux River data from

Table 5–5 are plotted on a normal probability distribution graph in Fig. 5–3. Note that the discharges are plotted on an arithmetic scale and the probabilities p and $1 - p$ are plotted to scales that are distorted so that a normal distribution would plot as a straight line. That the data do not plot as a straight line demonstrates the unsuitability of the normal distribution to fit this and, in fact, most river data. The straight line shown is based on Eq. 5–9. For the Big Sioux data this equation becomes

$$x = 13,844 + 14,505\,K$$

To plot the straight line, K may be read from Table 5–2 for selected values of p. By substituting the K values into Eq. 5–9, the corresponding values of x (or discharge in this case) are determined. The x values may be plotted versus p in Fig. 5–3 to

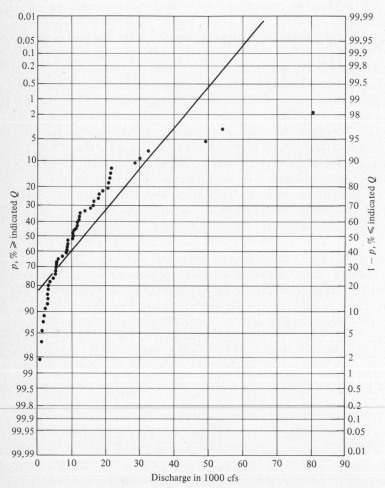

Figure 5–3 Normal distribution for the Big Sioux River at Akron, Iowa.

generate the line of best fit shown. The actual plotted points are not shown.

In the same way, a log-normal distribution will plot as a straight line on graph paper in which the probability scales are identical to those of the normal distribution, but the *x* values are plotted on a log scale. This has been done for the Big Sioux data in Fig. 5–4. The straight line shown is the best fit based on a log-normal distribution. Equation 5–13 becomes

$$\log x = 3.949 + 0.4380K$$

Here, values of *K* as a function of *p* are read from either Table 5–2 or Table 5–3 using the line of zero skew.

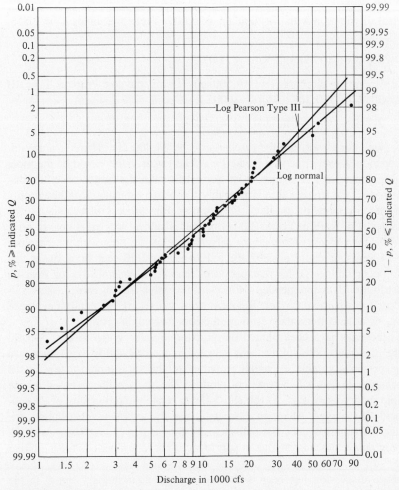

Figure 5–4 Log-normal and log Pearson Type III distributions for the Big Sioux River at Akron, Iowa.

The curved line in Fig. 5–4, representing the log Pearson Type III distribution, is obtained in exactly the same way, except that the values of K from Table 5–3 are based on the calculated coefficient of skew, $a = -0.368$. Either of the two lines shown is a marked improvement over the previous normal distribution. A somewhat philosophical, but nevertheless important, question can be raised at this point, namely, which of the two lines gives the better estimate of large discharges? Neither of the lines can be considered a perfect fit. The curved line based on the log Pearson Type III distribution fits the entire range of data better than the straight line of the log-normal distribution. However, the latter better represents the three largest flow rates, and these flows are in the area of greatest interest. If it were necessary to estimate the discharge with a return period of 200 years, which line should you select and what would be the resulting discharge? Although the lines could be extended to reach $t_p = 200$ years, we will instead return to the previous predictive equation. For the log-normal distribution with $a = 0$ and $p = 0.005$, Table 5–3 yields $K = 2.576$. Thus,

$$\log x = 3.949 + (0.4380)(2.576) = 5.0773$$

and the corresponding discharge is 119,500 cfs. Using the log Pearson Type III distribution with $a = -0.368$ and $p = 0.005$ leads to $K = 2.231$ and

$$\log x = 3.949 + (0.4380)(2.231) = 4.9262$$

After solving for x, we find the discharge to be 84,400 cfs. There is a considerable discrepancy between the two predictions. The larger value would be the more conservative estimate and it is also more consistent with the larger observed discharges. However, the lower value is probably the better estimate. It is based on a distribution that fits more of the data. Furthermore, the process of calculating the large discharges (Eq. 5–19 or Eq. 5–20) is one that is subject to a great deal of error in itself. While the process identified the largest flow as the flow with a recurrence interval of 54 years, it could possibly be a 100-year or 200-year flood event that happened to occur during the 53-year period of record. In general, the log Pearson Type III distribution is recommended. However, excessive and unrealistic skew values may occur for a particular river. Regional skew values are frequently developed to overcome, or at least minimize, estimation errors that might result.

Finally, the data have been plotted on an extreme value graph in Fig. 5–5. In this case, the recurrence interval has been selected as the probability axis rather than p, as in the previous graphs. To avoid crowding of the data points, a few of the smaller discharges in Table 5–5 were not plotted. The straight line is based on the extreme value theory using Eqs. 5–16b and 5–18 to relate the probability of exceedance to specific discharges. Although a better fit than the arithmetic distribution, the extreme value distribution does not satisfactorily fit the Big Sioux River data. As discussed previously, this may be due in part to incorrect return periods for the larger discharges.

This entire discussion has been with respect to U.S. customary units. In the United States, the hydrologic records throughout the years have generally been recorded and reported in these units. However, the analysis of flow rates in SI units would be identical in all respects.

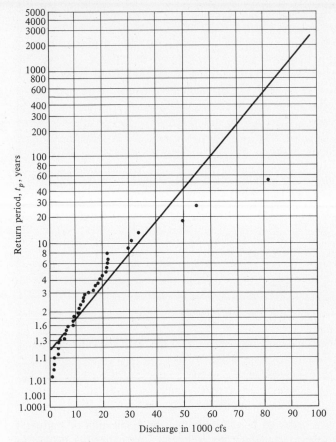

Figure 5–5 Extreme value distribution for the Big Sioux River at Akron, Iowa.

EXAMPLE 5–4

Use the four distribution methods to predict the 50-year flood on the Big Sioux at Akron, Iowa.

Solution

(1) Normal probability distribution: For $t_p = 50$ yr or $p = 0.02$, Table 5–2 yields $K = 2.054$. Substituting into Eq. 5–9,

$$x = 13{,}844 + (2.054)(14{,}505) = 43{,}600 \text{ cfs}$$

(2) Log-normal distribution: Using Table 5–2 or Table 5–3 with $a = 0$, we get $K = 2.054$. From Eq. 5–13

$\log x = 3.949 + (2.054)(0.4380) = 4.849$

Therefore,

$x = 10^{4.849} = 70,600$ cfs

(3) Log Pearson Type III distribution: Using Table 5–3 with $a = -0.368$, we get $K = 1.852$. From Eq. 5–13

$\log x = 3.949 + (1.852)(0.4380) = 4.760$

Therefore, $x = 57,600$ cfs

(4) Extreme value distribution: Equation 5–18 may be solved for the reduced variate b,

$b = -\ln[-\ln(1 - p)]$

$\quad = -\ln[-\ln(1 - 0.02)] = 3.902$

Thus, Eq. 5–16b becomes

$$3.902 = \frac{\pi}{\sqrt{6}\,(14,505)}\,[x - 13,844 + (0.45)(14,505)]$$

Solving, $x = 51,400$ cfs.

Note, that each of these results may alternatively be read from the plotted distribution functions.

The reader should have observed by this time that the application of statistical methods is not an exact science. Although the techniques are extremely helpful in the interpretation of hydrologic data and the prediction of design values, the engineer must remain aware of the many limitations involved. These different statistical methods can be applied to almost any type of hydrologic data. Not only are flood conditions important, but statistical methods are useful for the analysis of drought problems. When dealing with rainfall, the intensity of the precipitation as well as the overall quantity is important. Consequently, additional considerations are involved. For more information on these topics, reference should be made to an hydrology textbook.

 PROBLEMS

Section 5–2

5–1. With a single die, what is the probability of rolling a 1 six times in a row? What is the probability of not rolling a 1 during the course of six rolls of the die?

5–2. What is the probability of rolling a number equal to or greater than 3 on each of six rolls of a single die? What is the probability of rolling a number less than 3 on each of six rolls of the die?

5–3. A flood discharge has a recurrence interval of 50 years. (1) What is the probability of a flood of that magnitude or greater occurring during the next year? (2) What is the probability of it not occurring during the next year? (3) What is the probability of it occurring during the next 30 years? (4) What is the probability of it not occurring during the next 30 years? (5) What is the probability of it occurring during the next 50 years? (6) What is the probability of it occurring during the next 100 years?

5–4. Repeat Prob. 5–3 for a discharge with a recurrence interval of 30 years.

5–5. Repeat Prob. 5–3 for a discharge with a recurrence interval of 100 years.

5–6. Repeat Prob. 5–3 for a discharge with a recurrence interval of 10 years.

5–7. A discharge equal to or greater than 20,000 m^3/s has a 75 percent chance of occurring during a interval of 25 years. (1) What is the probability of it occurring during any one year? (2) What is the probability of it occurring during any 10 years? (3) What is the probability of it not occurring during any 50 years?

5–8. Repeat Prob. 5–7 for a flow rate that has a 40 percent chance of occurrence during an interval of 25 years.

5–9. Repeat Prob. 5–7 for a flow rate that has a 25 percent chance of occurrence during an interval of 100 years.

5–10. Assume that a stream is regulated by a reservoir so that all discharges fall between 200 and 700 cfs and any discharge within that range has an equal chance of occurrence. (1) What is the probability that the discharge at any instant will equal or exceed 350 cfs? (2) What is the probability that the discharge will fall within the range between 250 and 350 cfs? (3) What is the probability that the discharge will fall within the range between 150 and 350 cfs? (4) What is the probability that the discharge will equal 400 cfs?

5–11. A 40-year annual series of peak river discharges with the mean equal to 15,000 cfs and standard deviation equal to 7500 cfs approximates a normal distribution. (1) Determine the discharge with a return period of 10 years. (2) Determine the discharge with a return period of 20 years. (3) What is the return period that corresponds to a discharge of 20,000 cfs? (4) What is the probability that a discharge equal to or greater than 20,000 cfs will occur during a 5-year period?

5–12. Repeat Prob. 5–11 if the annual series data fit the extreme value theory.

5–13. A 50-year annual series of peak river discharges with a mean of 2000 m^3/s and standard deviation of 500 m^3/s is normally distributed. (1) What is the discharge with a recurrence interval of 50 years? (2) What is the discharge with a recurrence interval of 100 years? (3) Determine the probability that the 100-year discharge will occur in the next 50 years. (4) Determine the probability that the discharge will equal or exceed 3000 m^3/s during a 10-year period.

5–14. Repeat Prob. 5–13 using the extreme value distribution.

5–15. A 50-year annual series of peak discharges is described by the log-normal distribution. The geometric mean of the series is 60,000 cfs and the standard deviation of the log of the discharges is 0.43. (1) Determine the discharge with a recurrence interval of 25 years. (2) Determine the discharge with a recurrence interval of 100 years. (3) Find the recurrence interval associated with a discharge of 100,000 cfs. (4) What is the probability that a discharge of 100,000 cfs will be equaled or exceeded in the next 15 years?

5–16. Repeat Prob. 5–15 if the annual series is a log Pearson Type III distribution with a skew coefficient of (1) 0.2 and (2) 1.2.

5–17. Repeat Prob. 5–15 if the annual series is a log Pearson Type III distribution with a skew coefficient of (1) −0.2 and (2) −1.2.

5–18. A 60-year annual series of peak discharges in SI units fits a log-normal distribution. The mean of the log discharges is 3.0 and the standard deviation of the log discharges is 0.21. (1) What is the return period for a discharge of 2000 m^3/s? (2) What is the return period for a discharge of 3000 m^3/s? (3) Determine the probability that the discharge of 3000 m^3/s will occur in the next 100 years. (4) Find the discharge that has a return period of 50 years.

5–19. Repeat Prob. 5–18 using the log Pearson Type III distribution if the skew coefficient is (1) 0.6 and (2) 1.6.

5–20. Repeat Prob. 5–18 using the log Pearson Type III distribution if the skew coefficient is (1) 0.1 and (2) −0.1.

Section 5–3

The following four data sets are composed of the annual series discharges for four different rivers. They will be used in the subsequent problems.

	20-year record			30-year record	
Year	River 1 discharge (m^3/s)	River 2 discharge (cfs)	Year	River 3 discharge (m^3/s)	River 4 discharge (cfs)
1	5.20	12,000	1	46.7	1300
2	9.81	22,100	2	76.2	2370
3	2.30	6630	3	143.0	838
4	2.49	2710	4	44.2	1930
5	6.64	10,700	5	62.0	4000
6	15.0	16,200	6	23.3	2620
7	10.4	8620	7	48.0	3260
8	3.30	17,200	8	111.0	1440
9	6.92	46,100	9	78.7	2430
10	1.60	7630	10	229.0	990
11	4.48	4580	11	36.1	559
12	2.25	5190	12	19.4	968
13	8.81	4120	13	30.0	1750
14	4.07	8470	14	49.0	1830
15	4.88	11,900	15	36.4	1300
16	13.7	23,100	16	104.0	3910
17	6.30	8740	17	78.8	944
18	7.82	2500	18	18.1	1800

(continued)

	20-year record			30-year record	
Year	River 1 discharge (m^3/s)	River 2 discharge (cfs)	Year	River 3 discharge (m^3/s)	River 4 discharge (cfs)
19	5.05	4600	19	49.9	291
20	3.67	15400	20	43.0	1710
			21	139.0	2710
			22	47.4	1040
			23	27.1	2280
			24	47.1	597
			25	64.8	435
			26	78.2	2140
			27	34.8	2650
			28	25.2	3780
			29	33.1	2610
			30	20.0	1010

5–21. Array the discharges for River 1 and analyze the data as follows: (1) Determine the return period t_p and probability p associated with each value. (2) Calculate the arithmetic mean and standard deviation and the mean, standard deviation, and coefficient of skew for the logs of the discharges as required for the assigned portions of this problem. (3) Plot the discharges on normal probability paper and compare the results with the straight line of best fit based on the normal distribution. (4) Plot the discharges on log-normal paper and compare the results with the straight line of best fit based on the log-normal distribution. (5) Compare the plotted data in part (4) with the log Pearson Type III distribution by plotting this distribution on the graph using the calculated value of skew. (6) Plot the discharges on extreme value paper and compare the results with the straight line of best fit based on the extreme value distribution. (7) Which distribution best represents this data set?

5–22. For each of the assigned distributions in Prob. 5–21, determine (1) the return period for a discharge of 20 m^3/s and (2) the discharge with a return period of 100 years.

5–23. Repeat Prob. 5–21 using the discharges given for River 2.

5–24. For each of the assigned distributions in Prob. 5–23, determine (1) the return period for a discharge of 30,000 cfs and (2) the discharge with a return period of 10 years.

5–25. Repeat Prob. 5–21 using the discharges given for River 3.

5–26. For each of the assigned distributions in Prob. 5–25, determine (1) the return period for a discharge of 200 m^3/s and (2) the discharge with a return period of 200 years.

5–27. Repeat Prob. 5–21 using the discharges given for River 4.

5–28. For each of the assigned distributions in Prob. 5–27, determine (1) the return period for a discharge of 1600 cfs and (2) the discharge with a return period of 25 years.

5–29. The annual series of peak flow rates for a river is found to fit a log-normal distribution.

A discharge of 22 m³/s has a return period of 1.1 years, and a discharge of 200 m³/s has a return period of 35 years. Determine (1) the discharge with a return period of 100 years, (2) the return period for a discharge of 250 m³/s, and (3) the probability that a flow rate of 250 m³/s will be equaled or exceeded in the next 50 years.

5–30. Repeat Prob. 5–29 if the data fit an extreme value distribution.

5–31. The annual series of peak river discharge fits a log-normal distribution. A discharge of 90,000 cfs has a return period of two years, and a discharge of 600,000 cfs has a return period of 100 years. Estimate (1) the return period for a discharge of 800,000 cfs, (2) the discharge associated with a return period of 200 years, and (3) the discharge associated with a return period of 50 years.

References

Beard, L. R., *Statistical Methods in Hydrology,* U.S. Army Corps of Engineers, 1962.

Benjamin, J. R. and C. A. Cornell, *Probability, Statistics and Decision for Civil Engineers,* New York: McGraw-Hill, 1970.

Benson, M. A., "Evolution of Methods for Evaluating the Occurrence of Floods," U.S. Geology Survey *Water Supply Paper* 1580–A, 1962.

Chow, V. T., "Statistical and Probability Analysis of Hydrologic Data," Chapter 8, *Handbook of Applied Hydrology,* V. T. Chow (ed.), New York: McGraw-Hill, 1964.

Dalrymple, T., "Flood Frequency Analysis," U.S. Geological Survey *Water Supply Paper* 1543–A, 1960.

Eagleson, P. S., *Dynamic Hydrology,* New York: McGraw-Hill, 1970.

Hjelmfelt, A. T. and J. J. Cassidy, *Hydrology for Engineers and Planners,* Ames, IA: Iowa State University Press, 1975.

Linsley, R. K., M. A. Kohler, and J. L. H. Paulhus, *Hydrology for Engineers,* 3d ed. New York: McGraw-Hill, 1982.

Rodda, J. C., R. A. Downing, and F. M. Law, *Systematic Hydrology,* London: Newnes-Butterworth, 1976.

Viessman, W., T. E. Harbaugh, and J. W. Knapp, *Introduction to Hydrology,* New York: Intext Educational Publishers, 1972.

chapter 6

Pipelines

Figure 6–0 A section of the Trans-Alaska pipeline being lowered into a ditch. The pipeline system was designated the Outstanding Civil Engineering Achievement of 1978 by the American Society of Civil Engineers (courtesy of Alyeska Pipeline Service Company).

6

6-1 INTRODUCTION

This chapter will deal almost exclusively with pipes or conduits that are flowing full. Pipes flowing partially full will be discussed in Chapter 7 and culverts that may or may not flow full will be covered in Chapter 11. A free surface will not exist in the pipe itself, and the Froude number, which is important to open channel flows, will accordingly have no significance. The Reynolds number will be the significant dimensionless parameter, but its importance will also diminish at high Reynolds numbers as the flow becomes independent of viscous effects. An analysis of steady flow will be considered first. As usual, an understanding of the basic principles will be assumed, although they will usually be summarized or stated. Many different steady-flow applications will be considered. This will be followed by selected unsteady-flow problems. The subject of flow measurement in pipelines will be introduced but not covered extensively. The chapter will conclude with a detailed study of the hydraulic forces acting on a pipe. Considerable emphasis will be placed on water hammer problems.

6-2 HYDRAULICS OF STEADY FLOW IN CLOSED CONDUITS

The hydraulics of steady flow in closed conduits are governed by the continuity and energy equations. With reference to Fig. 6-1, the continuity equation states that the discharge remains constant from section to section, thus,

$$Q = V_1A_1 = V_2A_2 \tag{6-1}$$

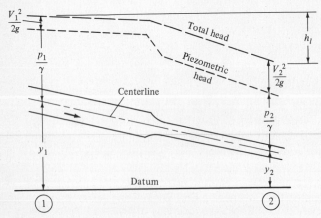

Figure 6-1 Definitional sketch for pipe hydraulics.

The energy equation, derived from the work-energy principle, may be stated as follows

$$\frac{p_1}{\gamma} + y_1 + \frac{V_1^2}{2g} = \frac{p_2}{\gamma} + y_2 + \frac{V_2^2}{2g} + h_l \tag{6–2}$$

The first three terms on either side of the equality sign are the *pressure head, elevation head,* and *velocity head,* respectively. The final term, h_l, is the *head loss* (or decrease in total head) due to friction.

Head Loss

The most general equation for head loss is the Darcy-Weisbach equation

$$h_l = f\frac{L}{D}\frac{V^2}{2g} \tag{6–3}$$

Here L is the pipe length, D is the diameter, and f is the Darcy-Weisbach resistance coefficient. This coefficient is a function of the pipe Reynolds number, VD/ν, and the relative roughness of the pipe, k/D, and may be obtained from the well-known Moody diagram[1] given in Fig. 6–2. Typical values of the absolute roughness k (in U.S. customary units) are given in Table 6–1.

Equation 6–3 can be used along with Fig. 6–2 to obtain a direct solution for the head loss. To determine the discharge given the other parameters requires an iterative solution, but assuming an f value based on the given value of k/D leads to rapid convergence. The design problem of estimating the necessary pipe size to deliver a specified flow rate under given conditions is best handled by assuming an f value and repeatedly solving for D until convergence of f values is achieved. A convenient approach to this problem is illustrated in Example 6–2.

The Moody diagram is in fact a graphic solution to

$$\frac{1}{\sqrt{f}} = -2\log\left(\frac{k}{3.7\,D} + \frac{2.51}{\mathrm{Re}\,\sqrt{f}}\right) \tag{6–4}$$

a complex equation obtained by Colebrook and White[2] that reasonably fits a broad range of smooth to rough pipe data. The resistance coefficient may be replaced using the Darcy-Weisbach equation (Eq. 6–3) to yield an explicit equation for V,

$$V = -2\sqrt{2gDh_l/L}\,\log\left(\frac{k}{3.7D} + \frac{2.51\,\nu}{D\sqrt{2gDh_l/L}}\right) \tag{6–5}$$

Alternatively, by further multiplying by the pipe area A, an explicit expression for the corresponding discharge results,

$$Q = -2A\sqrt{2gDh_l/L}\,\log\left(\frac{k}{3.7D} + \frac{2.51\,\nu}{D\sqrt{2gDh_l/L}}\right) \tag{6–6}$$

[1]L. F. Moody, "Friction Factors for Pipe Flow," *Trans.* ASME, Vol. 66, 1944.
[2]C. F. Colebrook, "Turbulent Flow in Pipes with Particular Reference to the Transition Region between the Smooth and Rough Pipe Laws," *Journal* Inst. Civil Engrs., London, Vol. 11, 1939.

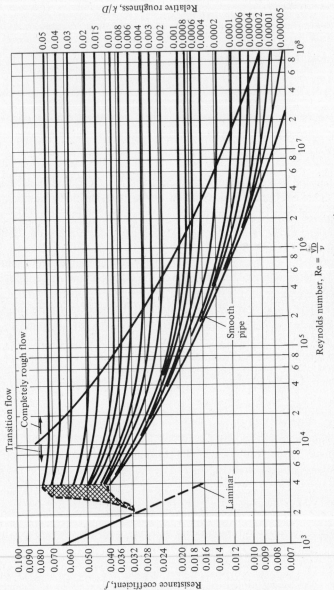

Figure 6–2 Moody resistance diagram for the Darcy-Weisbach equation.

Table 6–1 ABSOLUTE ROUGHNESS, k[1]

Material	k(ft)
Corrugated metal pipe	0.1–0.2
Riveted steel	0.003–0.03
Concrete	0.001–0.01
Wood stave	0.0006–0.003
Cast iron	0.00085
Galvanized iron	0.0005
Asphalted cast iron	0.0004
Commercial steel, wrought iron	0.00015
Drawn tubing	0.000005

[1]When using SI units, multiply k values by 0.3048.

This provides a direct solution to the discharge as opposed to the previously mentioned iterative solution. Although not attractive for hand calculations, it may be readily programmed and will therefore simplify certain types of computer algorithms.

The Darcy-Weisbach equation is valid for the laminar or turbulent pipe flow of any Newtonian fluid. A number of additional equations are also available for evaluating the head loss in a pipe due to friction. These tend to be more empirical in nature and are limited to the turbulent flow of water. This limitation is rarely restrictive in hydraulic engineering. Their main advantage is generally their ease of application. Although they are less accurate than the Darcy-Weisbach equation, many hydraulic engineering applications do not require or justify the higher accuracy anyway.

One such empirical equation is the *Hazen-Williams* equation:

$$V = 1.32 \, C_H R^{0.63} S^{0.54} \tag{6–7a}$$

or

$$V = 0.85 \, C_H R^{0.63} S^{0.54} \tag{6–7b}$$

for use with U.S. customary or SI units, respectively. In either case, C_H is a discharge coefficient with typical values given in Table 6–2, R is the hydraulic radius (equal to $D/4$ in a full pipe), and S is the slope of the total head or energy grade line, that is, the ratio of the head loss to the pipe length. There are both nomograph and tabular solutions of the Hazen-Williams equation that provide rapid solutions to head loss/discharge problems.

The Manning equation, to be discussed in more detail in conjunction with open-channel flow, may also be used to analyze the turbulent flow of water in pipes. In terms of average velocity, it may be written as

$$V = \frac{1.49}{n} R^{2/3} S^{1/2} \tag{6–8a}$$

Table 6–2 HAZEN-WILLIAMS COEFFICIENT FOR PIPES, C_H

Pipe description or material	C_H
Extremely smooth and straight	140
Asbestos-cement	140
Steel	
New, unlined	140–150
New, riveted	110
Old, riveted	95
Very smooth	130
Cast iron	
New, unlined	130
5-year-old	120
10-year-old	110
20-year-old	100
40-year-old	65–80
Smooth wood or masonry	120
Vitrified clay	110
Ordinary brick	100
Old pipe in bad condition	60–80

in U.S. customary units, and

$$V = \frac{1}{n}R^{2/3}S^{1/2} \tag{6–8b}$$

in SI units. Here R and S are as defined with respect to Eq. 6–7, and n is called the Manning n, a resistance coefficient with typical pipe values given in Table 6–3.

Table 6–3 MANNING n FOR PIPES

Pipe material	n
Brass, copper, or glass	0.009–0.013
Wood stave	0.010–0.013
Concrete	
Smooth	0.011–0.013
Rough	0.013–0.017
Vitrified or common clay	0.011–0.017
Cast iron	
Clean, coated	0.011–0.013
Clean, uncoated	0.012–0.015
Dirty or tuberculated	0.015–0.035
Brick (cement mortar)	0.012–0.017
Wrought iron, commercial steel	0.012–0.017
Riveted steel	0.013–0.017
Corrugated metal	0.020–0.024

EXAMPLE 6–1

Determine the head loss and pressure drop in 3000 ft of horizontal 3-in.-diameter galvanized iron pipe if the discharge is 0.5 cfs. Assume water at 60°F.

Solution

From Table 6–1, the absolute roughness $k = 0.0005$ ft. Thus the relative roughness is $k/D = 0.0005/0.25 = 0.002$. Also, $v = 1.217 \times 10^{-5}$ ft². The average velocity is

$$V = \frac{Q}{A} = \frac{0.5}{(\pi/4)(0.25)^2} = 10.19 \text{ ft/s}$$

and

$$\text{Re} = \frac{VD}{v} = \frac{(10.19)(0.25)}{1.217 \times 10^{-5}} = 2.09 \times 10^5$$

From the Moody diagram, $f = 0.024$ so that Eq. 6–3 yields

$$h_l = \frac{(0.024)(3000)(10.19)^2}{(0.25)(2)(32.2)} = 464.4 \text{ ft}$$

or

$$\Delta_p = \gamma h_l = (62.4)(464.4) = 29,000 \text{ psf} = 201 \text{ psi}$$

EXAMPLE 6–2

What diameter of cast iron pipe would be required to ensure that a discharge of 0.20 m³/s would not cause a head loss in excess of 0.01 m/100 m of pipe length? Assume a water temperature of 20°C.

Solution

In terms of the unknown diameter D, the relative roughness is

$$\frac{k}{D} = \frac{(0.00085)(0.3048)}{D} = \frac{2.59 \times 10^{-4}}{D}$$

and the Reynolds number is

$$Re = \frac{VD}{v} = \frac{4Q}{\pi Dv} = \frac{(4)(0.20)}{\pi D(1.007 \times 10^{-6})} = \frac{2.53 \times 10^5}{D}$$

The Darcy-Weisbach equation may be rearranged as follows:

$$h_l = f \frac{L}{D} \frac{V^2}{2g} = f \frac{L}{D} \frac{Q^2}{(\pi/4)^2 D^4 (2g)}$$

or

$$D = f^{1/5} \left[\frac{8LQ^2}{\pi^2 g h_l} \right]^{1/5}$$

Entering the given numerical values

$$D = f^{1/5} \left[\frac{(8)(100)(0.20)^2}{\pi^2 (9.81)(0.01)} \right]^{1/5} = 2.01 f^{1/5}$$

Assume a trial $f = 0.020$. Using this value, the trial diameter is

$$D = (2.01)(0.020)^{1/5} = 0.92 \text{ m}$$

This corresponds to $k/D = 0.00028$ and $Re = 1.75 \times 10^5$.

An improved resistance coefficient, $f = 0.017$, may now be read from the Moody diagram. Repeating,

$$D = (2.01)(0.017)^{1/5} = 0.90 \text{ m}$$

Thus, $k/D = 0.00029$ and $Re = 2.84 \times 10^5$. Again entering the Moody diagram, $f = 0.017$, which checks the previous value. Therefore, the required pipe diameter is $D = 0.90$ m.

Minor Losses

Minor losses are those energy losses in a pipeline that result from changes in either the magnitude of the velocity or the direction of the flow. The term minor is used because they are frequently small in magnitude relative to the energy or head loss due to friction. In fact, when a pipeline is long, the minor losses can often be ignored. However, in a short pipe, the so-called minor losses may exceed the head loss due to friction.

Pipe entrances, contractions, expansions, and valves all create zones of flow separation in which energy loss occurs. For turbulent flow, the loss is usually expressed as a loss coefficient C_L times a velocity head or difference in velocity heads.

At the submerged exit of a pipe into a reservoir, all of the kinetic energy at the exit is dissipated, which corresponds to a loss coefficient of unity.

Pipe bends and elbows set up a secondary flow that persists downstream but is eventually dissipated. The energy required to initiate the secondary flow is consequently an energy loss with respect to the main flow. This head loss will also be expressed by a coefficient times the velocity head of the main flow. Typical minor losses and loss coefficients are given in Figs. 6–3 through 6–5, and Table 6–4. The list is by no means all inclusive; for example, reducing elbows are not included, and recourse must be made to more extensive pipe handbooks. Certain head losses and the respective coefficients are examined further in the problem set.

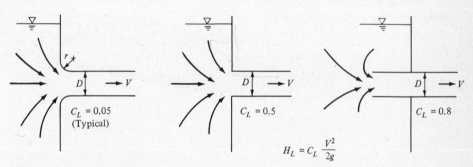

Figure 6–3 Pipe entrances.

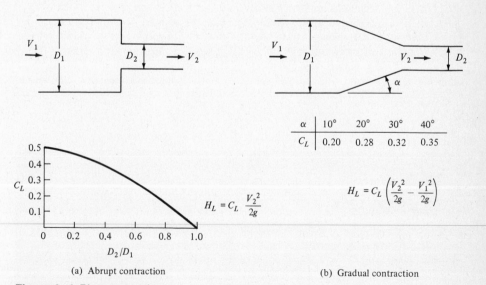

(a) Abrupt contraction (b) Gradual contraction

Figure 6–4 Pipe contractions.

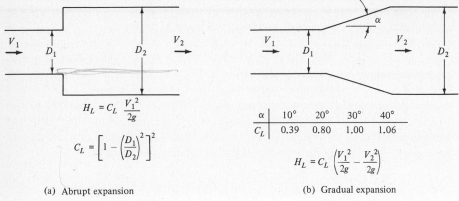

$$H_L = C_L \frac{V_1^2}{2g}$$

$$C_L = \left[1 - \left(\frac{D_1}{D_2}\right)^2 \right]^2$$

(a) Abrupt expansion

α	10°	20°	30°	40°
C_L	0.39	0.80	1.00	1.06

$$H_L = C_L \left(\frac{V_1^2}{2g} - \frac{V_2^2}{2g} \right)$$

(b) Gradual expansion

Figure 6–5 Pipe expansions.

Table 6–4 LOSS COEFFICIENTS IN APPURTENANCES, $H_L = C_L \dfrac{V^2}{2g}$

90 deg bend of radius r		Fittings		Valves	
$\dfrac{r}{D}$	C_L	Type	C_L	Type	C_L
1.0	0.40	Short-radius 90 deg elbow	0.9	Gate valve:	
1.5	0.32	Medium-radius 90 deg elbow	0.8	Fully open	0.2
2.0	0.27	Long-radius 90 deg elbow	0.6	¾ open	1.0
3.0	0.22	45 deg elbow	0.4	½ open	5.6
4.0	0.20	Tee, straight through	0.3	¼ open	24.
		Tee, through side outlet	1.8	Globe valve (open)	10.
				Swing check valve	2.5 (typical)

EXAMPLE 6–3

Consider a pipe system that consists of 2000 ft of 8-in. diameter commercial steel pipe followed by 1500 ft of 6-in. diameter cast iron pipe. The system leads from a reservoir with a water level 90 ft above the outlet of the pipe. Further, the 8-in. pipe has a square-edged entrance and contains a swing check

valve followed by a fully open gate valve. Following a gradual contraction (α = 20 deg), the 6-in. pipe contains six 45 deg elbows, and two 90 deg bends, one of 1-ft radius and the other of 2-ft radius. Assume that both pipes are fully rough (i.e., base the resistance coefficient on the relative roughness only) and estimate the discharge through the system.

Solution

Relative roughness values for the two pipes are

$$\left(\frac{k}{D}\right)_8 = \frac{0.00015}{8/12} = 0.00023$$

and

$$\left(\frac{k}{D}\right)_6 = \frac{0.00085}{6/12} = 0.0017$$

Assuming fully rough flow, Fig. 6–2 gives $f_8 = 0.014$ and $f_6 = 0.022$.
The appropriate loss coefficients are tabulated below:

Item	C_L
Entrance	0.5
Swing check valve	2.5
Gate valve (open)	0.2
Gradual contraction	0.28
45 deg elbow	0.4
90 deg bend, $r = 1$ ft	0.27
90 deg bend, $r = 2$ ft	0.20

Writing the energy equation between the reservoir surface and the outlet gives:

$$H_1 = H_2 + \text{friction losses} + \text{minor losses}$$

$$90 = \frac{V_6^2}{2g} + \left(f\frac{L}{D}\frac{V^2}{2g}\right)_8 + \left(f\frac{L}{D}\frac{V^2}{2g}\right)_6 + (0.5 + 2.5 + 0.2)\frac{V_8^2}{2g}$$

$$+ 0.28\left(\frac{V_6^2}{2g} - \frac{V_8^2}{2g}\right) + [(6)(0.4) + 0.27 + 0.20]\frac{V_6^2}{2g}$$

or

$$90 = \frac{V_6^2}{64.4} + \left[\frac{(0.014)(2000)}{8/12}\right]\frac{V_8^2}{64.4} + \left[\frac{(0.022)(1500)}{6/12}\right]\frac{V_6^2}{64.4}$$

$$+ 3.2\frac{V_8^2}{64.4} + 0.28\left(\frac{V_6^2}{64.4} - \frac{V_8^2}{64.4}\right) + 2.87\frac{V_6^2}{64.4}$$

Including continuity in the form of $V_6(6)^2 = V_8(8)^2$ and solving gives $V_6 = 8.29$ ft/s, or upon multiplying by the area, $Q = 1.63$ cfs.

Pipe Systems

Pipes in Series

One type of pipe system has already been considered in Example 6–3, namely pipes in series. Inspection of that example would indicate the two basic premises associated with pipes in series: first, continuity requires that the discharge is the same in both pipes, and second, the total head loss for the system equals the sum of the individual head losses of the respective pipes. These principles can be generalized for any number of pipes that are connected in series. It should be noted that if minor losses are included (and often they would not be), they must be referenced to the appropriate pipe size and velocity head. Similarly, the head loss due to friction must be evaluated separately for each pipe size. Any of the head-loss equations may be used. When the Darcy-Weisbach equation is used in the analysis of any type of pipe system, the resistance coefficients are frequently treated as constants (as illustrated in Example 6–3).

Pipes in Parallel

A second pipe system is composed of a number of pipes connected in parallel (Fig. 6–6). Although any number of pipes could be so connected, the development of the concepts will be based on just three pipes. In passing, note that the pipe sections upstream and downstream of the parallel pipes in Fig. 6–6 could be considered to be in series with the parallel pipe system of our immediate interest, and once the parallel pipe system is analyzed the remaining system could be treated as a number of pipes in series.

Returning to the three parallel pipes, continuity can be applied with the result

$$Q = Q_1 + Q_2 + Q_3 \tag{6–9}$$

Since each of the pipes must share common values of piezometric head at the junctions, it follows that the head loss must be identical in each of the parallel pipes. This gives a second relationship that may be expressed as

$$h_{l1} = h_{l2} = h_{l3} \tag{6–10}$$

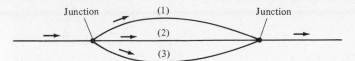

Figure 6–6 Three pipes in parallel.

If in a given problem the actual values of the piezometric head are known at the junctions (or the difference in piezometric head between the junctions is known), then the discharge in each of the parallel pipes can be obtained by treating it as a single line. Based on the selected resistance equation (Eqs. 6–3 through 6–8) and the known decrease in piezometric head or head loss, each discharge can be determined and the total discharge for the system will then be given by Eq. 6–9. The solution will involve an iterative approach only if the Darcy-Weisbach equation is used and the f values are not treated as constants.

However, the more typical problem will involve determining the flow through each of the pipes given the total discharge. In this event, the head loss itself is not known at the outset. The solution to this type of problem will be discussed using both the Darcy-Weisbach equation and the Manning equation.

If the Darcy-Weisbach equation is used, the constant head loss can be expressed in terms of the respective discharges according to

$$f_1 \frac{L_1 Q_1^2}{D_1^5 2g}\left(\frac{4}{\pi}\right)^2 = f_2 \frac{L_2 Q_2^2}{D_2^5 2g}\left(\frac{4}{\pi}\right)^2 = f_3 \frac{L_3 Q_3^2}{D_3^5 2g}\left(\frac{4}{\pi}\right)^2$$

and letting $C_i = L_i/D_i^5$ reduces the above to

$$C_1 f_1 Q_1^2 = C_2 f_2 Q_2^2 = C_3 f_3 Q_3^2 \qquad \text{(6–11)}$$

The discharge can now be obtained by solving Eqs. 6–9 and 6–11 simultaneously. Since the number of equations will increase at the same rate as the number of branches, the procedure is appropriate for any number of parallel pipes. Usually the resistance coefficients are treated as constants, a higher accuracy not being justified given the uncertainty associated with practical pipe problems. In the event that greater accuracy is desired, the f values would be assumed and the above procedure followed. After balancing Eqs. 6–9 and 6–11, the validity of the choice of f's can be checked with the Moody diagram. If the choices are incorrect, better f values would be selected based on the trial discharges and the entire process repeated until satisfactory accuracy is achieved.

In a similar fashion, but using the Manning equation (Eqs. 6–8a or 6–8b depending on the units used), the equality of the head loss between junctions gives

$$C_1 n_1^2 Q_1^2 = C_2 n_2^2 Q_2^2 = C_3 n_3^2 Q_3^2 \qquad \text{(6–12)}$$

where $C_i = L_i/D_i^{16/3}$. Here the slope S has been replaced by head loss/length and the hydraulic radius for a pipe, $D/4$, has been introduced. As before, the discharge in each branch may be determined by simultaneous solution of Eqs. 6–9 and 6–12. The procedure is illustrated in Example 6–4, which follows.

EXAMPLE 6–4

Three pipes are connected in parallel as in Fig. 6–6. Their characteristics are as follows:

Pipe no.	Diameter (m)	Length (m)	Manning n
1	0.20	1000	0.014
2	0.25	1200	0.014
3	0.30	1000	0.017

Determine the discharge in each of the pipes if the total flow rate is 0.80 m^3/s.

Solution

The values of C_i appropriate to the Manning equation are calculated as follows:

$$C_1 = \frac{1000}{(0.2)^{16/3}} = 5.34 \times 10^6$$

$$C_2 = \frac{1200}{(0.25)^{16/3}} = 1.95 \times 10^6$$

$$C_3 = \frac{1000}{(0.3)^{16/3}} = 6.15 \times 10^5$$

Applying Eq. 6–12, but using the square roots for convenience

$$(2312)(0.014)\, Q_1 = (1397)(0.014)Q_2 = (784)(0.017)Q_3$$

while continuity in the form of Eq. 6–9 gives

$$Q_1 + Q_2 + Q_3 = 0.80$$

Upon solving, $Q_1 = 0.157$ m^3/s or 19.6 percent of the flow, $Q_2 = 0.261$ m^3/s or 32.6 percent, and $Q_3 = 0.382$ m^3/s or 47.8 percent of the flow.

Branching Pipes

Another type of pipe system may be introduced by considering the three interconnected reservoirs of Fig. 6–7. Subsequently, we will generalize the analysis to include additional reservoirs and/or additional junctions. For the present consider the problem of estimating the discharge in each of the pipe sections.

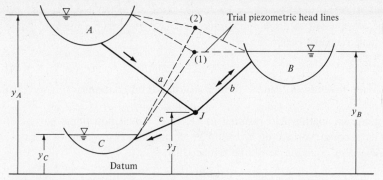

Figure 6–7 Three-branch reservoir system.

The necessary number of equations are obtained by utilizing energy and continuity concepts. However, it is difficult to incorporate the velocity head at the junction, a problem that is circumvented by assuming that it is a small quantity relative to the other energy terms and may justifiably be ignored. It is usual to make the same assumption with respect to possible minor losses. A second problem is that the flow direction in pipe *b* of Fig. 6–7 is not known at the outset. Our procedure will be to formulate the energy equation so as to determine the actual flow direction in pipe *b* and then proceed on the basis of the outcome of this first calculation. If the piezometric head (since the velocity head is neglected) at the junction is assumed equal to the elevation of reservoir surface *B*, as indicated by point 1 on Fig. 6–7, then no flow would occur in pipe *b*, and the continuity equation becomes $Q_a = Q_c$. In addition, the energy equations between *A* and *J* and *J* and *C* may be expressed as

$$y_A - h_J = h_{la} = f_a \frac{L_a}{D_a} \frac{V_a^2}{2g} = f_a \frac{L_a}{D_a^5} \left(\frac{4}{\pi}\right)^2 \frac{Q_a^2}{2g} \tag{6-13a}$$

and

$$h_J - y_C = h_{lc} = f_c \frac{L_c}{D_c} \frac{V_c^2}{2g} = f_c \frac{L_c}{D_c^5} \left(\frac{4}{\pi}\right)^2 \frac{Q_c^2}{2g} \tag{6-13b}$$

In the above, the Darcy-Weisbach equation has been chosen to estimate the head loss, and it will be left to the reader to formulate similar expressions using the Manning equation. In this first trial, the piezometric head at the junction is $h_J = y_B$. Since the various elevations are known, each of the equations may be solved for a discharge value. In the unlikely event that the two discharges are equal, the original assumption of piezometric head at the junction is correct. Otherwise, if $Q_a > Q_c$, then flow must be into reservoir *B* and the solution proceeds by assuming a new h_J at a higher elevation such as at point 2. Three energy equations are now required. In addition to Eqs. 6–13a and 6–13b, we may also write

$$h_J - h_B = h_{lb} = f_b \frac{L_b}{D_b^5} \left(\frac{4}{\pi}\right)^2 \frac{Q_b^2}{2g} \tag{6-13c}$$

Since a new piezometric head elevation has been assumed at the junction and the other elevations are known, a discharge can be obtained from each equation. These values can be tested against continuity

$$Q_a = Q_b + Q_c \tag{6-13d}$$

and if an inbalance remains, h_J can be further adjusted until continuity (Eq. 6–13d) is achieved.

If the original calculation had resulted in $Q_a < Q_c$, then flow must be out of reservoir B. The second estimate of h_J would be at point 2, below that of point 1, and the third head loss equation would be replaced by

$$h_B - h_J = h_{lb} = f_b \frac{L_b}{D_b^5} \left(\frac{4}{\pi} \right)^2 \frac{Q_b^2}{2g} \tag{6-13e}$$

The three values of discharge so obtained would be checked against

$$Q_a + Q_b = Q_c \tag{6-13f}$$

and h_J again adjusted until continuity is obtained. The procedure is demonstrated in Example 6–5 below.

EXAMPLE 6-5

Given three interconnected reservoirs with characteristics as follows:

Reservoir	Elevation (ft)	Pipe	Diameter (in.)	Length (ft)	f	$\dfrac{fL}{D^5}$
A	200	a	8	800	0.020	121.5
B	180	b	6	500	0.022	352.0
C	140	c	12	800	0.020	16.0

Determine the discharge in each pipe.

Solution

The head loss terms can be arranged to simplify repeated calculations by noting

$$h_l = f \frac{L}{D^5} \left(\frac{4}{\pi} \right)^2 \frac{Q^2}{2g} = (0.02517) \frac{fL}{D^5} Q^2$$

The flow direction in pipe b is obtained from Eqs. 6–13a and 6–13b by writing each and solving for the respective discharge.

$$200 - 180 = (121.5)(0.02517)Q_a{}^2$$

$$180 - 140 = (16.0)(0.02517)Q_c{}^2$$

yielding $Q_a = 2.56$ cfs and $Q_c = 9.97$ cfs. Since the discharge is less in pipe a, we can conclude that flow must be out of reservoir B. Trying $h_J = 150$ ft and repeating the two equations above yields

$$200 - 150 = (121.5)(0.02517)Q_a{}^2$$

$$150 - 140 = (16.0)(0.02517)Q_c{}^2$$

In addition, Eq. 6–13e gives

$$180 - 150 = (352.0)(0.02517)Q_b{}^2$$

and the three discharges become 4.04, 1.84, and 4.98 cfs in pipes a, b, and c, respectively. Comparison with Eq. 6–13f indicates that $Q_a + Q_b$ is greater than Q_c and thus the value of h_J must be increased somewhat. Further trials finally lead to a correct value of $h_J = 152.95$ ft and final discharges of $Q_a = 3.92$ cfs, $Q_b = 1.75$ cfs, and $Q_c = 5.67$ cfs.

The tediousness of the previous example encourages the use of a computer-based solution. The need becomes even more apparent as the number of reservoirs increases or more than one junction is included. A program that will handle any number of reservoirs with any reasonable number of junctions in either the U.S. customary or SI system of units is included as computer program 3 in Appendix E. In this program, the Darcy-Weisbach equation has again been chosen. Further, the restriction of constant f has been lifted, and rather than solving for the discharge as in Eqs. 6–13, the head loss has been entered directly into Eq. 6–5, the combined Darcy-Weisbach and Colebrook-White equation.

If the number of reservoirs are connected at more than one junction (see Fig. 6–8), then the elevation of the piezometric head at each junction must be estimated and continuity ultimately satisfied at each junction. The procedure, which is best accom-

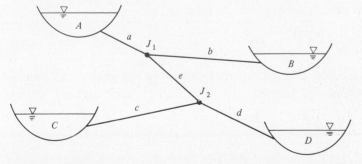

Figure 6–8 Reservoirs connected by branching pipes with multiple junctions.

plished using a computer, is to make initial estimates of the respective piezometric heads, at each junction, hold all but one constant, bring that one into balance, and then repeat the process for the remaining junctions. Since this estimate is closer than the initial estimates, convergence can be achieved by starting again at the first junction and repeating the entire process. Because of the time and difficulty involved, the solution will not be pursued herein.

Pipe Networks

The branching pipe analysis has been discussed in some detail because it represents a moderately important class of problems that may be solved by a straightforward albeit lengthy process. Another type of problem involves a pipe network such as that shown in Fig. 6–9. This is an enormously important problem because of its application to city water supply systems where a large number of such loops are encountered. The classic solution developed by Hardy Cross[3] is a relaxation method now solved rather easily by modern computer techniques. It will be considered here for its classic significance and because it remains the only reasonable hand calculation method. Other modern computer algorithms are available and presented in publications such as that by Roland Jeppson, which is referenced at the end of the chapter. These will not be discussed further as the entire subject of network analysis is often considered more in the domain of water supply rather than hydraulic engineering.

The essence of the Hardy Cross method lies in the two conditions that must be satisfied regardless of the number of loops involved. Namely, (1) the algebraic sum of the pressure drop around each loop must be zero and (2) because of continuity requirements, the net flow out of each junction must also equal zero. The procedure is an iterative one that can best be explained by example. But to summarize, an initial estimate is made of the flow in each pipe. While this should be the best possible estimate, the only necessary requirement is that the estimates satisfy continuity at each junction.

At the outset, the inflow to the system and the outflow or demand is assumed known. These are considered to apply at the junctions as in Fig. 6–9. If an inflow or outflow must be placed between existing junctions, that point can simply be treated as an additional junction. Since the diameter, length, and roughness of each pipe are known, the head loss therein may be expressed by

$$h_l = f\frac{L}{D}\frac{V^2}{2g} = f\frac{L}{D^5}\left(\frac{4}{\pi}\right)^2\frac{Q^2}{2g} = KQ^2 \tag{6–14}$$

The error in the estimated discharge of any pipe may be indicated as ΔQ, which when substituted into Eq. 6–14 gives

$$h_l = K(Q + \Delta Q)^2 = K[Q^2 + 2Q\Delta Q + (\Delta Q)^2]$$

[3]H. Cross, "Analysis of Flow in Networks of Conduits and Conductors," *Univ. of Illinois Engr. Exp. Sta. Bull. 286*, 1936.

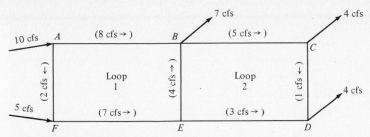

Figure 6–9 Pipe network.

Ignoring $(\Delta Q)^2$ on the basis that ΔQ is a small number, a reasonable assumption after a few iterations (if not initially) leads to

$$h_l = K\,(Q^2 + 2Q\Delta Q) \tag{6–15}$$

The head loss must be zero around any closed loop, and thus

$$\Sigma h_l = \Sigma(KQ^2) + 2\Sigma(KQ\Delta Q) = 0$$

Since continuity requires that the correction (or error) be the same for each pipe in the closed loop, this becomes

$$\Sigma h_l = \Sigma(KQ^2) + 2\Delta Q\Sigma(KQ) = 0$$

or upon solving for ΔQ

$$\Delta Q = -\frac{\Sigma(KQ^2)}{2\Sigma(KQ)} \tag{6–16a}$$

or alternatively

$$\Delta Q = -\frac{\Sigma h_l}{2\Sigma(h_l/Q)} \tag{6–16b}$$

In the above h_l is the calculated head loss in each pipe based on the estimated discharge Q. In application, consistency is important and the head loss in each of the loops will be treated in a clockwise direction. The above discharge correction ΔQ is applied to the first loop and the procedure is repeated successively loop by loop over the entire system until the correction is negligibly small and is within acceptable limits. To understand the procedure, the following example should be carefully scrutinized.

EXAMPLE 6–6

Estimate the flow in each pipe of Fig. 6–9 and the pressure at each of the junctions. The two inflows totaling 15 cfs are required to match the three demands shown at junctions *B*, *C*, and *D*. The pipe characteristics including the

Darcy-Weisbach resistance coefficients are given below (K values that are also tabulated here are calculated later in the solution).

Pipe	Length	Diameter	Resistance coefficient	Calculated
	(ft)	(in.)	f	K
AB	1000	12	0.014	0.352
BC	1000	12	0.014	0.352
AF	600	6	0.018	8.700
BE	600	8	0.016	1.835
CD	600	8	0.016	1.835
DE	1000	8	0.016	3.059
EF	1000	8	0.016	3.059

Finally, in determining the pressure at each junction, assume that each junction is at the same elevation and that the pressure at each of the outflow junctions cannot drop below 20 psi.

Solution

The K values of Eqs. 6–16 will be constant for each pipe since f values are given. If the respective pipe roughnesses were specified instead, the K values would change slightly as better estimates of the Q's were obtained. According to Eq. 6–14, K is calculated

$$K = f\frac{L}{D^5}\left(\frac{4}{\pi}\right)^2\frac{1}{2g} = \frac{8fL}{\pi^2 gD^5}$$

Thus, for pipe AB

$$K = \frac{(8)(0.014)(1000)}{\pi^2(3.22)(1)^5} = 0.352$$

This and the other K values are tabulated along with the given data above.

The required initial discharge estimate for each pipe is shown in parentheses in Fig. 6–9. Note that one value only is estimated for each loop, the rest being determined so as to satisfy continuity at each junction. The relaxation technique may now be applied alternately between loops 1 and 2. Commencing with loop 1:

Loop	Pipe	Q (cfs)	K	h_l (ft)	h_l/Q	New Q (cfs)
1	AB	8	0.352	22.53	2.82	9.96
	BE	−4	1.835	−29.37	7.34	−2.04

	EF	−7	3.059	−149.89	21.41	−5.04
	FA	−2	8.700	−34.80	17.40	−0.04
				$\Sigma = -191.53$	$\Sigma = 48.97$	

$$\Delta Q = -\frac{(-191.53)}{(2)(48.97)} = +1.96 \text{ cfs}$$

The h_l is obtained from Eq. 6–14 and the sign reflects the direction of the assumed Q. The algebraic calculations are as indicated. The correction ΔQ is finally calculated from Eq. 6–16b and the resulting new Q's are tabulated in the last column. Again, plus Q's are clockwise and negative Q's are counterclockwise. Proceeding to loop 2 and adjusting the original estimate in pipe *BE* on the basis of the new estimate from loop 1:

Loop	Pipe	Q (cfs)	K	h_l (ft)	h_l/Q	New Q (cfs)
2	BC	5	0.352	8.80	1.76	5.29
	CD	1	1.835	1.84	1.84	1.29
	DE	−3	3.059	−27.53	9.18	−2.71
	EB	2.04	1.835	7.64	3.74	2.33
				$\Sigma = -9.25$	$\Sigma = 16.52$	

$$\Delta Q = -\frac{(-9.25)}{(2)(16.52)} = +0.29 \text{ cfs}$$

The procedure is now repeated until ΔQ is acceptable.

Loop	Pipe	Q (cfs)	K	h_l (ft)	h_l/Q	New Q (cfs)
1	AB	9.96	0.352	34.92	3.51	11.08
	BE	−2.33	1.835	−9.96	4.28	−1.21
	EF	−5.04	3.059	−77.70	15.42	−3.92
	FA	−0.04	8.700	−0.01	0.35	1.08
				$\Sigma = -52.75$	$\Sigma = 23.56$	

$$\Delta Q = -\frac{(-52.75)}{(2)(23.56)} = +1.12 \text{ cfs}$$

Loop	Pipe	Q (cfs)	K	h_l (ft)	h_l/Q	New Q (cfs)
2	BC	5.29	0.352	9.85	1.86	5.52
	CD	1.29	1.835	3.05	2.37	1.52

DE	−2.71	3.059	−22.47	8.29	−2.48
EB	1.21	1.835	2.69	2.22	1.44
			Σ = −6.88	Σ = 14.74	

$$\Delta Q = -\frac{(-6.88)}{(2)(14.74)} = +0.23 \text{ cfs}$$

Loop	Pipe	Q (cfs)	K	h_l (ft)	h_l/Q	New Q (cfs)
1	AB	11.08	0.352	43.21	3.90	11.03
	BE	−1.44	1.835	−3.81	2.64	−1.49
	EF	−3.92	3.059	−47.01	11.99	−3.97
	FA	1.08	8.700	10.15	9.40	1.03
				Σ = 2.54	Σ = 27.93	

$$\Delta Q = -\frac{(2.54)}{(2)(27.93)} = -0.05 \text{ cfs}$$

Loop	Pipe	Q (cfs)	K	h_l (ft)	h_l/Q	New Q (cfs)
2	BC	5.52	0.352	10.73	1.94	5.51
	CD	1.52	1.835	4.24	2.79	1.51
	DE	−2.48	3.059	−18.81	7.59	−2.49
	EB	1.49	1.835	4.07	2.73	1.48
				Σ = 0.23	Σ = 15.05	

$$\Delta Q = -\frac{(0.23)}{(2)(15.05)} = -0.01 \text{ cfs}$$

Loop	Pipe	Q (cfs)	K	h_l (ft)	h_l/Q	New Q (cfs)
1	AB	11.03	0.352	42.82	3.88	11.03
	BE	−1.48	1.835	−4.02	2.72	−1.48
	EF	−3.79	3.059	−48.21	12.12	−3.79
	FA	1.03	8.700	9.23	8.96	1.03
				Σ = −0.18	Σ = 27.70	

$$\Delta Q = -\frac{(-0.18)}{(2)(27.70)} < 0.01 \text{ cfs}$$

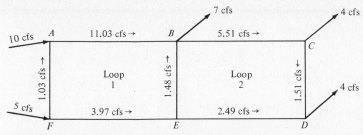

Figure 6–10 Solution to pipe network problem.

Figure 6–9 is redrawn above as Fig. 6–10 with the final discharges included. To finish the problem, the pressure drop in each pipe is tabulated along with the final discharges in the following table. The pressure drop between each junction is based on $\Delta p = \gamma h_l$, and the flow is in the direction from the first to the second letter of the pipe identifier.

Pipe	Q (cfs)	h_l (ft)	Δp (psi)
AB	11.03	42.82	18.56
BC	5.51	10.73	4.65
FA	1.03	9.23	4.00
EB	1.48	4.02	1.74
CD	1.51	4.24	1.84
FE	3.97	48.21	20.89
ED	2.49	18.81	8.15

To meet the condition that the pressure not drop below 20 psi at any of the outflow junctions, the pressure must equal or exceed that value at junction D. Thus, the pressure at the other junctions will be

Junction	Pressure (psi)
A	45.05
B	26.49
C	21.84
D	20.00
E	28.15
F	49.04

6–3 HYDRAULICS OF UNSTEADY FLOW IN PIPES

Unsteady flow will only be treated in an introductory manner. With the exception of the water hammer, which is considered later, unsteady flow will be restricted to those flows in which the head changes so slowly that the steady-state equations may be applied at any instant. Two major types of problems will be considered in this section. The first will involve flow from or between reservoirs when the head cannot be considered constant, and the second will deal with the establishment of flow from a reservoir before steady-state conditions are reached.

Varying Head Problems

A general case will be developed based on Fig. 6–11 and the approach of Featherstone and Nalluri (see reference at end of Chapter). The two reservoirs have constant horizontal cross section areas, A_1 and A_2, respectively. The connecting pipe has a length L, area A, and the discharge at any instant is Q. The reservoirs may also have prescribed, and not necessarily constant, inflows and outflows denoted by Q_1 and Q_2 as shown. At the instant shown, the reservoir levels or total heads are given by y_1 and y_2. The difference in total head or head loss H includes the friction and possibly minor losses. During a time period dt, the net result of inflow, outflow, and flow between the reservoirs is to cause a change in each of the reservoir levels and consequently in the difference in the total head values as well. These are shown in an assumed positive direction as dH_1 and dH_2. Thus, the change in total head is

$$dH = dH_1 - dH_2 \tag{6-17}$$

In addition, continuity may be expressed separately for each of the reservoirs with the result

$$Q_1 - Q = A_1 \frac{dH_1}{dt} \tag{6-18a}$$

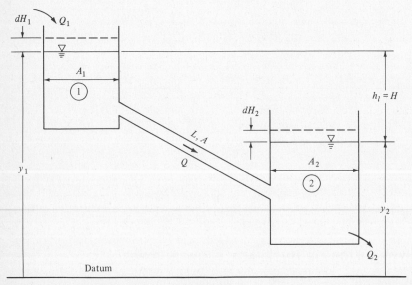

Figure 6–11 Reservoirs with varying heads.

$$Q - Q_2 = A_2\frac{dH_2}{dt} \tag{6-18b}$$

Since the change in the reservoir levels is assumed to be gradual, the head loss due to friction will be given by the Darcy-Weisbach equation and the minor losses, if any, as a sum of loss coefficient-velocity head products. The combined head loss between the two reservoirs is therefore

$$H = \left(f\frac{L}{D} + \Sigma C_{Li}\right)\frac{Q^2}{2gA^2}$$

Solving for the discharge yields

$$Q = A\sqrt{2g}\,\frac{H^{1/2}}{\left(f\dfrac{L}{D} + \Sigma C_{Li}\right)^{1/2}} = KH^{1/2} \tag{6-19}$$

Thus

$$dH = dH_1 - dH_2$$

$$= \left(\frac{Q_1 - Q}{A_1} - \frac{Q - Q_2}{A_2}\right) dt$$

$$= \left[\frac{Q_1}{A_1} - Q\left(\frac{1}{A_1} + \frac{1}{A_2}\right) + \frac{Q_2}{A_2}\right] dt$$

$$= \left[\frac{Q_1}{A_1} + \frac{Q_2}{A_2} - KH^{1/2}\left(\frac{1}{A_1} + \frac{1}{A_2}\right)\right] dt$$

$$dt = \frac{dH}{\left[\dfrac{Q_1}{A_1} + \dfrac{Q_2}{A_2} - KH^{1/2}\left(\dfrac{1}{A_1} + \dfrac{1}{A_2}\right)\right]} \tag{6-20}$$

Equation 6–20 represents the general case as stated at the outset. If inflow or outflow are zero, then the appropriate Q_1 and/or Q_2 terms are deleted. If the second reservoir is not present, as shown in Fig. 6–12, dropping A_2 and Q_2 leads to

$$dt = \frac{A_1\,dH}{Q_1 - KH^{1/2}} \tag{6-21}$$

In the application of this equation, it must be noted that H, the change in total head, has also been established as the head loss. Consequently, the exit velocity head, which elsewhere in the text has been considered as one of the energy terms, is now included in K as part of the energy loss. This will be illustrated in Example 6–7, wherein the exit velocity head is included by an additional $C_L = 1$.

In addition, if the unsteady flow out of the single reservoir is through an orifice rather than a pipe, as in Fig. 6–13, the pipe friction portion of the head loss may

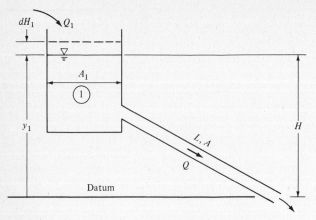

Figure 6–12 Reservoir discharging through pipe.

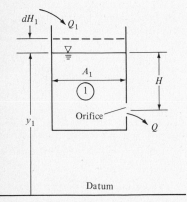

Figure 6–13 Reservoir discharging through orifice.

also be excluded. Although this last type of flow does not strictly belong in a chapter on pipelines, it is a logical special case covered by the differential equation for unsteady flow. A solution to a given problem depends upon integrating Eq. 6–20 or Eq. 6–21 to fit given conditions. This may be accomplished by closed-form integration or numerical integration, depending on the complexity of the problem. Several examples will illustrate the different cases.

EXAMPLE 6–7

The reservoir of Fig. 6–12 discharges 60°F water into the atmosphere through a 1-ft-diameter pipe with an absolute roughness height of 0.0002 ft and a length of 1000 ft. The reservoir has a cross-sectional area of 1000 ft². In addition to a square-edged entrance, the pipe contains six 90 deg medium radius elbows (not shown).

Determine the time required for the water level in the reservoir to drop from a height of 50 ft to 30 ft above the pipe outlet.

Solution

From Eq. 6–19, K is given by

$$K = \frac{A\sqrt{2g}}{\left[f\dfrac{L}{D} + \Sigma C_{Li} \right]^{1/2}}$$

The entrance has $C_L = 0.5$, each of the elbows has $C_L = 0.8$, and, as discussed above, since H is based on the outlet elevation, the exit velocity head must be included as a loss with $C_L = 1.0$. Thus

$$C_{Li} = 0.5 + (6)(0.8) + 1.0 = 6.3$$

In the event that f changes as the water surface falls, an average K will be calculated for the range in heads based on steady flow conditions. Initially assume fully rough flow, so that for $k/D = 0.0002$, $f = 0.014$. Then at $H = 50$ ft

$$50 = \left[\frac{(0.04)(1000)}{1} + 6.3 \right] \frac{V^2}{(2)(32.2)}$$

and $V = 12.6$ ft/s. Thus,

$$Re = \frac{(12.6)(1)}{1.217 \times 10^{-5}} = 1.04 \times 10^6$$

leading to an improved $f = 0.015$. Repeating

$$50 = \left[\frac{(0.015)(1000)}{1} + 6.3 \right] \frac{V^2}{(2)(32.2)}$$

yields a new $V = 12.3$ ft/s and

$$Re = \frac{(12.3)(1)}{1.217 \times 10^{-5}} = 1.01 \times 10^6$$

indicating that $f = 0.015$ is satisfactory. In the same manner, but using a head difference $H = 30$ ft, we would find that $f = 0.015$ is again satisfactory. Therefore, for the range of heads

$$K = \frac{(\pi/4)(1)^2\sqrt{(2)(32.2)}}{\left[\dfrac{(0.015)(1000)}{1} + 6.3 \right]^{1/2}} = 1.37$$

Integrating Eq. 6–21, with $Q_1 = 0$ (since there is no inflow specified for the reservoir) gives

$$t = \int_{50}^{30} \frac{1000 \, dH}{-1.37 \, H^{1/2}} = 732.2 \int_{30}^{50} H^{-1/2} \, dH$$

$$= (732.2)(2)(50^{1/2} - 30^{1/2})$$

$$= 2334 \text{ s} = 38.9 \text{ min.}$$

and 38.9 min. are required to lower the reservoir from 50 ft to 30 ft.

EXAMPLE 6–8

The reservoir of Fig. 6–13 consists of a tank with a constant diameter of 4 m. Water discharges from the tank through a well-rounded outlet that delivers a 10-cm-diameter jet. If there is a constant inflow of 0.02 m³/s into the tank, determine the time required for the water level to drop from 10 m to 2 m above the orifice.

Solution

The steady-state equation used to evaluate K (Eq. 6–19) may be compared with the conventional orifice equation

$$H = \frac{V^2}{2g} = \frac{Q^2}{2gA^2}$$

whereupon

$$K = A\sqrt{2g} = \left(\frac{\pi}{4}\right)(0.1)^2\sqrt{(2)(9.81)} = 0.0348$$

This may be substituted into Eq. 6–21 along with A_1 and Q_1 and reference to integral tables will provide a result that can then be evaluated between the limits of $H_1 = 10$ m and $H_2 = 2$ m. Alternatively, the following substitution provides a convenient solution:

$$h = kH^{1/2} - Q_1$$

then

$$H = \frac{1}{K^2}(Q_1 + h)^2$$

$$H = \frac{1}{K^2}(Q_1^2 + 2Q_1h + h^2)$$

$$dH = \frac{1}{K^2}(2Q_1 + 2h)\, dh$$

and Eq. 6–21 can be restated as

$$t = \int_{H_1}^{H_2} \frac{A_1\, dH}{Q_1 - KH^{1/2}} = \int_{H_1}^{H_2} \frac{(A_1/K^2)(2Q_1 + 2h)\, dh}{Q_1 - (h + Q_1)}$$

$$t = -\frac{2A_1}{K^2}\int_{H_1}^{H_2} \frac{(Q_1 + h)\, dh}{h} = -\frac{2A_1}{K^2}\left[\int_{H_1}^{H_2}\frac{Q_1\, dh}{h} + \int_{H_1}^{H_2} dh\right]$$

The integration limits can be obtained by noting that at $H = 10$ m

$$h = (0.0348)(10)^{1/2} - 0.02 = 0.0900$$

and similarly at $H = 2$ m

$$h = (0.0348)(2)^{1/2} - 0.02 = 0.0292$$

These limits may be introduced along with the other numerical values to get the required time.

$$t = -\frac{(2)(\pi/4)(4)^2}{(0.0348)^2}\left[\int_{0.0900}^{0.0292}\frac{Q_1}{h}\, dh + \int_{0.0900}^{0.0292} dh\right]$$

$$= 20{,}750\left[0.02\ln h + h\right]_{0.0292}^{0.0900} = 1729 \text{ s} = 28.8 \text{ min.}$$

EXAMPLE 6–9

In the two reservoir system of Fig. 6–11, reservoir 1 has a cross-sectional area of 10,000 ft² while that of reservoir 2 is 2000 ft². The connecting pipe has a diameter of 1 ft, a length of 20,000 ft, and an absolute roughness of 0.00015 ft. Minor losses can be neglected because of the pipe length. Initially, the level of reservoir 1 is 100 ft above that of reservoir 2. If there is no additional inflow or outflow to the system, determine the time required for the water level in reservoir 1 to drop by 10 ft.

Solution

The difference in total head represented in Fig. 6–11 is a relative quantity based on the two reservoir levels. In this problem, the time required for an absolute change (i.e., with respect to the one reservoir only) is to be determined. However, since there is no inflow or outflow, other than between the reservoirs, continuity can be used to find the absolute change in reservoir 2 that corresponds to the given 10 ft change in reservoir 1. That is

$$(10)(10,000) = (\Delta y)(2000)$$

and $\Delta y = 50$ ft. The difference in reservoir levels must therefore decrease from the initial 100-ft difference to a final value of $100 - 10 - 50 = 40$ ft.

As in the previous two examples, the value of f may be obtained at the limits of $H_1 = 100$ ft and $H_2 = 40$ ft. Based on steady flow at each of the limits, f values just below and just above 0.016 result, so that an average of $f = 0.016$ is selected. Then, with K given by

$$K = \frac{A\sqrt{2g}}{\left(f\dfrac{L}{D}\right)^{1/2}} = \frac{(\pi/4)(1)^2\sqrt{(2)(32.2)}}{\left[\dfrac{(0.016)(20,000)}{1}\right]^{1/2}} = 0.352$$

Eq. 6–20 becomes

$$dt = \frac{dH}{-0.352\, H^{1/2}\left(\dfrac{1}{10,000} + \dfrac{1}{2000}\right)}$$

$$t = -4735 \int_{100}^{40} \frac{dH}{H^{1/2}} = (4735)(2)H^{1/2}\,\Big|_{40}^{100}$$

$$t = 34,800 \text{ s} = 9.67 \text{ hr}$$

Flow Establishment

We have previously examined steady flow in a system such as that shown in Fig. 6–14. If the discharge is initially zero because of the closed downstream valve, then opening the valve results in a period of unsteady flow as the fluid is accelerated. Eventually, steady conditions are achieved, but it is this period of adjustment or flow establishment that is of interest in this section. In the development that follows, the reservoir level z and the resistance coefficient f will be assumed constant. Before the valve is open the hydrostatic conditions of Fig. 6–14a prevail. It will be assumed that at $t = 0$, the valve is instantaneously opened, however, the flow does not immediately reach the equilibrium conditions of Fig. 6–14c, but rather starts to accel-

THL–Total head line (or energy grade line)
PHL–Piezometric head line (or hydraulic grade line)

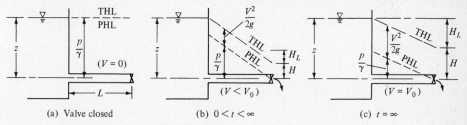

(a) Valve closed (b) $0 < t < \infty$ (c) $t = \infty$

Figure 6–14 Flow establishment in a pipe.

erate so that at some time t, the flow will resemble Fig. 6–14b. Then, at that instant the total head at the pipe exit will be

$$H = \frac{V^2}{2g}$$

The total head loss in the pipe at the same instant is

$$H_L = f\frac{L}{D}\frac{V^2}{2g} + \Sigma C_L \frac{V^2}{2g}$$

Thus at time t, the unbalance in total head available to accelerate the flow will be

$$\Delta H = z - (H + H_L)$$

$$= z - \left(1 + f\frac{L}{D} + \Sigma C_L\right)\frac{V^2}{2g}$$

or

$$\Delta H = z - K\frac{V^2}{2g} \qquad\qquad\qquad (6\text{–}22a)$$

where

$$K = 1 + f\frac{L}{D} + \Sigma C_L \qquad\qquad\qquad (6\text{–}22b)$$

has been introduced to simplify the notation.

The difference in head $(z - KV^2/2g)$ gives rise to an accelerative force $(z - KV^2/2g)\gamma A$, which can be equated to the product of the mass of fluid in the pipe ρAL and the acceleration. The latter quantity is simply the local acceleration dV/dt, since even under unsteady conditions the flow remains uniform. Since V is a function of time only, the total derivative is justified. Therefore,

$$\gamma A\left(z - K\frac{V^2}{2g}\right) = \rho AL\frac{dV}{dt}$$

or

$$2gz - KV^2 = 2L\frac{dV}{dt}$$

Rearranging

$$\int dt = 2L\int\frac{dV}{2gz - KV^2}$$

This expression integrates to

$$t = \frac{L}{\sqrt{2gzK}}\ln\frac{\sqrt{2gz} + \sqrt{K}V}{\sqrt{2gz} - \sqrt{K}V} + C \tag{6-23}$$

The constant of integration C must equal zero since $V = 0$ at $t = 0$. Letting V_0 be the equilibrium velocity, and comparing with the steady-state equation

$$z = \frac{V_0^2}{2g} + f\frac{L}{D}\frac{V_0^2}{2g} + \Sigma C_L\frac{V_0^2}{2g} = K\frac{V_0^2}{2g} \tag{6-24}$$

leads to the additional equations

$$t = \frac{L}{\sqrt{2gzK}}\ln\frac{V_0 + V}{V_0 - V} \tag{6-25}$$

and

$$t = \frac{L}{\sqrt{2gzK}}\ln\frac{1 + V/V_0}{1 - V/V_0} \tag{6-26}$$

The above implies that an infinite time is required to reach fully established steady flow. As a practical matter, the time necessary to reach nearly steady flow, say 99 percent of V_0, is easily calculated.

EXAMPLE 6–10

A valve at the outlet of the pipe system of Fig. 6–14 is suddenly opened. Determine the time required for the discharge to reach 99 percent of its ultimate value if the system has the following characteristics: $z = 50$ m, $L = 200$ m, $D = 1$ m, $f = 0.02$, and $\Sigma C_L = 14$.

Solution

Although the steady-state velocity and discharge can be readily determined from Eq. 6–24, they are not required for this solution. The coefficient K is obtained from Eq. 6–22b as

$$K = 1 + \frac{(0.02)(200)}{1} + 14 = 19$$

Then for $V/V_0 = 0.99$, Eq. 6–26 yields

$$t = \frac{200}{\sqrt{(2)(9.81)(50)(19)}} \ln \frac{1 + 0.99}{1 - 0.99} = 7.75 \text{ s}$$

6–4 FLOW MEASUREMENT IN PIPES

Velocity Measurement

Measurements typically fall into two major types, velocity measurements and discharge measurements. Velocity is generally determined using some form of *Pitot tube*, such as those shown in Fig. 6–15. They depend upon creating a pressure difference (often equal to the dynamic flow pressure, $\rho V^2/2$), which may then be related to the velocity using the Bernoulli equation with or without an additional correction factor. To illustrate, if the Pitot-static tube of Fig. 6–15a is inserted into a closed conduit flow with pressure p and velocity V, the nose opening will sense the combined ambient pressure plus dynamic pressure since the nose is a stagnation point. That is, according to the Bernoulli equation

$$p_{\text{st.pt.}} = p + \frac{\rho V^2}{2}$$

Further, if the geometry is so designed that the side openings are located at points where the pressure is exactly equal to the ambient pressure then they will of course sense that pressure. If the nose opening and the side openings (which would themselves be interconnected) are connected to a differential pressure meter, that meter will indicate the pressure difference $\Delta p = \rho V^2/2$. Upon solving for the velocity, this becomes

$$V = \sqrt{2\Delta p/\rho} \tag{6–27}$$

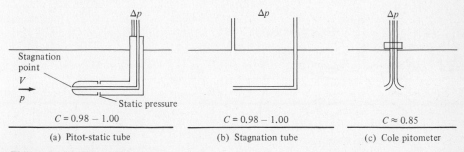

(a) Pitot-static tube   (b) Stagnation tube   (c) Cole pitometer

Figure 6–15 Pitot tubes.

To provide for different geometric shapes and the fact that the static pressure reading is not exactly equal to the ambient pressure, a correction factor C may be added leading to

$$V = C\sqrt{2\Delta p/\rho} \qquad\qquad\qquad\qquad (6\text{--}28)$$

Typical values of C are included on Fig. 6–15 for the different types of Pitot tubes, but high accuracy requires individual calibration. In addition to the conventional Pitot tube shown in Fig. 6–15a, part b contains a simpler *stagnation tube* in which the static or ambient pressure is obtained separately at the side of the pressure conduit. The advantage herein is that the stagnation tube can be made with a smaller diameter, thereby yielding a reading that more closely represents a point measurement. Part c, called a *Cole pitometer*, is designed for easy insertion into the pipe, whereupon one tube can be rotated so as to point in the downstream direction.

If the pipe discharge is of primary interest, the velocity may be measured at selected points across the conduit section and the discharge computed by graphic integration. If somewhat lower accuracy is acceptable, a single velocity measurement at the pipe centerline can be correlated with the discharge.

Discharge Measurement

Discharge in a pipe is most commonly measured using some form of differential head meter, such as the three included in Fig. 6–16. In each case the governing

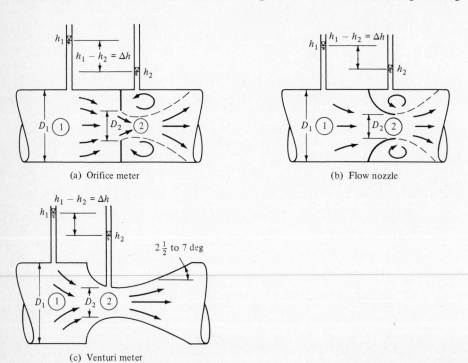

(a) Orifice meter   (b) Flow nozzle

(c) Venturi meter

Figure 6–16 Pipe flow meters.

equations may be obtained by judicious application of energy and continuity equations. For details refer to any textbook on fluid mechanics. The discharge may be expressed as a function of the pressure or piezometric head change created by the meter. With respect to the latter, the discharge is

$$Q = C_D A_2 \sqrt{2g(h_1 - h_2)} \tag{6–29}$$

Regardless of meter type, the area A_2 is the minimum opening and C_D is a discharge coefficient. The piezometric heads h_1 and h_2 are obtained at the two indicated sections using manometers (as shown) or pressure gauges.

The orifice meter is perhaps the most common meter since it may consist of no more than an orifice (usually machined) in a plate that is then inserted between the flanges in a pipe. Its biggest disadvantage is a rather large head loss. Depending upon where the pressure taps are located, there may be a significant variation in the magnitude of the discharge coefficient. An in-place calibration eliminates this problem and is recommended, in fact, for all meters when high accuracy is important. For a sharp-edged orifice, typical orifice meter coefficients are given in Table 6–5.

The flow nozzle reduces the head loss in exchange for a higher cost. On the other hand, the venturi meter almost eliminates the head loss but it is lengthy and most expensive. If the angle of divergence (Fig. 6–16c) exceeds approximately 7 deg, the flow will be unable to follow the diverging boundary. A zone of eddies will form and a low head loss will not be achieved. If actual calibration of a meter is not possible, then the venturi meter will usually provide the most reliable results.

Both the flow nozzle and the venturi meter may be subjected to a theoretical analysis, whereupon the discharge coefficient is determined to be

$$C_D = \frac{1}{[1 - (A_2/A_1)^2]^2} \tag{6–30}$$

On the average, Eq. 6–29, when used with Eq. 6–30, overestimates the discharge by about 2 percent since friction through the meter is ignored.

Some care is required in the installation and operation of any flow meter. There should be a stretch of 10 diameters of straight pipe ahead of the meter. Often,

Table 6–5 ORIFICE METER COEFFICIENT C_D

Orifice/pipe diameter ratio D_2/D_1	C_D
0.2	0.60
0.3	0.605
0.4	0.61
0.5	0.62
0.6	0.65
0.7	0.70
0.8	0.77

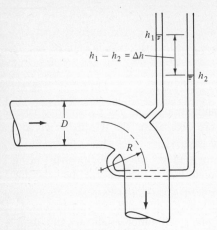

Figure 6–17 Elbow meter.

straightening vanes are also included to help suppress potential disturbances in the flow. A following straight stretch is also useful. It is sometimes desirable to locate the pressure taps on the side of the pipe rather than at the top or bottom to reduce air accumulation or plugging from rust or other foreign matter, both of which may be a problem in water lines.

Another useful type of meter is the elbow or bend meter, which takes advantage of the centrifugal effect created as the flow direction is changed. The resulting higher piezometric head at the outside of the bend and the lower head at the inside can be measured with pressure taps and manometer columns as shown in Fig. 6–17. Equation 6–29 again provides the discharge. Calibration is strongly recommended, but for a 90 deg elbow or bend of radius R the discharge coefficient may be approximated by

$$C_D = \sqrt{\frac{R}{2D}} \qquad\qquad (6\text{–}31)$$

EXAMPLE 6–11

A venturi with the characteristics shown in Fig. 6–18 is placed in a vertical pipeline. Determine the discharge of water if the upward flow results in $p_1 = 65$ kN/m^2 and $p_2 = 40$ kN/m^2. Correct for a discharge overestimation of 2 percent.

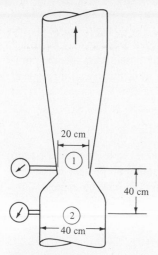

Figure 6–18 Venturi meter in vertical line.

Solution

Equation 6–29 may be rewritten as

$$Q = C_D A_2 \sqrt{2g} \left[\left(y_1 + \frac{p_1}{\gamma} \right) - \left(y_2 + \frac{p_2}{\gamma} \right) \right]^{1/2}$$

where

$$C_D = \frac{0.98}{[1 - (0.2/0.4)^4]^{1/2}} = 1.012$$

The 0.98 factor takes into account the friction through the meter. Upon substituting

$$Q = (1.012)\left(\frac{\pi}{4}\right)(0.2)^2[(2)(9.81)]^{1/2}\left[\left(\frac{65,000}{9810} \right) - \left(0.4 + \frac{40,000}{9810} \right) \right]^{1/2}$$

$$= 0.2064 \ \mathrm{m^3/s}$$

6–5 FORCES ON PIPELINES

A considerable number of forces may act on a pipeline. These include interior static and dynamic forces, as well as exterior forces or loadings. In addition, stresses result from factors such as temperature change. A number of these forces and stresses will

be considered here. The external forces on a pipeline involve a number of considerations depending upon the particular type of application. Most often, however, the
problem is one of soil mechanics and the solution to a given loading may be found
in a soil mechanics reference. The pipe structural analysis based on a specific combination of these stresses and using the appropriate thin-wall or thick-wall theory will
likewise not be included.

Internal Pressure

The internal pressure (which may be either static or dynamic) in a pipe gives rise to
a tangential or hoop stress as shown in Fig. 6–19a. Assuming a constant pressure p
and a unit pipe length, the equilibrium of forces in Fig. 6–19a may be expressed by

$$2\sigma t = 2rp$$

which immediately leads to

$$\sigma = \frac{pr}{t} \qquad\qquad (6\text{–}32)$$

where σ is the tangential stress and t is the wall thickness of the pipe. The modification of the analysis to include a hydrostatic pressure variation is left as an exercise.
The difference is usually negligible and therefore ignored. The tangential stress is
normally tensile, but a partial vacuum (i.e., subatmospheric pressure) will result in a
compressive wall stress that may be of concern, particularly in thin-walled conduits.

Longitudinal wall stresses are frequently zero, but may occur in a closed section
of pipe. The equilibrium analysis based on Fig. 6–19b yields

$$p\pi r^2 = \sigma t(2\pi r)$$

or

$$\sigma = \frac{pr}{2t} \qquad\qquad (6\text{–}33)$$

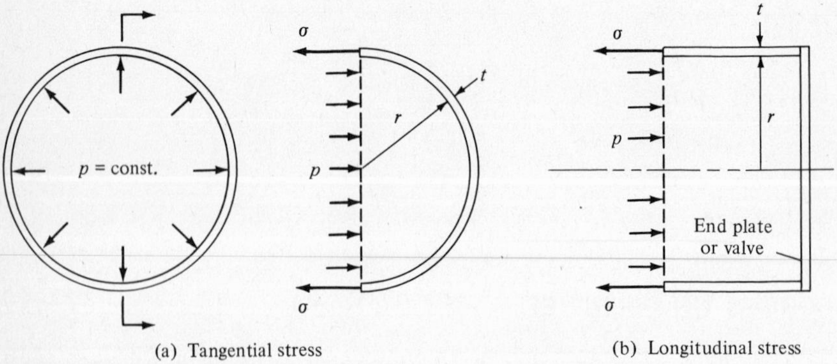

(a) Tangential stress (b) Longitudinal stress

Figure 6–19 Internal pipe pressure.

Thus, when the pressure causes a longitudinal stress, it is one-half of the tangential stress.

Thermal Stress

Thermal stresses rarely arise when pipe sections are connected by bell-and-spigot joints or couplings that allow considerable play. However, if a length of pipe is held rigidly at both ends, a significant stress may develop as the temperature changes. The analysis may be performed in two parts. If the pipe were unrestrained, then a temperature change ΔT would result in an elongation or shortening (as ΔT is positive or negative) of

$$\delta = \alpha L \Delta T$$

where α is the coefficient of thermal expansion of the pipe material and L is the pipe length. If the length change occurs, the stress fails to develop, but if a restraint prevents the change, then in the elastic range of material behavior

$$\sigma = E\epsilon = E\frac{\delta}{L}$$

Here, E is the modulus of elasticity and ϵ is the strain per unit length (δ/L). Upon combining the above equations, a relationship for the maximum thermal stress results,

$$\sigma = E\alpha \Delta T \qquad\qquad\qquad\qquad \text{(6–34)}$$

Typical values of α and E are included in Table 6–6 for both U.S. customary and SI units.

Table 6–6 PIPE MATERIAL CHARACTERISTICS

Material	Coefficient of thermal expansion		Modulus of elasticity E		Poisson ratio μ
	ft/ft°F $\times 10^6$	m/m°C $\times 10^6$	psi $\times 10^{-6}$	kN/m² $\times 10^{-6}$	
Aluminum	13.1	23.6	10.1	70	0.33
Brass	11.0	19.8	16	110	0.34
Cast iron	6.7	12.1	10	70	0.21
Concrete	5.5	10.0	4.5	30	0.15
Copper	9.4	16.9	16	110	0.34
Steel	6.5	11.7	30	207	0.30

Example 6–12

Determine the longitudinal stress that may be expected in a 500-ft-long steel pipe if a temperature decrease of 60°F occurs and the pipe is fully restrained. If some play is present at the pipe joints so that only one-half of the anticipated stress is actually reached, what length change must have occurred?

Solution

From Table 6–6, $\alpha = 6.5 \times 10^{-6}$ ft/ft°F and $E = 30 \times 10^6$ psi. Thus,

$$\sigma = \alpha E \Delta T = (6.5 \times 10^{-6})(30 \times 10^6)(60) = 11{,}700 \text{ psi (tension)}$$

Since

$$\delta = \alpha L \Delta T = (6.5 \times 10^{-6})(500)(60) = 0.20 \text{ ft}$$

one-half of the maximum stress would result when half of this length change occurs, or

$$\delta = \frac{0.20}{2} = 0.10 \text{ ft}$$

Forces due to Changes in Direction or Cross Section

Any change in pipe direction or cross-sectional area causes a change in the fluid momentum. This in turn gives rise to an additional force that acts on the pipe. By their very nature, these forces can usually be analyzed by the impulse-momentum principle.

Determination of the force on a bend or elbow normally requires that the momentum equation must be written in both coordinate directions. Care must be taken to sketch a proper control volume and include all of the forces acting on the fluid. As a first approximation, the fluid weight and elevation changes (unless they are large) are usually neglected. The procedure will be illustrated for the pipe contraction and accompanying control volume in Fig. 6–20. If these forces can not be elimi-

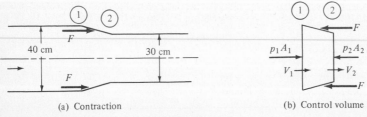

(a) Contraction   (b) Control volume

Figure 6–20 Force on a pipe contraction.

nated, the support structure must be designed to withstand them. Frequently, this requires that a massive thrust block be located at the transition or bend.

EXAMPLE 6–13

The pipe contraction of Fig. 6–20 reduces the diameter from 40 cm to 30 cm. Determine the force on the contraction if the water discharge is 0.75 m³/s and the upstream pressure is 300 kN/m².

Solution

The forces acting on the fluid are shown in their proper direction on the control volume in Fig. 6–20. Only the x direction needs to be considered in the solution. Section numbers 1 and 2 will be used to indicate the entrance and exit sections, respectively. As usual, the two velocities may be calculated by $V = Q/A$, whereupon $V_1 = 5.97$ m/s and $V_2 = 10.61$ m/s. The upstream pressure is given and the energy equation must be used to obtain the downstream pressure.

$$\frac{300,000}{9810} + \frac{(5.97)^2}{(2)(9.81)} = \frac{p_2}{9810} + \frac{(10.61)^2}{(2)(9.81)}$$

Upon solving, we find that $p_2 = 261,500$ N/m². The momentum equation in the x direction may be written as

$$p_1 A_1 - F - p_2 A_2 = \rho Q(V_2 - V_1)$$

Substituting the known information permits solving for the unknown force F:

$$(300,000)\left(\frac{\pi}{4}\right)(0.4)^2 - F - (261,500)\left(\frac{\pi}{4}\right)(0.3)^2$$
$$= (1000)(0.75)(10.61 - 5.97)$$

Thus, the force is 15,730 N. This is the force that the pipe exerts on the water. Therefore, the force that acts on the contraction is in the opposite direction or to the right.

Water Hammer

Water hammer is the result of a series of pressure waves that occur in a pipeline following a relatively sudden change in the velocity. The pressure differences may be quite large and very damaging to the pipe or its appurtenances. The phenomenon is frequently accompanied by a repeated banging sound and hence its name.

A primary cause of water hammer is a change in a valve setting. This is of considerable concern to the engineer because the design of a pipe system cannot always avoid accidental valve changes. Other causes include the starting or stopping of pumps, changes in the power demand of a turbine, pipe rupture, and entrapped air. The problems that result include damage to equipment and pipe failure due to excessively high or low pressure.

To introduce the problem, consider a steady pipe flow in which a downstream valve is suddenly closed (Fig. 6–21). The water just upstream of the valve will immediately stop. However, the water further upstream continues for a period of time and in the process compresses the water that is already at rest. A pressure rise will occur during the process, and as more and more of the water is stopped, the pressure wave or surge moves upstream. This phenomenon will be described in more detail during the subsequent analysis. We are again dealing with an unsteady flow, but it is considered separately in this particular section, because the nature of the problem is different than the previous unsteady flows. Even more to the point, the resulting forces are usually of great concern to the engineer.

There are two aspects that must be considered in the analysis of water hammer. First, the pressure wave is a compressible wave resulting from the compression (or rarefication) of the water and therefore travels through the water at the speed of sound, or *sonic velocity*. Second, the pressure change that results from the change in velocity must be evaluated.

The equation for the sonic velocity in an infinite fluid is derived in any textbook on fluid mechanics. The resulting equation is

$$c = \sqrt{K/\rho} \tag{6–35}$$

where c is the speed of sound in a fluid with modulus of compressibility or bulk modulus, K, and density, ρ. This equation would also apply to the speed of a compressive wave in a completely rigid pipe, since the pipe would not influence or interact with the fluid behavior. While we may find that Eq. 6–35 may serve on occasion as a first approximation, it will usually markedly overestimate the actual sonic veloc-

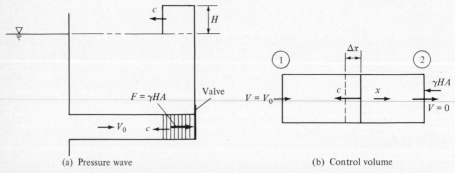

(a) Pressure wave   (b) Control volume

Figure 6–21 Definitional sketch for water hammer in a pipe.

ity, since the expansion of the pipe wall as the pressure increases moderates the compressive effects.

Following the treatment by Wylie and Streeter (see references at the end of Chapter 6), there are three distinct types of pipe supports for which the sonic velocity can be evaluated. They are as follows:

1. A pipe supported at the upstream end and otherwise free to move axially.

2. A pipe anchored against axial movement throughout its length.

3. A pipe anchored against axial movement throughout its length, but provided with expansion joints at regular intervals.

In type 1 constraint the pipe is allowed to move and the pressure increase leads to both axial and lateral stresses and strains. The second type of constraint eliminates axial movement, but axial stress still results from the Poisson effect. In type 3 constraint, axial movement is completely taken up by the available play at the joints so that no axial stress is created. In the analysis we will consider only type 3 but results will be included for the other two conditions.

The following derivation assumes an absence of friction effects, at least insofar as they affect the sonic velocity. Figure 6–21a shows a system with a valve at the downstream end. Sudden closure of the valve will result in a head increase H due to a pressure increase $\Delta p = \gamma H$. This increased head is shown relative to the horizontal line that would be the original total head line if completely frictionless flow were assumed. However, we will assume the more realistic condition that the original steady velocity V_0 is the actual velocity in the pipe based on friction and minor losses as appropriate. The increase in pressure due to water hammer must actually be added to the steady-flow pressure prevailing just before the valve was closed.

The sonic velocity will be obtained by considering the period of time L/c required for the pressure wave to travel the length of the pipe. To evaluate c, the volume of water entering the pipe must be related to the increased pipe volume due to pipe expansion and the reduced volume of water because of its own compression under the increased pressure. Since the rate at which the water continues to enter the pipe during the period L/c is $V_0 A$, the additional volume of water involved is

$$V_{added} = V_0 AL/c \qquad\qquad (6\text{–}36)$$

Because of the pipe joints, there will be no axial stress: however, the increased pressure will cause an additional lateral or hoop stress (above that of the previous steady-flow conditions). Referring to Fig. 6–19a and Eq. 6–32, the magnitude of this stress will be

$$\sigma = \frac{\Delta p D}{2t} = \frac{\gamma H D}{2t}$$

The corresponding lateral strain by Hooke's law is

$$\epsilon = \frac{\sigma}{E} = \frac{\gamma H D}{2tE}$$

where E is again the modulus of elasticity. This leads to an elongation of the diameter to $D + \epsilon D$. The increase in cross-sectional pipe area is

$$\Delta A = \frac{\pi}{4}(D + \epsilon D)^2 - \frac{\pi}{4}D^2$$

$$= \frac{\pi D^2 \epsilon}{2} = 2A\epsilon$$

where the small term involving ϵ^2 has been dropped. During the time period L/c, the pressure in the entire pipe may be assumed to reach Δp, thus the increase in pipe volume over the length L is

$$\Delta V_{\text{pipe}} = 2LA\epsilon = \frac{LA\gamma HD}{tE} \qquad (6\text{--}37)$$

Finally, the reduction in volume of the water within the pipe may be calculated using the defining equation for the modulus of compressibility of water:

$$K = \frac{dp}{d\rho/\rho} = -\frac{dp}{dV/V} \qquad (6\text{--}38)$$

The modulus of compressibility may be defined on the basis of the increase in pressure due to either the relative increase in density or the relative decrease in volume. Choosing the latter leads to a decrease in the volume of water

$$\Delta V_{\text{water}} = \frac{V\Delta p}{K} = \frac{LA\gamma H}{K} \qquad (6\text{--}39)$$

The volume of water added during the time L/c must equal the sum of the increased pipe volume and the reduced water volume. Consequently,

$$\frac{LAV_0}{c} = \frac{LA\gamma HD}{tE} + \frac{LA\gamma H}{K}$$

or when rearranged,

$$\frac{V_0}{c} = \frac{\gamma H}{K}\left(1 + \frac{K}{E}\frac{D}{t}C_0\right) \qquad (6\text{--}40)$$

The factor C_0 which has a value of unity for the type 3 constraint considered in the foregoing derivation, has been arbitrarily added to facilitate expressing the other two cases. They are more complex to analyze because of axial stresses and elongations. However, the results may be incorporated in Eq. 6–40 (and the subsequent equations) through C_0 by the following:

First case $\quad C_0 = \dfrac{5}{4} - \mu \qquad\qquad\qquad$ (6–41a)

Second case $\quad C_0 = 1 - \mu^2 \qquad\qquad\qquad$ (6–41b) and, as stated in the foregoing

Third case $C_0 = 1$ (6–41c)

where μ is Poisson's ratio, typical values of which are given in Table 6–6.

The sonic velocity is given in terms of the increase in head H in Eq. 6–40. We now need a second equation that will also relate the sonic velocity to the head. This can be obtained from the control volume of Fig. 6–21b. This control volume encompasses the wave front and extends in length from some section 1 where the velocity remains unchanged at V_0 to a section 2 behind the front where the velocity is zero and the pressure increase of γH has been realized. Within the control volume, the wave is moving to the left with a yet-unknown celerity c.

The momentum equation may be applied to the control volume, but the flow is unsteady so the net force acting on the control volume must equal the sum of the net rate of flow of momentum out plus the time rate of increase of momentum within the control volume. Thus, in the plus-x direction

$$-\gamma HA = (0 - \rho AV_0^2) - \rho AV_0 c$$

The final term reflects the decrease in momentum as more and more of the fluid is brought to rest. During a short time period Δt, the wave moves a distance Δx as shown. The mass of water stopped during this time is $c\Delta t A\rho$. Its velocity is changed by an amount $-V_0$ in the process and therefore the time rate of change of momentum is $-\rho AV_0 c$ as included in the above equation. Solving for H yields

$$H = \frac{cV_0}{g}\left(1 + \frac{V_0}{c}\right)$$ (6–42)

or if $c \gg V_0$, then

$$H = \frac{cV_0}{g}$$ (6–43)

which may also be expressed in terms of the pressure change as

$$\Delta p = \gamma H = \rho c V_0$$ (6–44)

The head H from Eq. 6–42 may now be substituted into Eq. 6–40, with the result that

$$c^2 = \frac{K/\rho}{\left(1 + \dfrac{V_0}{c}\right)\left[1 + \left(\dfrac{K}{E}\right)\left(\dfrac{D}{t}\right)C_0\right]}$$

Finally, if V_0/c is again assumed to be very small, the speed of sound in a given pipe becomes

$$c = \frac{\sqrt{K/\rho}}{\left[1 + \left(\dfrac{K}{E}\right)\left(\dfrac{D}{t}\right)C_0\right]^{1/2}}$$ (6–45)

Observe that if the pipe is assumed rigid (which is equivalent to an infinite value of the modulus of elasticity E), this equation simplifies to Eq. 6–35 as stated at the beginning of the section.

One effect of friction can be added at this point. The normal piezometric head line associated with the undisturbed steady flow conditions is the result of the pressure gradient that decreases in the flow direction and is required to overcome the frictional resistance. This prevails at the instant the valve is closed. Thus, the pressure wave that moves upstream continually encounters a higher and higher pressure. The water that is already at rest is at a nearly constant pressure, so that the higher pressure of the water being brought to rest causes the pressure to continually creep upward.

The head loss of the steady flow may be written

$$h_l = f \frac{L}{D} \frac{V_0^2}{2g}$$

and the increase in pressure head at the valve above that given by Eq. 6–43 can be shown to be

$$H_f = \frac{h_l}{\sqrt{2}} \tag{6–46}$$

Thus, the maximum increase in head due to water hammer is

$$H = \frac{cV_0}{g} + \frac{f}{\sqrt{2}} \frac{L}{D} \frac{V_0^2}{2g}$$

$$= \frac{V_0^2}{2g} \left(\frac{2c}{V_0} + \frac{fL}{\sqrt{2}D} \right) \tag{6–47}$$

The entire water hammer cycle will now be discussed. The process begun in Fig. 6–21 will be shown in its entirety in Fig. 6–22 for selected points along the length of the pipe. Assuming an instantaneous valve closure, a pressure rise occurs immediately at the valve (Eq. 6–44) with a corresponding increase in head (Eq. 6–43). The original undisturbed flow into the pipe continues and the pressure wave moves toward the reservoir, which it reaches at time L/c. At this point, all of the water in the pipe is at rest, but at a pressure above that in the reservoir. (Pressure at the valve will have continued to increase slightly as discussed previously in the development of Eq. 6–46.) This unbalance causes the water to flow out of the pipe in the opposite direction as the water expands back to approximately its original pressure. This relief pressure wave travels back down the pipe at the same celerity c, returning to the valve after a total elasped time $2L/c$.

However, the momentum of the excess water as it exits into the reservoir causes this upstream flow to continue, and a drop in pressure immediately results at the valve. If this pressure drop, which is essentially $-\Delta p$, would result in a pressure below the vapor pressure, then a water vapor cavity will form. Regardless this negative pressure wave travels from the valve to the reservoir during an additional time

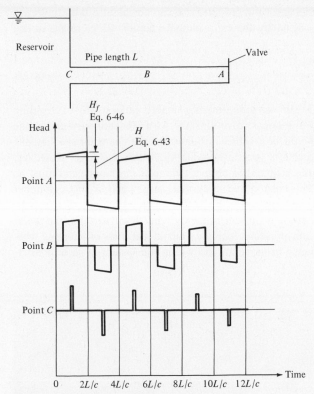

Figure 6–22 Effect of water hammer in a pipe.

L/c. The water is again at rest, but this time at a reduced pressure. The unbalance is now in the opposite direction and flow again occurs back into the pipe. The effect of this resumption of flow once again reaches the valve, this time after a total elasped time of $4L/c$. At this point, the flow is directed against the still-closed valve, a new pressure build-up begins, and the entire cycle is repeated. The process is eventually damped by the friction of the flow and the inelastic behavior of the pipe.

At points closer to the reservoir (points B and C), longer periods of neutral pressure occur as the pressure wave coming from the valve (and later from the reservoir) requires some time to reach the given point. The pressure increase and decrease near the reservoir occur for only a short time, resulting in the pressure spikes shown for point C.

All of the preceding discussion presupposes an instantaneous valve closure. The time of closure for the valve will not be covered in detail, but two aspects will be mentioned. If the time of closure is less than $2L/c$, then the valve is closed before the wave can reach the reservoir and return. Therefore, the pressure increase predicted by Eq. 6–44 will occur at the valve, but not at all upstream points. This can be called *rapid closure*. On the other hand, if the time of closure is greater than $2L/c$, so that *slow closure* occurs, then the maximum pressure increase will not be realized.

As a first approximation in this case, the pressure rise can be calculated by multiplying the pressure rise of Eq. 6–44 by the ratio of the quantity, $2L/c$, to the actual closure time, t_c. This gives

$$\Delta p = (\rho c V_0)\frac{2L/c}{t_c} = \frac{2LV_0\rho}{t_c} \tag{6–48}$$

Specifying a satisfactorily slow closure time is probably the best way of avoiding water hammer problems due to manual valve closure. As a precautionary guide, the calculated time of closure should be doubled to allow for the characteristic that the first half of the valve closure causes very little reduction in the discharge. However, automatic valves are sometimes required to protect hydraulic machinery and must be designed to close rapidly. In this case, a bypass system, surge tank, or surge chamber is required. These provide an alternative path for the water, at least until it can be brought to rest more slowly. If air is present in the system, it will cushion the water hammer waves and reduce the pressure increase, but at the risk of creating new problems because of its tendency to move around within the system in an unpredictable fashion.

EXAMPLE 6–14

The pipe system of Fig. 6–21 consists of 500 ft of 1-ft-diameter, ½-in.-thick steel pipe with a resistance coefficient $f = 0.018$. The water in the reservoir has a temperature of 70°F and a head of 25 ft. The pipe is fastened rigidly at the reservoir, but is otherwise free to move axially (type 1 constraint).

Ignore pressure creep and minor losses and determine the pressure rise at the valve if it is completely closed in (1) 0.2 s and (2) 2.0 s.

Solution

The steady-flow energy equation provides the water velocity,

$$25 = \frac{V_0^2}{2g} + \frac{(0.018)(500)}{1}\frac{V_0^2}{2g}$$

from which $V_0 = 12.69$ ft/s. At 70°F, the modulus of compressibility $K = 3.2 \times 10^5$ psi. From Table 6–6, $E = 30 \times 10^6$ psi and $\mu = 0.30$. Thus $C_0 = 1.25 - 0.30 = 0.95$ and from Eq. 6–45

$$c = \frac{\sqrt{(3.2 \times 10^5)(144)/1.94}}{\left[1 + \left(\dfrac{3.2 \times 10^5}{30 \times 10^6}\right)\left(\dfrac{1}{1/24}\right)(0.95)\right]^{1/2}} = 4371 \text{ ft/s}$$

Therefore

$$\Delta p = \rho c V_0 = (1.94)(4371)(12.69) = 107,600 \text{ psf} = 747 \text{ psi}$$

Also

$$\frac{2L}{c} = \frac{(2)(500)}{4371} = 0.23 \text{ s}$$

A valve closure time of 0.2 s is less than this value. Hence, in part (1) the pressure rise will be 747 psi. In part (2) the time of closure exceeds $2L/c$ and therefore the pressure rise (according to Eq. 6–48) will be only

$$\Delta p = \frac{(747)(0.23)}{2} = 86 \text{ psi}$$

PROBLEMS

Assume water as the fluid unless otherwise stated. Take the kinematic viscosity of water as 1.2×10^{-5} ft²/s or 1.11×10^{-6} m²/s unless a temperature is given or an instruction is stated to the contrary.

Section 6–2

6–1. Determine the head loss in a 1000-ft-long 1-ft-diameter steel ($k = 0.00015$ ft) pipe if the discharge is 10 cfs.

6–2. Determine the pressure drop in a 2000-m-long 0.5-m-diameter pipeline ($k = 0.0005$ m) when the discharge is 1.25 m³/s.

6–3. What is the pressure drop in a 500-ft-long, smooth, 3-in.-diameter pipe carrying oil (sp. gr. $= 0.90$ and $\nu = 10^{-4}$ ft²/s) if the average velocity is 3 ft/s?

6–4. Estimate the head loss in a 10-ft-diameter corrugated metal pipe if the length is 10 miles and the discharge is 250 cfs.

6–5. What is the head loss in a 100-m length of smooth 0.3-m-diameter pipe if the flow rate is 0.35 m³/s?

6–6. Determine the water discharge through a 0.5-m-diameter pipeline if the head loss is 1 m in a length of 100 m. Assume $k = 0.1$ mm. Evaluate the discharge using both the Moody diagram and the Colebrook-White equation.

6–7. What is the discharge through a 6-in.-diameter cast iron pipe if the head loss is 4 ft per 1000 ft. Compare the discharge using both the Moody diagram and the Colebrook-White equation.

6–8. Determine the water discharge through a smooth 1-ft-diameter pipeline if the pressure drop is 2 psi per 1000 ft.

6–9. Water is to be transmitted through a horizontal 2-ft-diameter steel pipeline. The discharge is 47 cfs. Determine the required spacing along the pipeline of pumps that are able to deliver 200 hp to the system. Note: $P\ (hp) = Q\ \gamma\ H_P/550$.

6–10. Determine the diameter of concrete pipe ($k = 0.004$ ft and $n = 0.013$) required to carry 100 cfs at a head loss of 1 ft/mile. Use both the Darcy-Weisbach equation and the Manning equation.

6–11. Determine the diameter of cast iron pipe ($k = 0.00085$ ft and $n = 0.014$) required to transmit 0.4 m³/s at a head loss of 1 m per 1000 m. Use both the Darcy-Weisbach equation and the Manning equation.

6–12. Select a value of k/D and verify that Eq. 6–4 corresponds to the appropriate curve on the Moody diagram.

6–13. Write a computer program based on Eq. 6–6 that solves for the pipe diameter given the other parameters. Test it against Prob. 6–10 or 6–11.

6–14. Repeat Prob. 6–1 using the Hazen-Williams equation.

6–15. Repeat Prob. 6–7 using the Hazen-Williams equation.

6–16. Repeat Prob. 6–8 using the Hazen-Williams equation.

6–17. Repeat Example 6–3 assuming that the gate valve is only one-quarter open.

6–18. The initial section of a pipe with a square-edged entrance 50 m below the surface of a reservoir has a diameter of 0.4 m and a length of 100 m. This is followed by a gradual contraction ($\alpha = 25$ deg) to 0.2-m pipe that has a length of 150 m. The first section of pipe contains an open gate valve and 4 medium-radius 90 deg elbows. The second pipe discharges directly into the atmosphere at an elevation 30 m below that of the reservoir. Both pipes are cast iron. Determine (1) the discharge and (2) the pressure just inside the first pipe.

6–19. Repeat Prob. 6–18 if the gate valve is one-quarter open.

6–20. Compare the head loss through a 20 deg contraction with the head loss through a 20 deg expansion. In each case the pipe diameters are 20 cm and 40 cm and the discharge is 0.45 m³/s. Explain the difference.

6–21. By writing the continuity, energy, and momentum equations between the upstream and downstream sections of an abrupt expansion, derive the equation for the head loss given in Fig. 6–5a. Hint: select a control volume that extends from the point of the expansion to a point further downstream.

6–22. Write a computer program to calculate the flow in any number of parallel pipes given the pipe characteristics and the total discharge. Use the Manning equation.

6–23. Repeat Prob. 6–22 using the Darcy-Weisbach equation. Input the respective f values as data.

6–24. Repeat Prob. 6–22 using the Darcy-Weisbach equation. Input the pipe roughness k values. Hint: use Eq. 6–6.

6–25. Three pipes are connected in parallel. Their characteristics are as follows:

Pipe no.	Diameter (m)	Length (m)	Manning n
1	0.50	5000	0.015
2	0.40	4000	0.016
3	0.30	3000	0.017

Determine the discharge in each pipe if the total discharge is 2.55 m³/s.

6–26. Three pipes are connected in parallel. Their characteristics are as follows:

Pipe no.	Diameter (ft)	Length (ft)	Roughness k (ft)
1	1.5	2000	0.0001
2	2.5	1450	0.0002
3	2.75	4500	0.00015

Determine the discharge through each pipe when the total discharge is 110 cfs. Verify your choice of f values as the pipes may not be fully rough. Adjust as necessary.

6–27. Four pipes are connected in parallel. Their characteristics are as follows:

Pipe no.	Diameter (ft)	Length (ft)	Roughness k (ft)
1	0.5	10,000	0.0002
2	1	10,000	0.0002
3	1.5	10,000	0.0003
4	2	10,000	0.0003

Determine the discharge through each pipe if the total discharge is 50 cfs. Assume that the pipe flow is fully rough in each case.

6–28. Three reservoirs are connected at a common junction. The reservoirs and connecting pipes have the following characteristics:

Reservoir	Elevation (ft)	Pipe	Diameter (in.)	Length (ft)	Roughness k (ft)
A	500	a	8	300	0.00019
B	400	b	8	1500	0.00019
C	200	c	12	1200	0.00019

Determine the direction and magnitude of the flow in each of the pipes. Assume that the flow is fully rough in each pipe.

6–29. Repeat Prob. 6–28 if a reservoir at D at an elevation of 150 ft is connected to the common junction by a pipe d with diameter of 10 in., length of 700 ft, and $k = 0.00017$ ft.

6–30. Repeat Prob. 6–28 if the roughness height k is replaced by Manning n values of 0.015, 0.014, and 0.013, respectively.

6–31. Redo Example 6–6 if the only inflow is 12 cfs at point A and the only outflow is at point D.

6–32. Redo Example 6–6 if the only inflow is 12 cfs at point A and the only outflow is at point B.

6–33. Redo Example 6–6 if each pipe has a Manning $n = 0.014$.

6–34. A two-loop pipe network has node designations identical to Fig. 6–9. Inflows of 0.4 m^3/s and 0.45 m^3/s enter at points A and B, respectively. Equal withdrawals are made at points C, F, and D. The pipe characteristics are as follow:

Pipe	Length (m)	Diameter (m)	Resistance coefficient f
AB	500	0.4	0.017
BC	400	0.5	0.016
AF	650	0.5	0.014
BE	750	0.35	0.015
CD	700	0.4	0.013
DE	550	0.5	0.016
EF	900	0.6	0.015

Determine the magnitude and direction of the flow in each pipe.

Section 6–3.

6–35. Two interconnected reservoirs (see Fig. 6–11) have surface areas of 10,000 ft^2. The connecting pipe has a diameter of 2 ft, a length of 40,000 ft, and a Darcy-Weisbach coefficient $f = 0.022$. The original difference in water surface elevation between the two reservoirs was 300 ft. Determine the time required for the difference to drop to 100 ft. Ignore minor losses.

6–36. Repeat Prob. 6–35 if 2 cfs are added continuously to the upper reservoir.

6–37. Two interconnected reservoirs (see Fig. 6–11) have surface areas of 4000 and 2000 ft^2, respectively. The connecting pipe has a diameter of 6 in., a length of 4000 ft, and a Darcy-Weisbach coefficient $f = 0.020$. The pipe also includes a square-edged entrance, a half-open gate valve, and four medium-radius elbows. The initial difference in water surface elevation between the two reservoirs was 100 ft. Determine the time required for the difference to drop to 50 ft.

6–38. Two interconnected reservoirs (see Fig. 6–11) have surface areas of 3000 and 1000 m^2, respectively. The connecting pipe has a diameter of 0.3 m, a length of 10,000 m, and a Darcy-Weisbach coefficient $f = 0.025$. The initial difference in water surface elevation between the two reservoirs was 120 m. Determine the time required for the difference to drop to (1) 110 m, (2) 100 m, and (3) 90 m. Ignore minor losses.

6–39. A 2000-m-long, 0.1-m-diameter pipe with $f = 0.030$ discharges water from a circular tank 8 m in diameter. The maximum and minimum operating water levels in the tank are 120 m and 105 m above the pipe outlet, respectively. The entrance to the pipe is square edged and the pipe includes a gate valve that is fully open. Determine the range of discharges and the time required for the water level to drop from the maximum to the minimum level.

6–40. Repeat Prob. 6–39 if water is continually added to the tank at the rate of 0.005 m³/s.

6–41. Repeat Prob. 6–39 if water is continually added to the tank at the rate of 0.009 m³/s.

6–42. A 10,000-ft-long, 1-ft-diameter cast iron pipe discharges water from a tank with a surface area of 15,000 ft². The initial water level in the tank is 200 ft above the outlet of the pipe. Determine the time required for the water level to fall (1) 50 ft and (2) 100 ft. Ignore minor losses.

6–43. Repeat Prob. 6–42 if water is added to the reservoir at the rate of 1 cfs.

6–44. Redo Example 6–8 if there is no inflow into the reservoir.

6–45. Redo Example 6–8 if the inflow of 0.02 m³/s continues for 10 min. then ends abruptly.

6–46. Redo Example 6–8 if the inflow of 0.02 m³/s starts when the head falls below a level at 6 m above the outlet.

6–47. Determine the time required for the pipe of Prob. 6–39 to reach 99 percent of its steady-flow discharge at both the maximum and minimum heads.

6–48. Determine the time required for the pipe of Prob. 6–39 to reach 98 percent of its steady-flow discharge at both the maximum and minimum heads.

6–49. Determine the time required for the pipe of Prob. 6–42 to reach 99 percent of its steady-flow discharge at the maximum head.

6–50. Determine the time required for the pipe system of Example 6–10 to reach (1) 20 percent and (2) 80 percent of its maximum flow rate.

Section 6–4

6–51. Determine the velocity if a Pitot tube indicates a pressure difference of 2 psi and the fluid is (1) water, (2) oil (sp.gr. = 0.85), and (3) air.

6–52. Repeat Prob. 6–51 if the pressure difference is 15 kN/m².

6–53. Determine the velocity of water if two manometers connected to a Pitot tube indicate a difference in piezometric head of (1) 1 ft and (2) 2 ft.

6–54. What pressure difference would be expected in a water velocity of 10 ft/s if a Pitot tube has a correction factor of (1) 1.0, (2) 0.98, and (3) 0.96.

6–55. Assuming average correction coefficients from Fig. 6–15, determine the pressure differences that each of the three types of Pitot tubes should be expected to indicate given a water velocity of 3 m/s.

6–56. An orifice meter with a diameter of 5 cm is placed in a 15-cm-diameter pipe. Estimate the discharge associated with a pressure difference of 18 kN/m^2.

6–57. A 4-in.-diameter pipe has a 3-in.-diameter orifice plate installed for flow measurement. What is the meter equation between the water discharge in cfs and the piezometric head differential in inches?

6–58. What is the discharge through the meter in Prob. 6–57 when the pressure difference is 2 psi?

6–59. Determine the discharge through the venturi meter of Example 6–11 if the pressures are as given but the meter is (1) horizontal and (2) inverted so that the flow direction is vertically downward?

6–60. Determine the relationship between the discharge Q and the piezometric head Δh in an elbow meter if the pipe diameter is 10 cm and the elbow has a radius of (1) 10 cm and (2) 20 cm.

Section 6–5

In each of the following assume that there is no longitudinal pipe stress unless stated to the contrary.

6–61. Determine the tangential stress due to a pressure of 150 psi in a pipe with a diameter of 16 in. and a wall thickness of (1) 0.75 in. and (2) 0.25 in.

6–62. If the allowable pipe stress is 10,000 psi, determine the minimum wall thickness necessary in a 2-ft-diameter pipe to withstand 175 psi.

6–63. What tangential stress will occur in a pipe with a diameter of 40 cm, a wall thickness of 1 cm, and an internal pressure of 900 kN/m^2?

6–64. Derive an equation equivalent to Eq. 6–32 for a situation where it is not adequate to work with the centerline pressure and the hydrostatic pressure distribution must be considered vertically across the pipe section.

6–65. What is the tangential stress in a 2-ft-diameter pipe with a wall thickness of 1 in. if the internal pressure is 2 psia?

6–66. A 1000-ft-long cast iron pipe is rigidly supported against movement. What stress will occur in the pipe if there is a temperature change of (1) +50°F and (2) −50°F?

6–67. What length change would occur in the pipe of Prob. 6–66 if movement were not restricted?

6–68. A gradual pipe contraction reduces the diameter from 4 ft to 3 ft. Determine the force on the contraction if the water discharge is 160 cfs and the upstream pressure is 50 psi.

6–69. Determine the total thrust on a 45 deg elbow in a 10-ft-diameter pipeline if the discharge is 3900 cfs and the pressure is 200 psi.

6–70. Determine the total thrust on a 90 deg reducing elbow that reduces the pipe diameter from 2 ft on the upstream end to 1.5 ft on the downstream end. The discharge is 24 cfs and the downstream pressure is 20 psi.

6–71. Determine the total thrust on a 45 deg reducing elbow that reduces the pipe diameter from 1 m to 0.5 m. The discharge is 3 m^3/s and the upstream pressure is 300 kN/m^2.

6–72. A 90 deg reducing elbow has upstream and downstream diameters of 40 and 25 cm, respectively. The corresponding pressures are 300 kN/m^2 and 200 kN/m^2. What is the total force on the elbow?

6–73. Determine the force on a ∪ pipe section located in a 4-ft-diameter pipeline. The discharge is 36 cfs and the pressure in the pipe is 25 psi.

6–74. A 4-ft-diameter pipe leads from a reservoir. The initial section of pipe is at an elevation of 1000 ft. It immediately slopes downward at an angle of 45 deg to an elevation of 300 ft, extends horizontally for 1000 ft, then slopes upward at 45 deg to an elevation of 800 ft. From there, it slopes downward at an angle of 45 deg until it reaches a turbine at an elevation of 100 ft. Just before the water enters the turbine at the lower end of the system, the pressure is 600 psi. Include the effects of friction and determine the thrust on each of the two 45 deg elbows at the 300-ft elevation, and the 90 deg elbow at the 800-ft elevation. The flow rate is 500 cfs and the resistance coefficient f = 0.011. Note that the reservoir elevation is not given. In addition, ignore any minor losses and the pressure changes through the elbows.

6–75. The horizontal pipe of Fig. 6–22 is supplied by a reservoir with a 400-ft head. The pipe has a diameter of 6 ft, a length of 4000 ft, a wall thickness of ¾ in., and a valve at the downstream end. The pipe material is steel (roughness k = 0.00015 ft, E = 30 × 10^6 psi, and μ = 0.3). For the water assume K = 300,000 psi and ν = 1.2 × 10^{-5} ft^2/s. (1) Calculate the steady flow velocity and for each of the three types of pipe constraints calculate the sonic velocity. (2) For each of the three types of constraints calculate the pressure rise at the valve if instantaneous closure occurs. Ignore the additional pressure creep due to friction.

6–76. If the average velocity in the pipe of Prob. 6–75 is 3 ft/s and type 3 constraint is assumed, calculate the maximum pressure at the valve and the stresses in the pipe if instantaneous closure occurs. Hint: At V_0 = 3 ft/s, the valve is not necessarily wide open. Use the energy equation to determine the pressure conditions at the valve under steady flow (ignore minor losses).

6–77. What is the pressure rise at the valve in Prob. 6–76 if the time of closure is (1) 1.5 s, (2) 3 s, and (3) 6 s?

6–78. The horizontal steel pipe of Fig. 6–22 is supplied from a reservoir with a head of 200 m. The pipe has a diameter of 20 cm, a length of 150 m, and a wall thickness of 0.25 cm. There is a valve at the downstream end. The pipe has a type 2 constraint. The water temperature is 30°C. If the valve is initially wide open, calculate the pressure rise at the valve if the time of closure is (1) 0.2 s, (2) 0.8 s, and (3) 5 s. Ignore minor losses and assume f = 0.014.

6–79. Repeat Prob. 6–78 if the pipe has a type 1 constraint.

6–80. Repeat Prob. 6–78 if the valve is initially partially closed so that the average velocity in the pipe is 3 m/s.

6–81. Repeat Prob. 6–78 if the valve is initially partially closed so that the average velocity in the pipe is 1 m/s.

6–82. Repeat Prob. 6–78 if the pipe is concrete (type 3 constraint) with a diameter of 0.6 m and a wall thickness of 3 cm. Use f = 0.020.

References

Brebbia, C. A. and A. J. Ferrante, *Computational Hydraulics,* London: Butterworths, 1983.

Featherstone, R. E. and C. Nalluri, *Civil Engineering Hydraulics,* London: Granada, 1982.

Jeppson, R. W., *Analysis of Flow in Pipe Networks,* Ann Arbor, MI.: Ann Arbor Science, 1976.

Prasuhn, A. L., *Fundamentals of Fluid Mechanics,* Englewood Cliffs, NJ.: Prentice-Hall, 1980.

Rouse, H. (ed.), *Engineering Hydraulics,* New York: Wiley, 1950.

Wylie, E. B. and V. L. Streeter, *Fluid Transients,* New York: McGraw-Hill, 1978.

chapter 7

Open Channel Hydraulics

Figure 7–0 November 5, 1913, Dedication day at the Los Angeles Cascades. The Owens River–Los Angeles Aqueduct, designated a National Historic Civil Engineering Landmark by the American Society of Civil Engineers, was the first major aqueduct built to conduct a large quantity of water to a major metropolitan area (courtesy of the Los Angeles Department of Water and Power).

7-1 INTRODUCTION

Open channel flow refers to any flow that occupies a defined channel and has a free surface. Because of the introductory nature of this book, major emphasis will be placed on flow in a rectangular channel. The simpler geometry leads to less mathematical complexity, yet illustrates all of the essential concepts. Other regularly used cross-sectional shapes such as trapezoidal and circular channels will be considered to a lesser degree. Some attention will also be given to the natural channel or river. The different flow conditions to be considered include steady uniform flow, unsteady flow (but only to a very limited extent), and steady nonuniform flow.

Uniform flow has been defined as flow with straight parallel streamlines. For a closed conduit flowing full, this will usually be satisfied if the conduit is straight and its cross section is constant along a portion of its length. For an open channel flow to be uniform, the same conditions must apply to the channel. That is, it must be straight and have a constant cross section. This is not sufficient, however, because the water surface must also be parallel to the channel slope. In simplest form, uniform open channel flow requires that $dy/dx = 0$, where y is the depth and x is the horizontal direction. This condition will be considered further in the following section. It should be apparent from the above that uniform flow will rarely occur in the natural channel.

Unsteady flow refers to conditions that change with time at a specific point or section. For example, if at any specified point in the flow the velocity is changing with time, the flow is unsteady. If at all points in the flow the velocity is constant with respect to time, the flow is steady. This is in contrast to nonuniform flow, in which the velocity changes from point to point in the flow direction. It is almost impossible to keep the water surface from changing with time in an unsteady open channel flow. Thus, an unsteady, uniform flow is most unlikely.

The following examples will illustrate these different conditions. Flow down a long rectangular chute at constant discharge would (after an initial length) be both steady and uniform. The opening of an upstream sluice gate in the same channel would produce an unsteady, nonuniform flow, at least until the channel adjusted to the new discharge. If at a specific section in the above chute there was a change in slope or cross-sectional size or shape, then the flow in the vicinity of the change would be nonuniform. In unsteady flow the reference point becomes important. To illustrate, consider a boat passing through otherwise still water. The water in the vicinity of the boat would appear as a steady flow to a passenger on the boat, whereas to a stationary observer the scene would appear as unsteady. It would be nonuniform in either case.

We will define the *depth* in an open channel flow as the vertical distance from the water surface to the lowest point in the channel bed. In a rectangular channel the depth will be constant across the entire section, while in a natural channel the depth at any section must be measured at the deepest point. In a uniform flow the depth will remain constant from section to section.

Because of the free surface, gravity may be expected to play an important role. Just as a pressure gradient was frequently responsible for flow in a closed conduit, gravity acting through the fluid weight causes the water to flow down a slope. Consequently, the Froude number will usually be the significant parameter.

7–2 PRESSURE DISTRIBUTION, RESISTANCE, AND THE MANNING EQUATION

Pressure Distribution

Uniform flow in an open channel requires that the water surface profile is parallel to the channel bottom, since only then will the streamlines be straight and parallel. This condition is shown in Fig. 7–1. Two related questions must be answered at the outset: for a slope angle θ as shown, what is the difference between the depth y_0 and the thickness d, and with this in mind what is the pressure at point 1? The thickness $d = y_0 \cos \theta$, thus for small angles, say up to about 5 deg, there is very little difference. Almost all open channel flows considered in this book will fall within this range, and the term depth will be used exclusively. It was stated in the review of fluid mechanics that the piezometric head is constant along a line normal to the flow direction in uniform flow. With reference to Fig. 7–1, the piezometric head at point 1 must therefore equal that at point 2:

$$h_1 = h_2$$

or

$$\frac{p_1}{\gamma} + y_1 = \frac{p_2}{\gamma} + y_2$$

therefore,

$$p_1 = \gamma(y_2 - y_1)$$

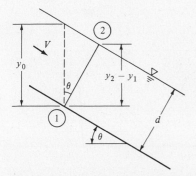

Figure 7–1 Definitional sketch of a steep channel.

since atmospheric pressure prevails at the surface and $p_2 = 0$. Now y_1 and y_2 are elevations measured vertically, as is the depth, and

$$y_2 - y_1 = d \cos \theta = y_0 \cos^2 \theta$$

The pressure at the bottom must therefore be

$$p_1 = y_0 \cos^2 \theta \qquad\qquad\qquad (7\text{--}1)$$

The pressure distribution remains linear, but depends on the slope, according to Eq. 7–1. For small slopes the angle may be ignored and the usual hydrostatic equation used.

Open Channel Resistance

A uniform flow with a more moderate slope is shown in Fig. 7–2. The channel cross section has been included and given a general irregular shape for which it is still possible to have a uniform flow, provided that along the entire channel all cross sections have that identical shape. Note that the velocity head has been added to locate the line of total head or energy grade line (EGL). The EGL must also be parallel to the water surface and channel bottom, since the average velocity will remain constant in a uniform flow.

Uniform flow represents an equilibrium between the driving force of gravity down the slope and the resistance due to friction around the perimeter of the channel. This equilibrium may be expressed for the region of length L shown. The water therein has a weight γLA, for which the down slope component is $\gamma LA \sin \theta$. This is resisted by the shear stress τ_0, which acts over the surface area PL. Here P, called the wetted perimeter, is the portion of the perimeter that resists the flow (overlying air drag being ignored). The hydrostatic pressure distributions on either end are equal in uniform flow and do not contribute to the analysis. Equating the two forces

$$\gamma LA \sin \theta = \tau_0 PL$$

or

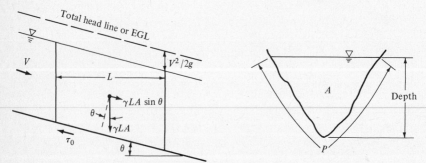

Figure 7–2 Open-channel resistance.

$$\tau_0 = \gamma \frac{A}{P} \sin \theta$$

The ratio of A/P is defined as the hydraulic radius R, a single-length dimension that represents the cross-sectional shape. This was introduced previously in conjunction with pipe hydraulics. The slope of a channel S_0 is defined as $\tan \theta$, but for the generally small angles involved we may write

$$S_0 = \tan \theta \approx \sin \theta$$

With the above substitutions, the resistance equation becomes

$$\tau_0 = \gamma R S_0 \tag{7-2}$$

Although the shear stress τ_0 has a constant magnitude around the circumference of a pipe flowing full, such is not the case in an open channel flow, and τ_0 in Eq. 7–2 must be treated as the average value around the wetted perimeter. If the flow is nonuniform, the analysis is greatly complicated by the necessary inclusion of the difference in the pressure distributions as well as a convective acceleration term. The derivation will not be examined here, but the result is the similar equation

$$\tau_0 = \gamma R S_f \tag{7-3}$$

The resistance equation for nonuniform flow differs in that the channel slope S_0 has been replaced by the slope of the EGL, S_f. This latter quantity is called the *friction slope*.

The square root of the quantity τ_0/ρ has dimensions of length/time, and for this reason is called the *shear velocity, u_**. From Eq. 7–2 we get

$$u_* = \sqrt{\tau_0/\rho} = \sqrt{gRS_0} \tag{7-4}$$

This is an important parameter that will show up again in Chapters 8 and 9.

The Manning Equation and Normal Depth

The second resistance equation (and one that will be used extensively) is the Manning equation, which was introduced previously in the context of pipe flow. It was originally proposed as an essentially empirical open channel equation appropriate to the turbulent flow of water. As formulated for uniform flow in the European metric system of units (or the present SI units), the Manning equation states that

$$Q = \frac{1}{n} A R^{2/3} S_0^{1/2} \tag{7-5a}$$

where Q is the discharge in m^3/s and n is the Manning n or resistance coefficient. Values of n for open channels are provided in Table 7–1. The remaining quantities, A, R, and the dimensionless S_0 are as defined previously. When the Manning equation was converted to U.S. customary units, it was felt that the Manning n values

Table 7–1 MANNING *n* FOR OPEN CHANNELS[1]

Surface	Manning *n*
Very smooth surface (glass, plastic, machined metal)	0.010
Planed timber	0.011
Unplaned wood	0.012–0.015
Smooth concrete	0.012–0.013
Unfinished concrete	0.013–0.016
Brickwork	0.014
Rubble masonry	0.017
Earth channels smooth, no weeds	0.020
Firm gravel	0.020
Earth channel with some stones and weeds	0.025
Earth channels in bad condition, winding natural streams	0.035
Mountain streams	0.040–0.050
Sand (flat bed), or gravel channels	
(d = median grain diameter[2], ft)	$0.034d^{1/6}$
(d = d-75 size[3], ft)	$0.031d^{1/6}$

[1]Reference should be made to Chow's *Open-Channel Hydraulics* for an extensive list of *n* values complete with photographs.
[2]The median grain diameter is the grain size for which 50 percent of the channel bed material is finer and 50 percent is coarser. In Chapter 8, this will be indicated as d_{50}. It should be noted that this equation, known as the Strickler equation, is less satisfactory for sand bed channels.
[3]The d-75 size (or d_{75}) is the size for which 75 percent of the bed material is finer.

should remain unchanged. Consequently, the numerical value had to be converted and the equation takes the form.[1]

$$Q = \frac{1.49}{n}AR^{2/3}S_0^{1/2} \tag{7–5b}$$

in U.S. customary units.

The Manning equation is often applied to nonuniform flows by the simple expedient of replacing the channel slope S_0 by the friction slope S_f. This works satisfactorily and is consistent with the previous theoretical treatment of the resistance equation (see Eqs. 7–2 and 7–3).

The depth associated with uniform flow is called the *normal depth*. For a given channel and discharge there will be only one normal depth, which is usually obtained by using the Manning equation as illustrated in Example 7–1. An iterative solution is required to calculate the normal depth, and the reader is urged to develop a computer program to avoid the repetitive computations. The calculation of the normal

[1]For an interesting and thorough discussion, refer to page 98 of *Open-Channel Hydraulics* by Chow, referenced at the end of the chapter. In a word, the conversion factor is based on $(3.28)^{1/3} = 1.486$, where 3.28 is the conversion from feet to meters. Based on the accuracy with which Manning *n* values are known, a rounding to 1.49 is generally accepted.

depth is basic to the analysis of open channel flows and is required even for nonuniform flow.

EXAMPLE 7–1

Determine the normal depth in a trapezoidal channel with side slopes of 1.5 on 1, a bottom width of 25 ft, and a channel slope of 0.00088, if the discharge is 1510 cfs and the Manning n is 0.017 (Fig. 7–3).

Solution

The Manning equation in U.S. customary units may be written

$$Q = \frac{1.49}{n} A R^{2/3} S_0^{1/2} = \frac{1.49}{n} \frac{A^{5/3}}{P^{2/3}} S_0^{1/2} \qquad R = \frac{A}{P}$$

where for the trapezoidal shape

$$A = 25 y_n + 1.5 y_n^2$$

and

$$P = 25 + 2 y_n \sqrt{1 + (1.5)^2}$$

It is convenient to separate the known from the unknown quantities, which leads to

$$\frac{Qn}{1.49 S_0^{1/2}} \left[= \frac{(1510)(0.017)}{(1.49)(0.00088)^{1/2}} = 580.8 \right] = \frac{(25 + 1.5\, y_n^2)^{5/3}}{(25 + 2 y_n \sqrt{3.25})^{2/3}}$$

Solving iteratively, the normal depth is found to be $y_n = 6.22$ ft.

The relationship between the normal depth and discharge expressed by the Manning equation exists in a natural channel as well. The usual method of describing the cross section of a river is through the use of a number of coordinate points. Various schemes are available to compute area, wetted perimeter, and hydraulic radius from

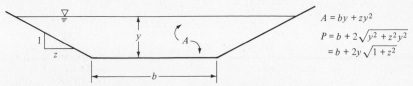

$$A = by + zy^2$$
$$P = b + 2\sqrt{y^2 + z^2 y^2}$$
$$= b + 2y\sqrt{1 + z^2}$$

Figure 7–3 Definitional sketch of a trapezoidal channel.

the sets of *x, y* coordinates. This will be discussed in Section 7–8. However, there is one aspect of the natural channel that remains to be covered in this section. A typical river consists of a main channel and one or more floodplain or overbank sections (Fig. 7–4). Only one overbank section is shown, but the procedures that follow are easily generalized for additional overbank sections.

In most rivers the normal water surface or *stage* is contained within the main channel. Only the higher flood stage discharges spill onto the floodplain. Assuming this condition for the present, we are still interested in applying the Manning equation, but the cross-sectional shape is now too extreme to be well represented by the single length dimension of the hydraulic radius. Further, the average velocities in the overbank sections are generally much less than the average velocity in the main channel. Thus, for purpose of analysis, the cross section is divided into subsections consisting of the main channel and one or more overbank sections. It will be assumed in the process that the water surface elevation is constant across the entire section and that the same value of slope (which for a river must be the friction slope) applies to all subsections.

The subsections are somewhat arbitrarily divided by the dashed line in Fig. 7–4 into regions 1 and 2. Applying the Manning equation to each section while recognizing the common friction slope leads to

$$Q = Q_1 + Q_2$$

$$= \frac{1.49}{n_1} A_1 R_1^{2/3} S_f^{1/2} + \frac{1.49}{n_2} A_2 R_2^{2/3} S_f^{1/2}$$

or

$$Q = 1.49 S_f^{1/2} \left(\frac{A_1^{5/3}}{n_1 P_1^{2/3}} + \frac{A_2^{5/3}}{n_2 P_2^{2/3}} \right) \tag{7–6}$$

The actual area is used for each subsection, but the wetted perimeter is based on only the channel boundary as shown. This ignores any shear stress between the regions and introduces some error if V_1 is much larger than V_2, as it is inclined to be. Only the U.S. customary version is included in Eq. 7–6, and the 1.49 must be replaced with unity when working with SI units.

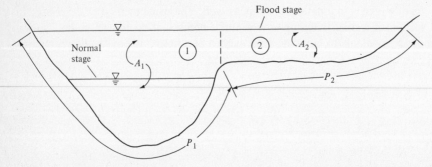

Figure 7–4 River with floodplain.

Example 7–2

A cross section of a river in flood stage is approximated by the three rectangles of Fig. 7–5. Estimate the discharge if the main channel has a sand bed with a median grain size of 1 mm and the two overbank portions are heavily overgrown with estimated n values of 0.065. The friction slope is 0.0005.

Solution

Using the Strickler equation from Table 7–1, n for the main channel is

$$n = 0.034d^{1/6}$$

$$= (0.034)\left(\frac{1}{304.8}\right)^{1/6} = 0.013$$

assuming that the bed of the stream is flat. Substituting into Eq. 7–5 (but allowing for three subsections) we get

$$Q = (0.0005)^{1/2}\left[\frac{(180)^{5/3}}{(0.013)(39)^{2/3}} + \frac{(200)^{5/3}}{(0.065)(102)^{2/3}} + \frac{(160)^{5/3}}{(0.065)(161)^{2/3}}\right]$$

$$= 1020 \text{ m}^3/\text{s}$$

This example is potentially in considerable error. The Strickler equation is not particularly good for sand channels. Under the given flow conditions, transport of bed material by the river is likely and the bed will not necessarily remain flat. These complications will be considered more carefully in the following chapter. In actual fact, a Manning $n = 0.20$ is probably more reasonable for the main channel.

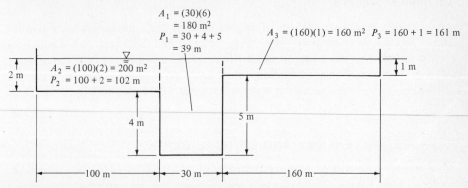

Figure 7–5 River cross section for Example 7–2.

Hydraulically Efficient Cross Sections

The most efficient cross section is one that delivers the greatest discharge for a given area. For a given discharge this corresponds to the minimum cross-sectional area and minimum wetted perimeter. The circle is the most efficient geometric shape in this respect, and an open channel with a semicircular cross section is equivalent. Other geometric shapes are less efficient, but it can be shown that maximum hydraulic efficiency is achieved when the cross section is closest to a semicircle. This is frequently associated with a hydraulic radius, $R = y/2$.

Accordingly, the most efficient rectangular channel should have a width $b = 2y$. This may be proven for a rectangular channel by the following argument. For values of Q and S and n, the Manning equation may be written

$$A = CP^{2/5}$$

where C is known (thereby demonstrating that the wetted perimeter P is a minimum when A is a minimum). For a rectangular channel of depth y and bottom width b, we have $A = by$ and $P = b + 2y$. Thus,

$$A = (P - 2y)y = CP^{2/5}$$

Differentiating with respect to y in order to obtain the minimum P yields

$$\left(\frac{dP}{dy} - 2\right)y + (P - 2y) = \frac{2}{5} CP^{-3/5} \frac{dP}{dy}$$

Upon setting $dP/dy = 0$, we get $P = 4y$ or the predicted $b = 2y$.

Without going through the details, the most efficient trapezoidal channel is a half-hexagon, the channel therefore containing 60 deg side slopes. Although hydraulic efficiency is an attractive goal, other factors such as construction costs and channel material may be more important. For materials such as wood or poured-in-place concrete, the semicircular cross section is often not practical. An unlined trapezoidal channel cut through alluvial material must have side slopes less than the angle of repose (see Section 8–4).

Channel Freeboard

In the design of any channel, an allowance must be made for factors such as wave height and design uncertainty. This safety factor, called *freeboard*, varies with conditions and no best value can be stated. However, an estimate can be obtained using the U.S. Bureau of Reclamation design chart given in Fig. 7–6. These curves provide the height of the banks above the water surface and, in the case of a lined canal, the height of the lining above the water surface.

7–3 SPECIFIC ENERGY AND CRITICAL DEPTH

In this section we will develop some aspects of the energy equation that are unique to open channel flows. These concepts will be introduced through an analysis of one type of transition and resulting nonuniform flow in a rectangular channel. Many of

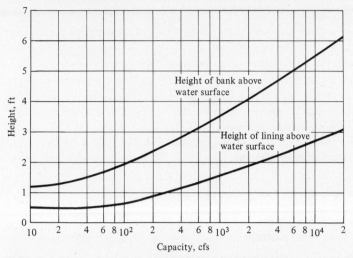

Figure 7–6 Recommended freeboard and height of lining (U.S. Bureau of Reclamation).

these concepts are important throughout the remainder of the chapter. The transitions of interest may involve changes in the elevation of the channel bottom, changes in the channel width, or both. The important characteristic is that the change must occur over such a short length of channel that friction may be ignored. As a result, the channel slope can also be ignored, at least as a first approximation. The channel will be shown horizontally for clarity, but for most slopes the results of this analysis can be applied to a sloping channel as well.

Starting with the upward change in channel bottom Δz depicted in Fig. 7–7, let us examine the problem of determining the downstream conditions of depth and velocity given the upstream depth and either the discharge or upstream velocity. The channel is assumed rectangular with width b and the flow is steady. Therefore, the remaining upstream quantities including the total head can be calculated immediately. Since friction is ignored, the EGL is horizontal as shown.

The energy equation (see Section 2–4) may be written between the upstream and downstream channel sections. With a datum chosen at the elevation of the lower level we get

$$y_1 + \frac{V_1^2}{2g} = (y_2 + \Delta z) + \frac{V_2^2}{2g}$$

The unknown quantities are y_2 and V_2, but the velocity can be readily eliminated. As will frequently be the case, the discharge Q in a rectangular channel can advantageously be replaced by the discharge per unit width, q,

$$q = Q/b = Vy \tag{7–7}$$

and continuity may be expressed by

$$q = V_1 y_1 = V_2 y_2 \tag{7–8}$$

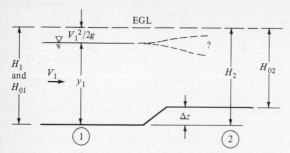

Figure 7–7 Transition in an open channel.

This may be substituted into the energy equation to yield

$$y_1 + \frac{q^2}{2gy_1^2} - \Delta z = y_2 + \frac{q^2}{2gy_2^2} \tag{7-9}$$

leaving only the unknown y_2 on the right-hand side. This may be solved by iteration but a difficulty remains: there will usually be two positive real roots, only one of which is correct, and we have no basis on which to make the correct decision. There is also the possibility of no positive real roots to compound the problem. We therefore need to look further for a relationship that will provide some additional insight into the behavior of a free surface flow.

The solution to the above problems lies in the introduction of a new energy relationship called the *specific energy* or *specific head*. Deceptively simple, the specific energy, denoted by H_0, is the energy (or head) as measured relative to the channel bottom rather than a horizontal datum. Thus, for any channel we may write

$$H_0 = y + \frac{V^2}{2g} \tag{7-10}$$

If the channel is rectangular, then through the introduction of Eq. 7–7 this becomes

$$H_0 = y + \frac{q^2}{2gy^2} \tag{7-11}$$

The quantities H_{01} and H_{02} are included on Fig. 7–7. Equation 7–9 may now be expressed in terms of the specific energy.

$$H_{01} - \Delta z = H_{02} \tag{7-12}$$

We see that unlike the actual energy equation, the specific energy is not constant if the elevation of the channel bottom changes. However, the specific energy will be constant if the bottom remains horizontal.

Specific Energy Diagram, Critical Depth, and Froude Number

To apply this new concept, it is instructive to plot a graph or *specific energy diagram* of H_0 versus depth y for a channel of constant width such as that in Fig. 7–7. Consider Eq. 7–11, and note that q remains unchanged as the flow passes through the transition. At the limit as y goes to zero, H_0 approaches infinity, and as y gets very large, y and H_0 become equal. The entire graph is shown in Fig. 7–8. The lower limb of the diagram represents the smaller depths, while the larger depths compose the upper limb, which is asymptotic to the 45° line shown.

The minimum point on the curve is defined as the *critical depth* y_c and the corresponding specific energy is H_{0min}. For any value of H_0 greater than H_{0min} there are two possible depths. They are for the present called y_A and y_B. At either depth, the horizontal distance from the vertical axis to the curve consists of two parts. The distance from the vertical axis to the 45 deg line must equal the depth, and from Eq. 7–10 the remaining distance must be the velocity head, $V^2/2g$. From the relative line lengths it is apparent that y_A and the entire lower limb are relatively shallow, rapid flows, while y_B and the upper limb are relatively deep but slower flows. The first case, with $y_A < y_c$, is called a *supercritical* flow and the second, with $y_B > y_c$, is a *subcritical* flow.

We may determine the critical depth and the corresponding minimum point on the specific energy diagram by differentiating H_0 with respect to the depth and equating the result to zero. Differentiating

$$\frac{dH_0}{dy} = 1 - \frac{q^2}{gy^3} \tag{7–13}$$

Setting the derivative equal to zero and solving for the depth, which is now the critical depth, gives

$$y_c = \sqrt[3]{\frac{q^2}{g}} \tag{7–14}$$

Combining Eq. 7–14 with $q = V_c y_c$ we get

$$q^2 = gy_c^3 = V_c^2 y_c^2$$

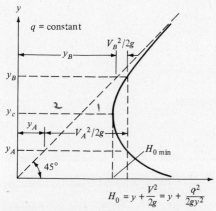

$$H_0 = y + \frac{V^2}{2g} = y + \frac{q^2}{2gy^2}$$

Figure 7–8 Specific energy diagram.

and

$$y_c = \frac{V_c^2}{g} = 2\left(\frac{V_c^2}{2g}\right)$$

which expresses the critical depth in terms of the critical velocity. Writing Eq. 7–10 at the minimum point therefore yields

$$H_{0min} = y_c + \frac{V_c^2}{2g} = \frac{3}{2}y_c \qquad (7\text{–}15)$$

which provides an additional relationship involving the critical depth. But be aware that unlike Eq. 7–14, which can always be used to determine the critical depth in a rectangular channel, Eq. 7–15 only applies to H_{0min}. It cannot be used, for instance, to obtain the critical depth from a given value of H_0.

The last aspect to be introduced before we apply the specific energy diagram to flow transition problems is the Froude Number, Fr, which was previously introduced in Section 2–5. If we define Fr in terms of the average velocity and the depth we immediately get

$$\text{Fr} = \frac{V}{\sqrt{gy}} \qquad (7\text{–}16)$$

The Fr at critical depth must be

$$Fr_c = \frac{V_c}{\sqrt{gy_c}}$$

but from the above, $y_c = V_c^2/g$ and therefore $Fr_c = 1$. With reference to Eq. 7–16 and Fig. 7–8, if $y > y_c$, then $V < V_c$ and Fr < 1. Conversely, if $y < y_c$, then $V > V_c$ and Fr > 1. The different flow regimes may be summarized as follows:

Flow type	Depth	Velocity	Froude number
Subcritical flow	$y > y_c$	$V < V_c$	Fr < 1
Critical flow	$y = y_c$	$V = V_c$	Fr = 1
Supercritical flow	$y < y_c$	$V > V_c$	Fr > 1

Applications of the Specific Energy Diagram

Transition with Change in Bed Elevation

Consider again the transition shown in Fig. 7–7. The channel bottom increases in elevation due to the upward step. The specific energy, H_0, must therefore decrease, which means that a solution will be obtained by moving to the left (i.e., in the direction of decreasing H_0) on the specific energy diagram. To use this information we must determine whether we are on the upper or lower limb. The Froude number could be calculated for the upstream conditions, but it is best to calculate y_c (Eq. 7–14) and compare with the value of y_1.

It is assumed in Fig. 7–9, which is a repeat of the transition in Fig. 7–7, that the flow is subcritical. With this assumption, y_1 and H_{01} can be identified on the specific energy diagram of Fig. 7–9a. Moving to the left a distance Δz locates H_{02} and y_2. In passing through the transition, the depth follows the curve from point 1 to point 2. Hence, the depth decreases through the transition, and in the process the water surface elevation must decrease as well. The latter effect is less obvious but must occur, because the velocity head (the horizontal distance from the 45 deg line to the curve) increases in going from section 1 to 2. Since EGL is horizontal, the water surface must drop. The resulting profile is shown in Fig. 7–9b.

Completing the solution still requires solving a cubic equation, but the range of the downstream depth will be known (see Example 7–3) so that the iterative solution will converge on the correct root. Equation 7–12 can be used, and the only unknown can be expressed in terms of y_2,

$$H_{01} - \Delta z = H_{02} = y_2 + \frac{q^2}{2gy_2^2} \qquad (7\text{--}17)$$

which may now be solved for the downstream depth.

If at the outset it is found that the upstream flow is supercritical, then y_1 must lie on the lower limb as shown in Fig. 7–10a. The specific energy must still decrease as shown, leading to an increase in the depth. The complete profile is included in Fig. 7–10b. If the channel bottom drops rather than rises, the solutions differ only in that the specific energy increases rather than decreases through the transition. Equations 7–12 and 7–17 apply to the channel drop as well, provided that the drop in elevation is treated as $-\Delta z$.

Consider the transition and specific energy diagram of Fig. 7–9 again. What can be expected to happen if Δz is increased to a greater and greater extent? As Δz increases, H_{02} must continue to decrease, and as this occurs y_2 decreases as well. The limit is reached for this subcritical flow when y_2 equals the critical depth, at which point the transition becomes a *choke*. A further increase in Δz results in the

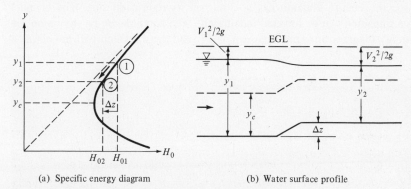

(a) Specific energy diagram   (b) Water surface profile

Figure 7–9 Subcritical transition with upward step.

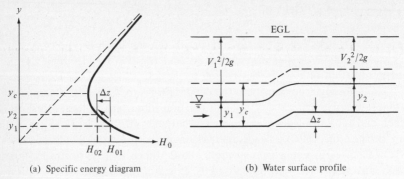

(a) Specific energy diagram   (b) Water surface profile

Figure 7–10 Supercritical transition with upward step.

impossible situation where H_{02} is less than $H_{0 \min}$. It is impossible in the sense that the solution appears to be off the curve, and, in fact, there would be no positive solution to Eq. 7–17. What has happened is that the upstream flow now has insufficient energy to pass through the transition at the specified discharge.

The flow will not cease, but rather adjusts itself to either a lower discharge or an increased specific energy. Assuming the discharge remains unchanged because of some upstream source, the subcritical flow under consideration will increase both its upstream depth and specific energy by means of a gentle swell or a series of small waves that form and travel upstream. The new upstream depth will be such that the flow can just pass the transition. Thus, y_2 will equal y_c, H_{02} will equal $H_{0 \min}$, and H_{01} will exceed $H_{0 \min}$ by the magnitude of Δz. The transition is called a choke since the critical depth prevails regardless of the increase in upstream energy.

The supercritical flow in Fig. 7–10 behaves in a somewhat different fashion as Δz increases. A choke occurs when the minimum specific energy is reached: however, the adjustment that follows a further increase in Δz is in the form of a surge (the surge or wall of water is discussed in the following section) that moves upstream. When equilibrium is finally achieved, the supercritical flow will have been replaced by the identical subcritical flow discussed in the previous case, and the transition will continue to act as a choke.

The specific energy diagram has no limit on its right-hand side. The channel bottom in a transition can be lowered indefinitely. The solution can be shown on the specific energy diagram or found from Eq. 7–17, and the transition will not act as a choke.

EXAMPLE 7–3

Determine the downstream depth in a horizontal rectangular channel in which the bottom rises 0.5 ft, if the steady flow discharge is 300 cfs, the width is 12 ft, and the upstream depth is 4 ft.

Solution

The discharge per unit width is

$$q = Q/b = 300/12 = 25 \text{ cfs/ft}$$

and the critical depth is

$$y_c = \sqrt[3]{\frac{q^2}{g}} = \sqrt[3]{\frac{(25)^2}{32.2}} = 2.69 \text{ ft}$$

The upstream depth is therefore subcritical and the solution will follow that of Fig. 7–9. The upstream specific energy is

$$H_{01} = y_1 + \frac{q^2}{2gy_1^2} = 4 + \frac{(25)^2}{(2)(32.2)(4)^2} = 4.61 \text{ ft}$$

Applying Eq. 7–17,

$$4.61 - 0.5 = 4.11 = y_2 + \frac{(25)^2}{(2)(32.2)y_2^2}$$

We can determine from Fig. 7–9a that the depth and water surface both decrease. Thus, y_2 must be greater than y_c and less than $y_1 - \Delta z$, or in terms of the numerical values, 2.69 ft$<y_2<$3.5 ft. Solving the cubic equation by iteration within this range yields $y_2 = 3.09$ ft.

EXAMPLE 7–4

Repeat Example 7–3 if the increase in the channel bottom is 1 ft, but all other quantities remain unchanged.

Solution

From the previous example $y_c = 2.69$ ft, $H_{01} = 4.61$ ft, and the specified upstream conditions are subcritical. Using Eq. 7–12

$$H_{02} = H_{01} - \Delta z = 4.61 - 1 = 3.61 \text{ ft}$$

But $H_{0\min} = 1.5y_c = 4.03$ ft, which is greater than the calculated value of H_{02}. The upstream flow must adjust to the choke conditions. The downstream depth will remain critical, that is $y_2 = 2.69$ ft, and the downstream specific energy is 4.03 ft. The new upstream conditions must be based on

$$H_{01} = H_{02} + \Delta z = 4.03 + 1 = 5.03 \text{ ft}$$

The cubic equation for H_{01} is

$$H_{01} = 5.03 = y_1 + \frac{(25)^2}{(2)(32.2)y_1{}^2}$$

and solving this by iteration, $y_1 = 4.57$ ft. This upstream depth has just enough energy to pass the given discharge through the transition.

Transition with a Change in Width

A second type of transition involves either a contraction or expansion in the width of the rectangular channel. Initially, we will assume that the channel bottom remains flat so that the specific energy remains constant through the transition. The flow is still steady, but the change in width results in a change in the unit discharge. In particular, if the upstream and downstream widths are b_1 and b_2, respectively, the unit discharges are $q_1 = Q/b_1$ and $q_2 = Q/b_2$.

The specific energy diagram as considered heretofore was based on a constant value of q, a situation that no longer exists. We can, however, draw a specific energy diagram for each value of q. This has been done in Fig. 7–11a for the contraction shown in Fig. 7–11b (ignore the dashed curves labeled q_3 and q_4 until later). The q_2 curve is to the right of the q_1 curve because $q_2 > q_1$. To verify this examine Eq. 7–11. At any depth y the larger the magnitude of q the larger will be H_0. In addition, Eq. 7–14 immediately demonstrates that y_{c2} is greater than y_{c1} as shown.

As before, it must be determined whether the upstream flow is subcritical or supercritical. To illustrate the subsequent analysis, let us assume that y_1 is less than y_c, so that the flow is supercritical. The depth y_1 and specific energy H_{01} can now be located on the q_1 curve. Since the specific energy is constant, the solution must lie

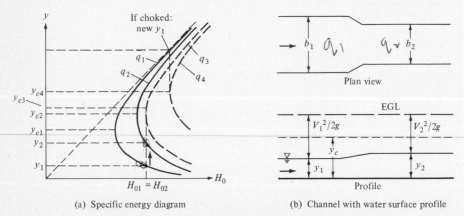

(a) Specific energy diagram   (b) Channel with water surface profile

Figure 7–11 Supercritical transition with width contraction.

on the vertical line and the water surface will rise from y_1 to y_2 as shown. The cubic equation that once again results is arrived at through the following calculation. From the given upstream conditions, H_{01} can be determined and set equal to H_{02}. The downstream depth may then be found from

$$H_{01} = H_{02} = y_2 + \frac{q_2^2}{2gy_2^2} \tag{7–18}$$

by iteration. Note that the subscript on q is now important.

If the flow were subcritical, then the upper limbs of Fig. 7–11 would be used. In the case of an expansion, the same procedures are followed, except that the two discharge curves are reversed. If there is a step in the channel bottom as well as a width change, the two procedures are combined. The solution would proceed from the q_1 curve to the q_2 curve, but in the process the specific energy would be changed by the magnitude of Δz. If downstream conditions (rather than the upstream ones) are specified, the reader will have little difficulty in modifying the foregoing approaches as the situation demands.

Continually decreasing the downstream width in the channel of Fig. 7–11 will eventually lead to the creation of a choke. The upstream adjustment of the flow will take place with either a swell or surge, as was discussed in conjunction with the increase in bottom elevation. In Fig. 7–11 the specific energy diagram associated with the discharge that results in a choke is labeled q_3 (drawn with a dashed line). A further contraction to say q_4 (also shown dashed) requires an increase in the upstream energy. The flow adjustment increases the depth to provide this energy. The new upstream depth will be at the point on the q_1 curve (indicated as "new y_1") vertically above the minimum point on the q_4 curve. Now the adjusted flow can just pass through the transition at the critical depth y_{c4}.

EXAMPLE 7–5

A rectangular channel contracts from a width of 3 m to 2.5 m in a short transition section. If the discharge is 6.5 m³/s and the upstream depth is 0.4 m, determine the downstream depth.

Solution

The unit discharges are

$$q_1 = 6.5/3 = 2.167 \text{ m}^3/\text{s/m}$$

and

$$q_2 = 6.5/2.5 = 2.6 \text{ m}^3/\text{s/m}$$

The corresponding critical depths are

$$y_{c1} = \sqrt[3]{\frac{(2.167)^2}{9.81}} = 0.782 \text{ m}$$

and

$$y_{c2} = \sqrt[3]{\frac{(2.6)^2}{9.81}} = 0.883 \text{ m}$$

The upstream flow is supercritical and the solution will be identical to that of Fig. 7–11. The upstream specific energy is

$$H_{01} = 0.4 + \frac{(2.167)^2}{(2)(9.81)(0.4)^2} = 1.896 \text{ m}$$

From Eq. 7–18,

$$H_{01} = H_{02} = 1.896 = y_2 + \frac{(2.6)^2}{(2)(9.81)y_2{}^2}$$

yielding $y_2 = 0.496$ m.

One-dimensional flow has been assumed throughout this section. While quite valid for subcritical flows, it is likely to prove an oversimplification in many super-critical transitions such as the preceding example. In particular, oblique standing waves may form within and downstream of a change in cross section. For a complete analysis of supercritical transitions, the reader should consult the open channel references at the end of the chapter.

EXAMPLE 7–6

Repeat Example 7–5 if the channel bottom has a downward step of 0.3 m. Assume that all of the other characteristics remain unchanged.

Solution

The computations will follow those of Example 7–5 through the calculation of the upstream specific energy. From Eq. 7–12, the downstream specific energy is

$$H_{02} = H_{01} - \Delta z = 1.896 - (-0.3) = 2.196 \text{ m}$$

Equation 7–18 may now be used to obtain the downstream depth,

$$2.196 = y_2 + \frac{(2.6)^2}{(2)(9.81)y_2^2}$$

and upon solving, $y_2 = 0.443$ m. It is left to the reader to sketch the solution on a specific energy diagram.

Nonrectangular Channels

The analysis of channels with nonrectangular cross sections essentially follows the procedures for rectangular channels. The main differences are that the discharge per unit width can no longer be defined, and the iterative solution that is required to solve for the unknown depth is more complicated. A general cross section has been chosen for the definitional sketch of Fig. 7–12, but the most common nonrectangular cross sections are probably trapezoidal and circular. With reference to this sketch, the total head may be written as

$$H = z + y + \frac{V^2}{2g} = z + y + \frac{Q^2}{2gA^2} \tag{7-19}$$

and the specific energy is

$$H_0 = y + \frac{V^2}{2g} = y + \frac{Q^2}{2gA^2} \tag{7-20}$$

where A is the cross-sectional area and y is the depth as shown. A specific energy diagram similar to that of the rectangular channel can be drawn, since Q is a constant and A can be expressed as a function of y. Differentiating with respect to the depth (similarly to Eq. 7–13 for a rectangular channel) leads to

$$\frac{dH_0}{dy} = 1 - \frac{Q^2}{gA^3}\frac{dA}{dy} \tag{7-21}$$

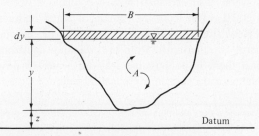

Figure 7–12 Nonrectangular channel.

The derivative dA/dy can be evaluated by noting in Fig. 7–12 that the differential area dA (shown shaded) associated with a small increase in depth dy can be expressed by $dA = B\,dy$ where B is the top width. Thus Eq. 7–21 becomes

$$\frac{dH_0}{dy} = 1 - \frac{Q^2 B}{gA^3} \tag{7–22}$$

If the above derivative is set equal to zero the equation for critical depth should result. This is

$$\frac{A^3}{B} = \frac{Q^2}{g} \tag{7–23}$$

The right-hand side contains the quantities that are generally known at the outset, and both A and B on the left-hand side can be expressed in terms of the depth, although this is a rather time-consuming computation if applied to the cross section of a river. The equation can be solved by iteration for the critical depth. Note that for a rectangular channel, Eq. 7–23 simplifies to Eq. 7–14.

If the quantity A/B which is the average depth, is taken as the representative length, the Froude number may be written

$$\text{Fr} = \frac{V}{\sqrt{gA/B}} \tag{7–24}$$

which reduces to Eq. 7–16 for a rectangular channel. The critical conditions of Eq. 7–23 may be rearranged as follows

$$\frac{Q^2/A^2}{g(A/B)} = \frac{V^2}{g(A/B)} = \text{Fr}_c^2 = 1$$

to demonstrate that this definition of Fr is also consistent with respect to the minimum point on the specific energy diagram.

The one-dimensional analysis of transition problems may now be completed for any cross section. It is recommended that a computer program be used to make the computations.

7–4 HYDRAULIC JUMP AND SURGES

The *hydraulic jump* is shown in Fig. 7–13. It is a steady, nonuniform phenomenon that occurs in open channel flows when a supercritical flow encounters a deeper subcritical flow. Supercritical flow could result from the release of water under a sluice gate, whereas the subcritical downstream depth might simply be the normal depth for the channel at that discharge as determined from the Manning equation. We will see in the next section that there is no possible way for water to flow smoothly from a supercritical to a subcritical depth in a prismatic channel.[2] Conse-

---

[2] The channel transitions of Section 7–3 do provide a possible solution. If a contraction forces the depth to critical, and this is followed by an expansion, then a smooth transition from supercritical to subcritical may occur. However, this requires a change in the cross-sectional geometry.

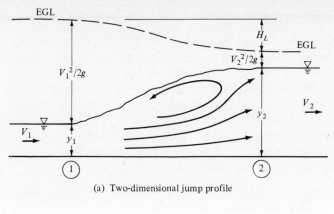

(a) Two-dimensional jump profile

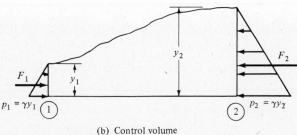

(b) Control volume

Figure 7–13 Hydraulic jump.

quently, the relatively abrupt increase in depth associated with a hydraulic jump provides the only solution.

As the supercritical flow encounters the slower-moving water, it tends to flow under it and then spread upward. This creates a large eddy or roller as shown, which in the more intense jumps dissipates a significant amount of energy. This feature of the hydraulic jump makes it very desirable below a spillway or sluice gate. If the high-velocity (and high-energy) flow below a dam can be converted to a lower-velocity (and lower-energy) flow before the water is returned to the natural channel, the potential for erosion problems is greatly reduced. This will be considered further after an analysis of the jump itself.

The *surge,* which is a moving hydraulic jump, will be the only unsteady phenomenon to receive serious attention in this chapter. Its analysis, which is similar to that of the hydraulic jump, will be considered subsequently as well. There are many other important unsteady flow situations. The passing of a flood hydrograph discussed in Chapter 3 is one example. The normal action of the tide in rivers and estuaries represents another unsteady flow. Any time a control device such as a sluice gate is adjusted so as to increase or decrease the discharge, there is a period of unsteady flow as the flow adjusts to the new conditions.

Hydraulic Jump

The hydraulic jump will be analyzed with respect to the rectangular channel. For this purpose a two-dimensional hydraulic jump is shown in Fig. 7–13a. The channel width is b, and the upstream depth and velocity (or discharge) will be assumed known. We are interested first in determining the *conjugate* or downstream depth.[3] Because of the unknown amount of energy dissipation, the energy equation cannot be applied at this point. The greater downstream depth results in a change in momentum, and the momentum equation (Eq. 2–13) will be written between sections 1 and 2 on the control volume shown in Fig. 7–13b. The forces in the flow direction consist of the hydrostatic forces on either end, and we may write

$$\frac{\gamma y_1^2 b}{2} - \frac{\gamma y_2^2 b}{2} = \rho Q(V_2 - V_1)$$

In addition continuity requires,

$$Q = V_1 y_1 b = V_2 y_2 b$$

The two equations may be combined by expressing both Q and V_2 in terms of the upstream velocity V_1:

$$\frac{\gamma y_1^2 b}{2} - \frac{\gamma y_2^2 b}{2} = \rho V_1 y_1 b \left(\frac{V_1 y_2}{y_2} - V_1 \right)$$

Upon rearranging,

$$y_1^2 - y_2^2 = \frac{2 y_1 V_1^2}{g} \left(\frac{y_1}{y_2} - 1 \right)$$

$$y_1 + y_2 = 2 \frac{y_1}{y_2} \frac{V_1^2}{g}$$

and finally

$$\frac{y_2}{y_1} \left(\frac{y_2}{y_1} + 1 \right) = \frac{2 V_1^2}{g y_1} = 2\,\mathrm{Fr}_1^2$$

which relates the depth ratio to the upstream Froude number. This is a quadratic equation in y_2/y_1, which may be solved by either completing the square or using the quadratic formula to get

$$\frac{y_2}{y_1} = \frac{1}{2} \left(\sqrt{1 + 8\,\mathrm{Fr}_1^2} - 1 \right) \tag{7–25}$$

Given the upstream conditions, this equation may be used to find the conjugate (or required) downstream depth. Equation 7–25 is plotted on Fig. 7–14.

---

[3]Some books use the term *sequent* rather than conjugate depth. Conjugate depth will be used exclusively hereafter.

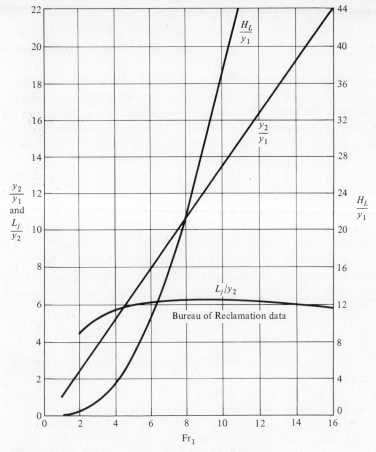

Figure 7–14 Characteristics of the hydraulic jump.

If the channel cross section is not rectangular, the same concepts of continuity and momentum may still be applied. The hydrostatic forces will have to be found using Eq. 2–6, and the area will be a more complicated function. An iterative solution will be required, because the downstream depth will be included within the area terms similar to the general equation for critical depth discussed previously (Eq. 7–23).

There are times when the downstream conditions are known and the upstream depth is required. Rather than use Eq. 7–25, it is more convenient to return to the hydraulic jump derivation. If instead of eliminating Q and V_2 as was done previously, V_1 and Q are both expressed in terms of the downstream velocity V_2, the subsequent derivation leads to

$$\frac{y_1}{y_2} = \frac{1}{2}\left(\sqrt{1 + 8\,\mathrm{Fr}_2{}^2} - 1\right) \tag{7-26}$$

where the downstream Froude number is defined by

$$Fr_2 = \frac{V_2}{\sqrt{gy_2}}$$

Determination of y_2 (or y_1 if Eq. 7–26 is used) permits calculation of the unknown velocity as well. The energy equation may now be used to evaluate the head loss or loss of energy in the jump. Using the channel bottom of Fig. 7–13 as the datum, the head loss may be written as the difference in total head between the two sections. That is,

$$H_L = H_1 - H_2 = \left(y_1 + \frac{V_1^2}{2g} \right) - \left(y_2 + \frac{V_2^2}{2g} \right) \tag{7–27}$$

The ratio of head loss to upstream depth can be expressed as a function of the upstream Froude number by using continuity to eliminate V_2 and then substituting Eq. 7–25 for the depth ratio. The resulting relationship is plotted on Fig. 7–14.

Some algebraic manipulation is required, but by use of both continuity and momentum the velocity heads in Eq. 7–27 can both be eliminated. This exercise, which is left to the reader, leads to the much simpler equation for head loss,

$$H_L = \frac{(y_2 - y_1)^3}{4y_1y_2} \tag{7–28}$$

The length of a hydraulic jump is a necessary factor in the design of a hydraulic structure involving a jump. Recourse is usually made to experimental studies such as those made by the U.S. Bureau of Reclamation, which are included in Fig. 7–14. The leading edge of the jump is fairly easy to identify because of the nature of the abrupt roller, but the effective downstream end is far more difficult. It is desirable that the length encompasses essentially all of the energy dissipation, which includes the roller as well as the downstream region in which the high-velocity flow spreads out and becomes a uniform lower-velocity flow.

It should be stressed that for given upstream conditions there is only one conjugate depth, y_2 as given by Eq. 7–25, for which the jump will occur. If the correct conditions are not present at the desired location, the jump will locate at a different location where compatible depths are available. This process involves channel friction and will be covered in detail in the following section.

To overcome the difficulty of keeping the jump where it is intended below a dam or other hydraulic structure, a *stilling basin* such as that shown in Fig. 7–15 is frequently used. There are many standard forms that consist of baffle blocks, sills, as well as other devices, all of which contribute to the change in flow momentum and, as a consequence, stabilize the location of the jump. Much of the design at this point is based on experiments, but Example 7–8 will illustrate how baffle blocks may be incorporated into the analysis. More details on stilling basins are available in the references at the end of the chapter. For design information the reader is particularly directed to the U.S. Bureau of Reclamation procedures as described in French's book.

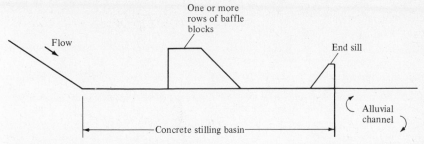

Figure 7–15 Sketch of stilling basin.

EXAMPLE 7–7

Determine the downstream depth required for a jump to occur in a 20-ft-wide rectangular channel if the upstream depth is 2 ft and the discharge is (1) 200 cfs and (2) 640 cfs. What is the head loss in each case?

Solution

For the discharge of 200 cfs

$$V_1 = \frac{200}{(2)(20)} = 5 \text{ ft/s}$$

and

$$Fr_1 = \frac{5}{\sqrt{(32.2)(2)}} = 0.62$$

This value is less than unity and a hydraulic jump cannot occur. Repeating for $Q = 640$ cfs,

$$V_1 = \frac{640}{(2)(20)} = 16 \text{ ft/s}$$

and

$$Fr_1 = \frac{16}{\sqrt{(32.2)(2)}} = 1.99$$

Applying Eq. 7–25 (or Fig. 7–14)

$$\frac{y_2}{2} = \frac{1}{2}\left(\sqrt{1 + (8)(1.99)^2} - 1\right)$$

which yields a depth $y_2 = 4.72$ ft.
In the second case the head loss in the jump will be

$$H_L = \frac{(y_2 - y_1)^3}{4y_1y_2} = \frac{(4.72 - 2)^3}{(4)(2)(4.72)} = 0.53 \text{ ft}$$

EXAMPLE 7–8

Assume that a stilling basin such as that of Fig. 7–15 contains a series of 30 baffle blocks in a 40-m-wide channel and that a downstream sill is not present. The upstream depth and velocity are 1.5 m and 12 m/s, respectively. If each block exerts a thrust of 60 kN, determine the downstream depth and compare the result with the solution if there were no baffle blocks.

Solution

The upstream Froude number is

$$\text{Fr}_1 = \frac{12}{\sqrt{(9.81)(1.5)}} = 3.13$$

If no baffle blocks are present, the downstream depth may be obtained directly from

$$\frac{y_2}{1.5} = \frac{1}{2}\left(\sqrt{1 + (8)(3.13)^2} - 1 \right)$$

for which the result is $y_2 = 5.93$ m.

Inclusion of the baffle blocks requires that we return to the momentum analysis itself. A control volume like that of Fig. 7–13, but with modification to allow for the presence of the blocks, may be used. The reader should sketch this control volume. The momentum equation may be written as follows:

$$\frac{(9810)(1.5)^2(40)}{2} - \frac{(9810)y_2^2(40)}{2} - (60,000)(30)$$

$$= (1000)(12)(40)(1.5)(V_2 - 12)$$

where the third term incorporates the force exerted by the 30 blocks. The unknown velocity V_2 may be eliminated by introducing the continuity equation,

$$(12)(1.5) = V_2y_2$$

However, we are left with a cubic equation in y_2. Solving by iteration, $y_2 = 4.85$ m. Note that the force on each baffle block, here given as 60 kN, can be estimated by using the drag force equation $F = C_D A \rho V^2 / 2$. The drag coefficient C_D might well be in the range of 1 to 2 and A is the projected area of the block. Is the given drag force reasonable?

Surges

As stated previously, the surge is the unsteady counterpart of the hydraulic jump. A surge may travel either upstream against the current or downstream overtaking the existing flow. In either case the surge takes the same form as the hydraulic jump, possessing a relatively abrupt wave front and a similar roller action. There are also negative surges, for example, downstream of a gate that has suddenly been partially closed. We will consider only the positive types that result in an increase in depth.

Surges can occur when a flash flood results in an unusually large amount of water converging on a river or stream. In the event of a dam failure a devastating surge may result. On a more regular basis, rivers in regions of extreme tides, such as the Bay of Fundy in Canada and the Severn Estuary in Great Britain, experience an upriver surge as the tidal water is rapidly funneled into the rivers. This type of surge, known as a *tidal bore,* occurs twice a day and may have a surge height of several feet. In a regulated channel, rapid or partial closure of a gate may cause a surge to move upstream, while the rapid opening of an upstream gate may result in a downstream surge.

The surge may be analyzed by the hydraulic jump equations if it is made stationary through the use of relative motion. This is achieved if the reference frame is translated with the surge velocity. Assume that a steady uniform flow from left to right with depth y_1 and velocity V_1 exists in Fig. 7–16a. Due to some downstream condition a surge begins to move upstream with a velocity V_s, and at a given instant is located as shown in Fig. 7–16a. All velocities, including the surge velocity, are to be interpreted as relative to a fixed reference frame. After the surge passes, the resulting flow has a depth y_2 and a velocity V_2. Although shown in the downstream direction, the actual direction of V_2 is unknown and can be in either direction, or the velocity may be zero.

The surge may be made stationary for purposes of analysis by adding the velocity V_s in the opposite direction to all points in the flow field. This has been done in Fig. 7–16b, which is now a steady flow with a stationary hydraulic jump. The jump equations may now be used provided that the upstream and downstream velocities are taken as $V_s + V_1$ and $V_s + V_2$, respectively. Because an additional unknown that was not present in the hydraulic jump analysis has been added, one more piece of information must be provided in the statement of the problem. For example, if the surge of Fig. 7–16 were the result of a rapid closure of a downstream gate, then we could set V_2 equal to zero. In some cases it may be difficult to keep track of the

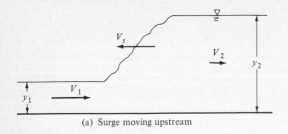

(a) Surge moving upstream

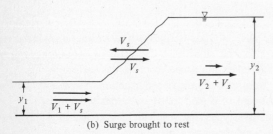

(b) Surge brought to rest

Figure 7–16 Definitional sketch for surge analysis.

signs on the various velocity terms and it may be preferable to draw the steady flow control volume and write the momentum equation directly rather than use the hydraulic jump equations.

EXAMPLE 7–9

A surge of the type shown in Fig. 7–16 occurs in a wide (assumed rectangular) river. The presurge steady flow has a depth of 1.2 m and a velocity of 1.5 m/s. If the incoming tide results in the formation of a surge with a height of 0.7 m (i.e., $y_2 = 1.9$ m), determine the surge velocity and the magnitude and direction of the velocity after the surge has passed.

Solution

The conversion to a steady flow is shown in Fig. 7–16b. Rather than use Eq. 7–25, in this case it is more convenient to use the unnumbered equation immediately preceding it. This equation becomes

$$\frac{y_2}{y_1}\left(\frac{y_2}{y_1} + 1\right) = \frac{2(V_1 + V_s)^2}{gy_1}$$

Substituting,

$$\left(\frac{1.9}{1.2}\right)\left(\frac{1.9}{1.2} + 1\right) = \frac{(2)(1.5 + V_s)^2}{.\,(9.81)(1.2)}$$

which may be solved directly to obtain a surge velocity $V_s = 3.41$ m/s.

To obtain the final velocity, the steady flow continuity equation may be written as

$$(V_1 + V_s)(y_1) = (V_2 + V_s)(y_2)$$

or

$$(1.5 + 3.41)(1.2) = (V_2 + 3.41)(1.9)$$

This yields $V_2 = -0.31$ m/s. Therefore, after the surge passes the resulting flow is in the upstream direction (driven by the incoming tide) with a velocity as indicated of 0.31 m/s.

7–5 GRADUALLY VARIED FLOW

The term *gradually varied flow* is used to describe a particular type of steady non-uniform flow. As the name implies, the changes in the depth and velocity occur gradually over a considerable length of channel and the nonuniformity of the flow is not pronounced. One consequence of this is that even though the flow is slightly nonuniform, a hydrostatic pressure variation with depth may be assumed with little error. A second consequence resulting from the usually long length of channel involved is that friction cannot be ignored.

We will assume that we are dealing with relatively long lengths of prismatic channel. However, no restriction will be placed on that shape for the present. It may be anticipated under these circumstances that normal depth would generally occur provided that the channel has a constant downward slope in the flow direction. This depth would be determined using the Manning equation (Eqs. 7–5), with the slope being that of the channel slope. A transition region must occur if the depth is forced away from normal depth for any reason. To illustrate, suppose that at some point in the channel the slope changes abruptly from one value to another. The two sections must have different normal depths. The water surface profile in the length over which this change occurs would form one type of gradually varied flow transition. A horizontal stretch in a channel of otherwise constant slope poses an additional difficulty, inasmuch as normal depth cannot occur in a horizontal channel. To analyze all of the different types of transitions that might be encountered, it is necessary to determine how many and what sort of water surface profiles are in fact possible.

Water Surface Profiles

We will proceed using Fig. 7–17 as a definitional sketch. To identify the possible profiles we will first develop the differential equation that governs the flow. Sections

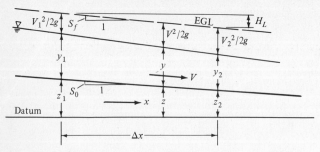

Figure 7–17 Definitional sketch of gradually varied flow.

1 and 2 have been identified for later use, but for the present consider an arbitrary section where the total head is

$$H = z + y + \frac{V^2}{2g} \tag{7–29}$$

The total head, as always, is measured from a horizontal datum. However, note that the depth is measured from the channel bottom, which is generally not horizontal. The velocity is the average velocity at the section, and both the velocity and depth are functions of x only.

The channel slope is S_0 and the slope of the energy grade line, or friction slope, is S_f. The former will generally be given and the latter may be expressed in terms of the Manning equation. Because the EGL always slopes downward in the flow direction and the channel slope usually does, the downward slope is defined as positive. This in itself should cause no confusion, but the relationship of the slopes to the terms in Fig. 7–17 may require some thought. The channel elevation z, the depth y, and the total head H will all be considered as positive upward. Thus,

$$S_0 = -\frac{dz}{dx} \tag{7–30}$$

and

$$S_f = -\frac{dH}{dx} \tag{7–31}$$

A positive value of dz/dx would imply that z is increasing in the flow direction; in other words, this would be an uphill channel. If the positive value of dz/dx were substituted into Eq. 7–30, the equation would indicate a negative value for S_0 consistent with the sign convention stated previously for the slope.

The total head given in Eq. 7–29 may be differentiated with respect to x to get

$$\frac{dH}{dx} = \frac{d}{dx}\left(z + y + \frac{V^2}{2g} \right) = -S_f$$

or

$$\frac{d}{dx}\left(y + \frac{V^2}{2g}\right) = S_0 - S_f$$

The quantity within the parentheses is the specific head H_0, thus

$$\frac{dH_0}{dx} = \frac{dH_0}{dy}\frac{dy}{dx} = S_0 - S_f$$

But from either Eq. 7–13 or Eq. 7–22, depending on whether the cross section is rectangular or not, we may write

$$\frac{dH_0}{dy} = 1 - \mathrm{Fr}^2$$

With the introduction of this expression we get the governing equation for gradually varied flow,

$$\frac{dy}{dx}(1 - \mathrm{Fr}^2) = S_0 - S_f \qquad (7\text{--}32)$$

Before considering this equation, we can establish the actual number of possible water surface profiles by first classifying channels into five types on the basis of slope. A channel has a *mild* slope if the normal depth is greater than the critical depth. Uniform flow on a mild slope must therefore be subcritical, but as we will see, nonuniform water surface profiles may be either subcritical or supercritical. The normal and critical depths serve to further divide the channel into three distinct regions as shown in Fig. 7–18; a region $y > y_n$, a region between the normal and critical depths $y_c < y < y_n$, and a region $y < y_c$. The first two of these regions are subcritical while the final region is supercritical.

In Fig. 7–18 and the four figures that follow for the remaining channel classifications, the normal depth will be indicated by a series of long dashes and the critical depth by a series of short dashes. Also shown in these five figures, as will be explained shortly, are the possible profiles and to the right of each, a physical example.

Continuing the classification, a *steep* channel occurs when the normal depth is less than the critical depth. As shown in Fig. 7–19, uniform flow would be super-

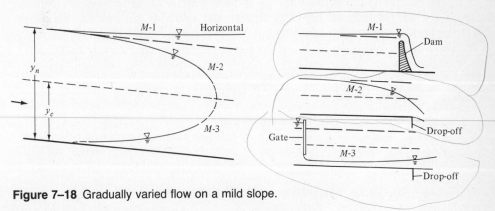

Figure 7–18 Gradually varied flow on a mild slope.

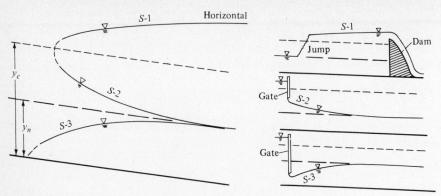

Figure 7–19 Gradually varied flow on a steep slope.

critical and there are two regions of supercritical flow and one region of subcritical flow. The limiting condition between the steep and mild classifications, called the *critical* slope, occurs when the normal depth just equals the critical depth as indicated by the alternating sets of long and short dashes in Fig. 7–20. There can be no profile between y_n and y_c and only two regions are possible.

As the slope in a channel becomes less and less, the normal depth increases accordingly. A *horizontal* channel with a slope of zero may be considered to represent the lower limit of the mild slopes. As shown in Fig. 7–21, the normal depth approaches infinity and consequently only two regions, one subcritical and one supercritical, are of concern here. The *adverse* slope of Fig. 7–22 differs from the horizontal only in being somewhat more extreme, as the slope is now negative. Two regions can be identified in this situation as well. Uniform flow is not possible in either the horizontal or adverse channel.

We may summarize the preceding discussion by noting that five separate classifications have been made, and twelve different regions of nonuniform flow have resulted. Equation 7–32 may now be used to determine the actual form of the profile in each of the twelve regions. The first water surface profiles will be developed very thoroughly, but the development of some of the remaining profiles will be left largely to the reader. As can be seen in Figs. 7–18 through 7–22, the profiles are identified by a letter that indicates the channel slope and a number that indicates the region. The profile in the region above both y_n and y_c is labeled 1, the region between those two depths is called region 2, and the region below both depths is region 3.

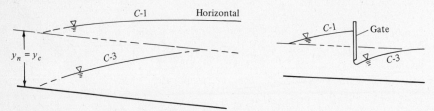

Figure 7–20 Gradually varied flow on a critical slope.

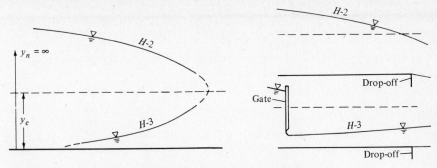

Figure 7–21 Gradually varied flow on a horizontal slope.

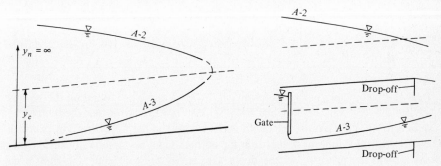

Figure 7–22 Gradually varied flow on an adverse slope.

One additional general statement is required at the outset of the analysis. When the Manning equation is written with S_0, the appropriate depth is the normal depth. If the depth is not equal to the normal depth, then the flow is nonuniform and the slope is the friction slope associated with the actual depth. Depending on the channel geometry, the depth may be related to the area and hydraulic radius in a rather complicated fashion: nevertheless, if $y > y_n$ then $S_f < S_0$ and conversely, if $y < y_n$ then $S_f > S_0$.

Commencing with the upper region of the mild slope, we wish to establish the form of the *M–1* curve. In this region $y > y_n$, Fr<1 (since $y > y_c$ and hence is subcritical), and $S_f < S_0$. According to Eq. 7–32, the derivative dy/dx must therefore be positive, which requires that the depth y increase in the x direction. As the depth approaches the normal depth, S_f approaches S_0 and the right-hand side of Eq. 7–32 approaches zero. Since Fr does not approach 1, it remains that dy/dx must approach zero. In other words, the *M–1* profile approaches the normal depth asymptotically. Finally, as the depth gets large both Fr and S_f approach zero leaving $dy/dx = S_0$. Since the depth is measured from the channel bottom that drops away at the rate S_0, the water surface must approach a horizontal line at this limit. The result of these three conditions is the *M–1* curve given in Fig. 7–18. A summary of the foregoing curve characteristics, along with those of the remaining 11 profiles, is included in Table 7–2.

TABLE 7–2 SUMMARY OF WATER SURFACE PROFILES

Channel	Depth	Froude number	Slope	dy/dx	Control upstream/ downstream	Curve type
Mild	$y > y_n > y_c$	Fr < 1	$S_f < S_0$	+	DS	$M\text{--}1$
Mild	$y_n > y > y_c$	Fr < 1	$S_f > S_0$	−	DS	$M\text{--}2$
Mild	$y_n > y_c > y$	Fr > 1	$S_f > S_0$	+	US	$M\text{--}3$
Steep	$y > y_c > y_n$	Fr < 1	$S_f < S_0$	+	DS	$S\text{--}1$
Steep	$y_c > y > y_n$	Fr > 1	$S_f < S_0$	−	US	$S\text{--}2$
Steep	$y_c > y_n > y$	Fr > 1	$S_f > S_0$	+	US	$S\text{--}3$
Critical	$y > y_n = y_c$	Fr < 1	$S_f < S_0$	+	DS	$C\text{--}1$
Critical	$y_n = y_c > y$	Fr > 1	$S_f > S_0$	+	US	$C\text{--}3$
Horizontal	$y > y_c$	Fr < 1	$S_f > S_0 = 0$	−	DS	$H\text{--}2$
Horizontal	$y_c > y$	Fr > 1	$S_f > S_0 = 0$	+	US	$H\text{--}3$
Adverse	$y > y_c$	Fr < 1	$S_f > S_0$	−	DS	$A\text{--}2$
Adverse	$y_c > y$	Fr > 1	$S_f > S_0$	+	US	$A\text{--}3$

Relevent parameters needed to establish the M–2 curve are found on the second line of Table 7–2. Since Fr < 1 and $S_f > S_0$, Eq. 7–32 requires that dy/dx must be less than zero; in other words, the depth must decrease in the flow direction. As with the M–1 curve, the depth must approach the normal depth asymptotically since dy/dx goes to zero in the process. As the depth approaches critical depth, the right-hand side of Eq. 7–32 remains finite while Fr approaches unity. This implies that dy/dx must become infinite. The large degree of curvature that occurs as the critical depth is approached violates the basic assumption of nearly uniform flow, and the vertical water surface profile will not actually occur. The completed profile may now be drawn in Fig. 7–18, with the final portion near the critical depth shown by a dashed line.

In the final region involving a mild slope, the fact that Fr > 1 and $S_f > S_0$ means that the depth will once again increase in the flow direction. As before, the profile will tend toward the vertical as critical depth is approached. As the depth goes to zero, a limit that cannot be physically reached, both Fr and S_f go to infinity. For Eq. 7–32 to remain in balance as this occurs, dy/dx must maintain some positive value. These three conditions determine the form of the M–3 profile in Fig. 7–18.

A few general observations may be made, which can be verified by continuing through the rest of the profiles using the above approach. First, when the flow approaches or leaves the normal depth, it must do so asymptotically. Second, as the profile approaches the critical depth, the theoretical result is a vertical surface whereas the actual slope will have some finite value. Third, if the water surface profile approaches the bottom of the channel, it must do so at some positive angle.

An example of each of the mild profiles is included in Fig. 7–18. The first is a dam, but any downstream obstruction on a mild slope backs up the flow, causing an M–1 curve in the process. A further increase in the height of the dam would raise the M–1 curve and it would extend still further upstream. Because of the nature of the M–1 curve, the water surface profiles were originally called backwater profiles. In all figures, it should be noted that the vertical scale is greatly exaggerated relative to the horizontal.

If a channel with a mild slope ends in a dropoff, the relatively slow water cannot maintain its normal depth to the end of the channel and the M–2 curve results. Theoretically, the flow should pass through critical at the brink, but again the large amount of curvature coupled with a nonhydrostatic pressure distribution at the dropoff alters the actual profile, and it will pass through critical depth a distance upstream of the brink equal to three to four times the critical depth. The actual depth at the brink will be approximately $0.7y_c$. On the other hand, the relatively high-velocity supercritical flow associated with the M–3 profile will simply shoot off the end as shown.

In the first two examples, both of which were subcritical, the location and form of the profile are entirely determined by the downstream end. We may say that these profiles are based on a downstream control. As defined in Chapter 3, a control section is one at which there is a unique relationship between the discharge and the depth. Expanding on this point, a downstream control exerts its influence in the upstream direction and ''controls'' the geometry of the profile. In the case of the supercritical flow, the dropoff at the downstream end has no effect on the upstream profile. Instead, the upstream sluice gate locates the profile, a condition called upstream control. The upstream control, therefore exerts its influence in the downstream direction. The subject of control is an important one that will be discussed in more detail later. Note, however, that the location of the control is included in Table 7–2 for each of the profiles.

The steep profiles, which are analyzed in the same manner as the mild profiles, are given in Fig. 7–19 along with one example in each case. It is a useful exercise to verify the form of each of these profiles. The examples should be studied with respect to the concept of control. What happens, for instance, to the water surface profiles as the two sluice gates are raised or lowered?

The critical slope (Fig. 7–20) represents the limit between the mild and steep slopes. The occurrence of critical slope is very rare and the two curves are included only for completeness. The form of the curves should fall between those of the two more important profiles, and since the M and S curves are alternately concave up and down in regions 1 and 3, the C curves tend toward the horizontal. The summary in Table 7–2 may be used as previously. As a water surface profile approaches critical depth it tends to the vertical, whereas a profile approaching normal depth does so asymptotically. An inconsistency results in the case where $y_c = y_n$: this inconsistency may be resolved by examination of Eq. 7–32, which suggests that dy/dx must maintain some positive value. The actual form of the C curves will depend on the shape of the channel cross section.

The horizontal and adverse profiles follow directly from Table 7–2 and Eq. 7–32. The two sets have qualitatively similar profiles that differ in degree only. In the horizontal case $S_0 = 0$ and in the adverse channel $S_0 < 0$.

Computation of Profiles—The Direct Step Method

The general form of the twelve possible profiles has now been established. Before attempting to combine them in a channel with changing slopes, gates, and other features, we will develop the means to calculate the actual profile. There are many procedures available to analyze the profiles, although some of them are restricted to rectangular channels. Details may be found in the references to open channel flow found at the end of this chapter. One type of solution is based on the integration of Eq. 7–32 over the distance x. The various integration methods differ in how they handle the shape of the cross section and in the determination of the friction slope. They usually represent some degree of approximation but are relatively easy to apply because of the availability of tabulated integral functions. The integration methods will not be discussed further because step or difference computations may be done far more easily and accurately on a computer or even a programmable calculator.

The procedure that follows, called the *direct step* method, may be used for all of the profiles in any regular constant-shape channel. The necessary equation may be obtained by writing the energy equation between sections 1 and 2 of Fig. 7–17. This gives

$$z_1 + y_1 + \frac{V_1^2}{2g} = z_2 + y_2 + \frac{V_2^2}{2g} + H_L \tag{7-33}$$

The head loss between the two sections, H_L, and the elevation difference, $z_1 - z_2$, may be replaced in Eq. 7–33 by introducing Δx, the horizontal distance between the two sections, as follows

$$\frac{H_L}{\Delta x} = S_f \quad \text{and} \quad \frac{z_1 - z_2}{\Delta x} = S_0$$

Upon substituting and rearranging

$$S_0 \Delta x - S_f \Delta x = \left(y_2 + \frac{V_2^2}{2g} \right) - \left(y_1 + \frac{V_1^2}{2g} \right)$$

or

$$\Delta x = \frac{\left(y_1 + \frac{V_1^2}{2g} \right) - \left(y_2 + \frac{V_2^2}{2g} \right)}{S_f - S_0} \tag{7-34}$$

Equation 7–34 may be used to calculate the water surface profile using a step-by-step sequence. A starting point of known depth is required. This depth, y_1, will usually be obtained at a control. The calculations will then proceed in the direction of the effect of the control. A small incremental change in depth is assumed, which gives y_2, and from this the horizontal step Δx from y_1 to y_2 is calculated. Once this point is established, the new depth y_2 becomes the starting depth y_1 for the next step and the process is repeated until the entire curve is completed. The smaller the increment in y, the more accurate is the procedure, because the process assumes linear characteristics between the two sections.

For both the known and assumed depths, the corresponding velocities may be calculated since the flow is steady. The channel slope will be known and the friction slope may be evaluated from the appropriate form of the Manning equation. If U.S. customary units are used, then

$$S_f = \left(\frac{n_{AV} V_{AV}}{1.49\, R_{AV}^{2/3}} \right)^2 \qquad (7\text{–}35a)$$

similarly, in SI units

$$S_f = \left(\frac{n_{AV} V_{AV}}{R_{AV}^{2/3}} \right)^2 \qquad (7\text{–}35b)$$

In either case, and average value of n, V, and R based on the values at sections 1 and 2 are used. Usually n will not change from section to section.

Each calculation gives the distance from the known depth to the assumed depth. The sum of the Δx values gives the location of each depth from the control or starting depth. Since the form and type of profile are known, the sign on the Δx terms need be of no concern. A tabulated solution illustrating the procedure will follow in Example 7–10.

EXAMPLE 7–10

A discharge of 800 cfs occurs in a long 20-ft-wide rectangular channel that has a slope of 0.0005 and ends in an abrupt drop-off. The Manning n is 0.018. Calculate the water surface profile from the dropoff to a point at which the depth has reached 99 percent of the normal depth.

Solution

Both the normal and critical depths are required to determine the type of curve involved. From the Manning equation,

$$Q = 800 = \frac{1.49 A R^{2/3} S_0^{1/2}}{n} = \frac{(1.49)(20 y_n)^{5/3}(0.0005)^{1/2}}{(0.018)(20 + 2y_n)^{2/3}}$$

Solving by iteration, $y_n = 8.01$ ft. For a discharge per unit width $q = 800/20 = 40$ cfs/ft, the critical depth is

$$y_c = \sqrt[3]{\frac{(40)^2}{32.2}} = 3.68 \text{ ft}$$

Thus, the profile will be the M–2 curve shown in Fig. 7–23. The depth at the brink will be $(0.7)(3.68) = 2.58$ ft. The profile will cross the critical depth line approxi-

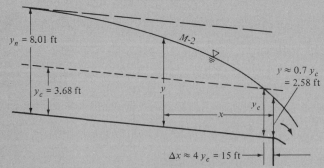

Figure 7–23 M-2 curve for Example 7–10.

mately $(4)(3.68) = 15$ ft upstream of the brink. Because of the asymptotic nature of the $M-2$ curve as it approaches the normal depth, it will only be extended until the depth reaches $(0.99)(8.01) = 7.93$ ft.

The computations, which should be self-explanatory, are given in Table 7–3. The friction slope is calculated using Eq. 7–35a. The final column gives the results as measured from the brink. This includes the 15 ft distance out to where the depth reaches critical depth and the profile computations actually begin.

Table 7–3 COMPUTATION OF *M*–2 CURVE FOR EXAMPLE 7–11

y	A	R	V	$y + \dfrac{V^2}{2g}$	$\Delta\left(y + \dfrac{V^2}{2g}\right)$	R_{AV}	V_{AV}	S_f	Δx	$x = \Sigma\Delta x$
(ft)	(ft^2)	(ft)	(ft/s)	(ft)	(ft)	(ft)	(ft/s)		(ft)	(ft)
(1)	(2)	(3)	(4)	(5)	(6)	(7)	(8)	(9)	(10)	(11)
3.68	73.6	2.68	10.87	5.515	—	—	—	—	—	15
4.68	93.6	3.19	8.55	5.815	0.300	2.94	9.71	3.27×10^{-3}	108	123
5.68	113.6	3.62	7.04	6.450	0.635	3.41	7.79	1.72×10^{-3}	518	641
6.68	133.6	4.00	5.99	7.237	0.787	3.81	6.52	1.04×10^{-3}	1449	2090
7.68	153.6	4.34	5.21	8.101	0.864	4.17	5.60	6.82×10^{-4}	4750	6840
7.93	158.6	4.42	5.04	8.324	0.233	4.38	5.13	5.36×10^{-4}	6212	13052

The calculations in Table 7–3 are subject to some error because of the number of significant figures carried through the calculations. The sixth column, which is the difference in specific head between two sections, requires subtracting two relatively large numbers to get a relatively small difference, and the final calculation of Δx requires the difference between S_f and S_0. Both of these calculations can result in the loss of significant figures. Consequently, care should be taken to ensure that a loss of accuracy does not occur. To illustrate this effect as well as that of step size, the profile was also calculated using a computer program in which the size of the vertical step was varied. The results, which are tabulated in Table 7–4, may be compared with those of Table 7–3. The second column is a repeat of the results from Table 7–3, and the next four columns are the computer results using step sizes of 1, 0.5, 0.2, and 0.01 ft, respectively. The calculations with the smaller step sizes generate more points in the profile, but for comparison purposes only those corresponding to the original y values are tabulated here.

Table 7–4 SENSITIVITY ANALYSIS OF PROFILE CALCULATIONS

y (ft)	x (ft) Table 7–3 results $\Delta y = 1$ ft	x (ft) Computer results $\Delta y = 1$ ft	x (ft) Computer results $\Delta y = 0.5$ ft	x (ft) Computer results $\Delta y = 0.2$ ft	x (ft) Computer results $\Delta y = 0.01$ ft
(1)	(2)	(3)	(4)	(5)	(6)
2.58	0	0	0	0	0
3.68	15	15	15	15	15
4.68	123	122	137	142	143
5.68	641	635	681	694	697
6.68	2090	2077	2213	2255	2263
7.68	6840	6771	7680	8068	8160
7.93	13052	12742	13651	14716	15196

The third column uses the same step size as the original example and demonstrates the effect of rounding errors. The remainder of the table illustrates the significance of the step size. The smaller the step size, the more accurate the resultant water surface profile will be. A high accuracy is not required in most cases since we are dealing with very flat curves over most of their length. It is significant to note that a measureable effect of the drawdown due to the drop-off extends for a long distance upstream.

Combined Water Surface Profiles

The water surface profiles are the result of an obstruction to the flow or some change or changes to the channel itself, such as its slope. The transition or transitions that result may involve one or more distinct profiles, as illustrated in Figs. 7–24 through

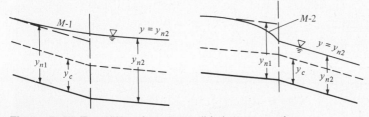

Figure 7–24 Transitions from one mild slope to another.

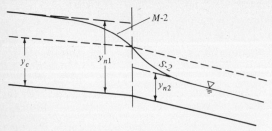

Figure 7–25 Transition from mild to steep slope.

7–26. The change in slope can lead to situations in which the smooth water surface profile is not in itself adequate and a hydraulic jump will occur, as in Figs. 7–27 and 7–28. In addition, structures placed within the channel force a transition curve. The sluice gates placed in the two channels of Fig. 7–29 illustrate this situation. The transitions studied in Section 7–3 may also result in possibly lengthy, gradually varied flow profiles in addition to the local change in water surface considered previously. As some of these examples are discussed, we will see that the solution to the transition requires one or more of the profiles contained in Figs. 7–18 through 7–22. Furthermore, an understanding of the concept of control is usually necessary to locate the curve.

In Fig. 7–24 all of the slopes are mild, and subcritical flow may be expected throughout. There is no profile in Fig. 7–18 that the water surface can follow so that it might become supercritical. In both sketches, the required downstream control is provided by the downstream normal depth. This downstream control is a friction control because it is related to the channel friction through the normal depth, y_n, and the Manning equation. From the break in grade onward, the flow must continue at the downstream normal depth. Upstream, the appropriate profile occurs. Once the

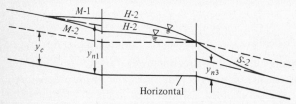

Figure 7–26 Transition from mild to horizontal to steep slope.

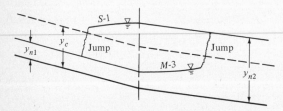

Figure 7–27 Transition from steep to mild slope.

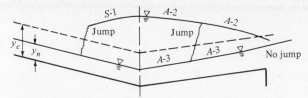

Figure 7–28 Transition from steep to adverse slope.

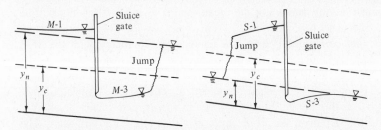

Figure 7–29 Profiles caused by a sluice gate.

profile is analyzed in this fashion the actual curve may be calculated by the direct step method, starting at the break in grade with a depth equal to y_{n2} and proceeding upstream. As a final point, the location of the M–1 or the M–2 curve at the grade change cannot be other than as shown. There is no water surface profile that smoothly approaches the normal depth on a mild slope, and the profile can only leave the normal depth by M–1 or M–2 curve.

A transition from one steep slope to another would have the opposite behavior. The transition would occur in the downstream section and the upstream normal depth would serve as an upstream friction control. There is no profile that smoothly leaves the normal depth on a steep slope.

It can be rigorously proven that the transition curve in Fig. 7–25 must pass through critical depth at the grade break. It can be demonstrated more readily, however, by simply noting that upstream the flow must be subcritical and therefore have a downstream control, while the downstream flow is supercritical and must have an upstream control. Thus, the flow must be critical at the break, and this provides the control for both directions. An even simpler, but equally convincing, argument is that no other combination of curves from Figs. 7–18 and 7–19 can be arranged to connect the two normal depths.

Figure 7–26 is similar to Fig. 7–25 in that a subcritical portion is followed by a supercritical section. The control will be at the grade change following the horizontal section. In this case two distinct situations can occur. Depending on the length and other characteristics of the horizontal section, the H–2 curve as it projects upstream may fall either above or below the normal depth on the mild slope. The M curve that precedes it will depend on which case occurs, and cannot be predicted without calculating the H–2 curve first. It is not to be expected that the H–2 curve will exactly match the upstream normal depth because its location depends entirely on the downstream conditions.

A transition from a steep to a mild channel is shown in 7–27. Consideration of the M and S profiles reveals that the transition cannot be accomplished by the water surface profiles alone. The solution requires the presence of a hydraulic jump, but two possibilities exist as shown. To determine at the outset which case is correct for a given problem, the jump equation, Eq. 7–25, may be applied at the break in grade, with y_1 in the jump equation taken as the upstream normal depth, y_{n1}. The conjugate depth y_2 from Eq. 7–25 can be compared with the downstream normal depth. If $y_2 > y_{n2}$, the jump must be downstream of the break, and if $y_2 < y_{n2}$, the jump will be upstream. This is because y_2 decreases as y_1 increases in Eq. 7–25 for constant discharge and channel conditions. When this tendency is compared with the M–3 profile, which increases in depth in the downstream direction, we see that if the conjugate depth at the break is greater than y_{n2}, then as we proceed downstream on the M–3 curve, a point must be reached for which y_{n2} equals the conjugate depth. The jump will occur at this point. When the jump is downstream of the break, Eq. 7–26 will provide the depth on the M–3 curve at which the jump will start. The profile may be calculated from the break in grade downstream to this point using the direct step method.

Returning to the comparison of the conjugate depth at the break with the downstream normal depth, if $y_2 < y_{n2}$, a downstream jump would never reach y_{n2} and an upstream jump is the only possible solution. The S–1 curve can be projected upstream until the depth decreases to the previously calculated conjugate depth, thereby locating the jump.

The situation is somewhat more complicated in Fig. 7–28. A jump may occur in either section, but since the adverse section terminates with a drop-off, there is also the possibility that no jump will occur. There is no easy solution to the water surface profile in Fig. 7–28. The no-jump case can be examined first by plotting the A–3 curve from the grade change to the drop-off. If the profile does not get close to the critical depth, this will be the correct solution. Otherwise, the A–2 profile can be extended upstream to the grade change and a jump at the break compared with the depth of the A–2 profile at that point. This will locate the jump with respect to the break in grade just as with Fig. 7–27. If the jump is upstream, it will be located exactly as it was for the steep to mild channel. If it is downstream, then the jump must be fitted between the A–3 and the A–2 curves using Eq. 7–25 to calculate conjugate depths for various points along the A–3 curve.

Figure 7–29 illustrates the effect of a gate on the flow and the water surface profiles that result. On the mild slope, the flow will be at normal depth until the gate is lowered into the flow. Immediately thereafter, it will act as a control on the upstream section and an M–1 curve will form. The more the gate is lowered into the flow, the farther the M–1 curve will extend upstream. However, the gate has little effect on the downstream region until the position of the gate drops below the critical depth, since only then does it act as an upstream control on the resulting supercritical flow. Until that point is reached, there will be an eddy behind the gate and some energy loss, but the flow will remain at nearly normal depth.

On the steep channel, the instant the gate enters the flow (which will be at normal depth) a surge will form, move upstream some distance, and become stationary as shown. The gate again acts as a control on both the upstream subcritical flow and the downstream supercritical flow. The hydraulics of the sluice gate itself will be considered in Section 7–9 on flow measurement.

7–6 ADDITIONAL TRANSITION CONSIDERATIONS

The analysis of the short transition using the concept of specific energy in Section 7–3 ignored both friction loss and the possible head loss resulting from eddies. In the short length involved, the effect of friction may well be negligible, but the head loss due to the contraction or expansion requires further consideration. The contraction and expansion losses that will be discussed in this section apply not only to the short transition but to the natural river as well.

An important aspect of open channel flow that has received little attention thus far pertains to the ability of the water to get into the channel at the rate that has been assumed. To illustrate this type of problem one entrance condition will be examined. A third topic that will receive brief consideration in this section is the transition in direction, or the flow around a bend.

Flow Contraction and Expansion

Head loss due to the formation of eddies may occur whenever there is a contraction or expansion of a channel. These are similar to the minor losses due to contractions and expansions in pipes that were covered in Section 6–2. An additional consideration in the channel transition is whether the flow is subcritical or supercritical. If the flow is supercritical, oblique standing waves form in the transition, increasing the design requirements. Because of their complexity, supercritical flows, which fortunately occur with lower frequency than subcritical flows, will not be considered here.

There are a variety of theoretical and experimental equations for head loss in subcritical open channel flow. As with minor losses in pipes, these usually involve the product of a coefficient and some type of velocity head term. For a given form of transition the contraction will have a considerably smaller head loss than the corresponding expansion. One expression for expansions is

$$H_L = C_L \left(\frac{V_1 - V_2}{2g} \right)^2 \tag{7–36}$$

where V_1 and V_2 are the upstream and downstream velocities, respectively. The abrupt expansion of Fig. 7–30a has an expansion coefficient C_L approaching 0.9 to 1.0 whereas C_L reduces to 0.3 for the tapered expansion of Fig. 7–30b.

The contraction loss may be expressed by

$$H_L = C_L \frac{V^2}{2g} \tag{7–37}$$

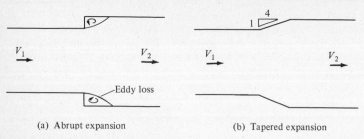

| (a) Abrupt expansion | (b) Tapered expansion |

Figure 7–30 Channel expansions.

where V is the velocity in the contracted section. The abrupt contraction that is the counterpart to Fig. 7–30a has a loss coefficient ranging in value up to approximately 0.23. Rounding or tapering the contraction reduces C_L considerably.

These head losses must be considered in conjunction with the changes in elevation of the water surface considered previously in Section 7–3 and 7–5. For example, assume that the channel shown in Fig. 7–9 has a mild slope. Normal depth y_2 will prevail downstream of the transition. A drop in the water surface will occur as shown: however, the upstream specific energy will be greater by the amount of head loss (from Eq. 7–37) and the upstream depth y_1 will be greater as well. An M–1 profile will extend upstream of the transition until the depth decreases to the upstream normal depth. Since the channel has a constant width, this depth will be the same as y_2 downstream of the transition.

The natural river has a cross section that changes constantly from section to section. This, in effect, acts as a series of alternating expansions and contractions. The head loss in either case may be written in terms of the difference in velocity heads,

$$H_L = C_L \left| \frac{V_1^2}{2g} - \frac{V_2^2}{2g} \right| \tag{7–38}$$

where V_1 and V_2 are the average velocities at any two adjacent sections. Typical values for the coefficients are given in Table 7–5. The use of these coefficients will be discussed further when river hydraulics are examined in Section 7–8.

Table 7–5 RIVER HEAD LOSS COEFFICIENT, C_L

Characteristic of cross sections	Coefficient	
	Expansion	Contraction
Gradual transition (the usual case)	0.3	0.1
Abrupt transition (the most extreme case)	0.8	0.6

Entrance Hydraulics

A question of some interest concerning a channel and a specified discharge is: how does that actual flow rate get into the channel? This may involve an upstream control structure that regulates the flow through gates or other devices as discussed in Section 7–9. The entrance condition to be considered here is a rectangular channel that connects directly with a reservoir, as sketched in Fig. 7–31. If the channel has a steep slope, then the entrance to the reservoir will be the control. The critical depth at the entrance can be determined as two-thirds of the specific energy H_0 at that point (Eq. 7–15). The discharge can then be determined from y_c using Eq. 7–14. An S–2 curve would follow as the depth decreases to normal depth for that discharge. In this case, the discharge is determined by the head in the reservoir and the channel slope has no bearing on the discharge.

If the channel slope is mild, both the upstream head and the downstream friction control enter into the problem. The discharge can be obtained by the following iterative procedure. A trial depth greater than y_c can be substituted into the Manning equation to get a trial Q from whence the velocity and velocity head may be calculated. This velocity head may be added to the trial depth to get a value of specific head that may be compared with the given value based on the reservoir elevation. If they are not the same, the procedure can be repeated with a new depth until satisfactory convergence is achieved. The discharge associated with the correct value of H_0 will be the desired discharge.

Whether the channel is steep or mild can be determined at the outset by substituting y_c into the Manning equation. If the resulting Q is less than the Q given by Eq. 7–14 (which is the maximum possible discharge for the channel under the given head), then the slope is mild and the iterative procedure must be used. If the discharge based on the Manning equation (but using the value of y_c for the depth) is greater than that given by Eq. 7–14, the channel is steep and the discharge is the value found from Eq. 7–14 as discussed previously. The entire process is illustrated in Example 7–11.

It has been assumed in this discussion that there is no energy loss as the water enters the channel. If a loss occurs, the procedure must be modified so that the specific head at the entrance is less than the reservoir head by the amount of the head loss.

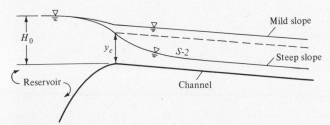

Figure 7–31 Entrance to rectangular channel.

EXAMPLE 7–11

The reservoir of Fig. 7–31 has a head of 4 m above the bed of the 10-m-wide rectangular channel that leads from it. The Manning n is 0.016 and the channel slope is 0.0006. Determine the discharge in the channel.

Solution

To determine whether the channel slope is mild or steep, the flow rate is determined at the critical depth,

$$y_c = 2/3 \, H_0 = (2/3)(4) = 2.667 \text{ m}$$

From Eq. 7–14

$$q^2 = gy_c^3 = (9.81)(2.667)^3 = 186.1$$

from which

$$Q = (10)(186.1)^{1/2} = 136.4 \text{ m}^3/\text{s}$$

This is the maximum possible flow rate in the channel. It will be the actual discharge if the slope is critical or steep.

The discharge given by the Manning equation for the critical depth is

$$Q = \frac{[(10)(2.667)]^{5/3}(0.0006)^{1/2}}{(0.016)[10 + (2)(2.667)]^{2/3}} = 59.1 \text{ m}^3/\text{s}$$

Since this is less than the 136.4 m³/s obtained previously, the slope will be mild. Note in passing that in the unlikely event that the two calculated discharges were identical, the channel would have a critical slope.

It remains to determine the channel depth that will have a specific head equal to 4 m. This depth must be greater than y_c and less than H_0. The tabulation below illustrates the procedure. The discharge is found for each trial depth using the Manning equation as above, and H_0 is the sum of the trial depth and the resulting velocity head.

Trial depth (m)	Q (m³/s)	V (m/s)	$\dfrac{V^2}{2g}$ (m)	H_0 (m)
3.5	86.7	2.477	0.313	3.81
3.8	97.2	2.558	0.333	4.13
3.7	93.7	2.532	0.327	4.03
3.67	92.6	2.524	0.325	4.00

The final depth of 3.67 m results in a specific head equal to that provided by the reservoir and the resulting discharge is 92.6 m³/s.

Flow in a Bend

As flow passes around the bend in Fig. 7–32, the inertia of the water tries to keep the water flowing in a straight path. The water near the surface, which has a velocity somewhat greater than the average velocity, possesses greater inertia than that near the bottom, which has a velocity less than the average. This greater inertia directs the higher-velocity water toward the outer channel bank while the slower water moves inward to satisfy continuity. This results in a spiral flow that may persist for some distance downstream. In an alluvial channel, such as Fig. 7–32b, this accounts for the typical erosion pattern on the outer bank and the tendency for deposition to occur at the inner bank.

The presence of a bend also causes tilting of the water surface as shown. This can be analyzed using the rectangular cross section of Fig. 7–32c and a unit length of channel. It will be assumed that the width B is much less than the radius of curvature r, that the surface slope is linear, and that the average velocity represents the flow. According to Newton's second law, the imbalance of hydrostatic forces shown must equal the product of the mass of water and the radial acceleration, V^2/r. This may be written

$$\frac{\gamma y_1^2}{2} - \frac{\gamma y_2^2}{2} = \frac{\rho B(y_1 + y_2)}{2} \frac{V^2}{r}$$

Dividing through by $(y_1 + y_2)$ and rearranging

$$y_1 - y_2 = \frac{V^2 B}{gr} \tag{7–39}$$

This gives an approximate value of the difference between the inner and outer water surfaces for any shape cross section (provided that the top width B is used). The actual tilt may be as much as 20 percent greater due to the inaccuracy of the assumptions.

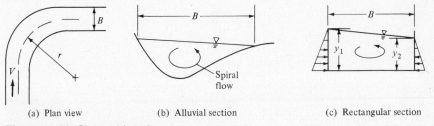

(a) Plan view   (b) Alluvial section   (c) Rectangular section

Figure 7–32 Channel bends.

If the flow is supercritical, standing waves of a height of about the same magnitude as the tilt may be expected. In the design of such a bend, superelevation of the bed and spiral transition sections are sometimes used to reduce the height of the waves.

7-7 FREE SURFACE FLOW IN PIPES

Most of the concepts discussed in this chapter are appropriate to open channel pipe flow. This includes critical and normal depth, specific energy, the hydraulic jump, and water surface profiles. However, some of the possible flow conditions are rare and others are of little interest in many cases. The analysis of the various free-surface pipe problems proceeds in essentially the same manner as in any other channel, with the additional consideration that a normal depth greater than the pipe diameter will result in a profile or jump that ends up as a full pipe. Once the pipe becomes full, the resulting pressure flow must be analyzed using the concepts of Chapter 6.

A partly full pipe is shown in Fig. 7–33. The geometric properties may be calculated as follows:

$$z = \frac{D}{2} - y \tag{7-40}$$

$$\theta = \cos^{-1}\left(\frac{z}{r}\right) = \cos^{-1}\left(\frac{2z}{D}\right) \tag{7-41}$$

$$A = r^2\theta - (r \sin \theta)(r \cos \theta)$$

$$= \frac{D^2}{4}\left(\theta - \frac{\sin 2\theta}{2}\right) \tag{7-42}$$

$$P = r(2\theta) = D\theta \tag{7-43}$$

and

$$R = A/P \tag{7-44}$$

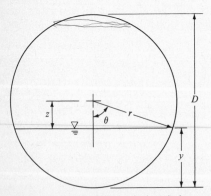

Figure 7–33 Definitional sketch for partially full pipe.

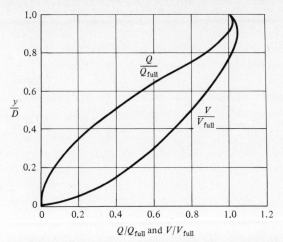

Figure 7–34 Hydraulic properties of circular pipes flowing partially full.

where θ is in radians. These geometric properties provide a theoretical means to determine profiles, normal depth, and so forth. They are not convenient, particularly in iterative solutions, and a computer program is very desirable if many calculations are to be made.

The relationship between normal depth and discharge or velocity can be obtained more readily by determining the full pipe Q or V based on the pipe slope. Figure 7–34, which was developed originally by Camp[4] may then be used to find the depth for a specified discharge or velocity graphically or, alternatively, the discharge or velocity for a given depth, as the case may be. This approach is frequently used in the design of storm and sanitary sewers. An approximate allowance is made in Fig. 7–34 for the variation of Manning n with depth. Consequently, if Eqs. 7–40 through 7–44 are used with the Manning equation they will not match Fig. 7–34, unless the same variation of n with depth is included.

Because of the rapid convergence of the pipe boundary as y exceed $0.8D$, the wetted perimeter increases faster with depth than does the cross-sectional area. Consequently, R begins to decrease as y approaches D. As indicated in Fig. 7–34, the maximum discharge occurs when $y = 0.94D$. This is not a practical design criterion, however, because disturbances and waves in the flow will bring the water surface into contact with the top of the pipe and it will likely flow full instead.

7–8 THE RIVER CHANNEL

The natural river has already been considered to some extent throughout this chapter. Many of the equations, such as Eq. 7–23 for critical depth, are applicable to any cross section. It is apparent that the calculations are quite tedious without the use of a computer. In this section, the emphasis will be on the calculation of a water surface

---

[4]T. R. Camp, "Design of Sewers to Facilitate Flow," *Sewage Works Journal* vol. 18, pp. 1–16. 1946.

profile (similar to the profiles of Section 7–5). Only the more common subcritical flow will be considered, although computations in a supercritical river flow are quite similar, except that the calculations proceed in the downstream direction. Further, overbank flow will be excluded, except for a brief reference at the conclusion.

The direct step method is not suitable for a river because the cross section changes continually and the necessary geometric data will only be available at selected cross sections. Therefore, we must use a computational technique that calculates the water surface elevation from one section to another, rather than the distance between sections, as was the case in the direct step method.

Geometric Properties

It will be assumed that the geometric data is provided in the form of x,y coordinates as given in Fig. 7–35. The x coordinates are measured from some arbitrary zero and increase when going from the left bank to the right bank. The y coordinates are elevations that are relative to sea level or any arbitrary datum. The geometric properties that must be calculated for a given water surface elevation are A, P, R, and

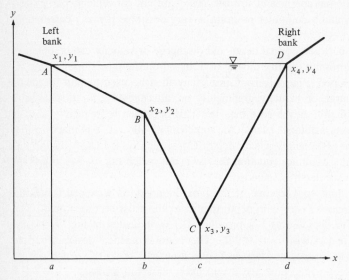

(a) Simplified cross section looking downstream

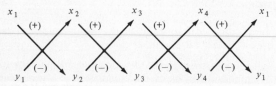

(b) Schematic for computation of area

Figure 7–35 River cross section by coordinates.

latter may be added to WS to get the total head H at the downstream section. In addition, the friction slope S_f may be calculated at this section using the Manning equation. The calculations are tabulated for typical values of A and P in Table 7–6.

Proceeding to the next section in the upstream direction, a value of WS is estimated. For lack of better information, this may be done as follows:

$$WS_2 = WS_1 + S_{f1}\Delta x \tag{7–48}$$

where the subscripts 1 and 2 refer to the downstream and upstream sections, respectively, and Δx is the distance between them. Using the trial WS, the total head H_{2a} may be found by adding the velocity head as before. In addition, a second calculation of H_2, H_{2b}, may be made by calculating S_{f2}, multiplying the average value of S_f over the reach by Δx, and adding the product to the downstream H. If H_{2a} equals H_{2b}, the assumed WS was correct. If not, a revised value must be chosen and the process repeated.

A best correction to WS_2 may be made by defining the difference in H_2 as

$$\Delta H_2 = H_{2b} - H_{2a}$$

It can be shown that the correction ΔWS, which should be algebraically added to the previous value of WS, is

$$\Delta WS = \frac{\Delta H_2}{1 - \text{Fr}_2{}^3 + \dfrac{3S_{f2}\,\Delta x}{2R_2}} \tag{7–49}$$

This will usually result in convergence of the H_2 values, but it can be repeated as necessary. The entire process is repeated for the next upstream section, and so on. Note in the correction that Fr is best calculated using Eq. 7–24 directly. Generally, B will be known and should be used. In the calculations for Table 7–6, B has been equated to P, a reasonable assumption in a wide river, and Eq. 7–24 becomes

$$\text{Fr} = \frac{V}{\sqrt{gA/B}} \approx \frac{V}{\sqrt{gA/P}} = \frac{V}{\sqrt{gR}}$$

Table 7–6 Water Surface Profile Computations in a River*

Section number	x	WS	A	$\dfrac{V^2}{2g}$	H_a	P	R	S_f	Average S_f	H_b
(ft)	(ft)	(ft)	(ft²)	(ft)	(ft)	(ft)	(ft)	($\times 10^3$)	($\times 10^3$)	(ft)
1	—	4.00	180	0.39	4.39	65	2.77	1.809	—	4.39
2	1000	5.81	207	0.29	6.10	71	2.92	1.276	1.543	5.93
2	1000	5.69	199	0.32	6.01	70	2.84	1.430	1.620	6.01 (ok)
3	1500	8.12	175	0.41	8.53	63	2.78	1.903	1.667	8.51
3	1500	8.11	174	0.42	8.53	63	2.76	1.944	1.687	8.54 (ok)

*River data: $Q = 900$ cfs, $n = 0.025$.

the top width B. To illustrate the calculations a simple cross section has been chose in Fig. 7–35a, but any reasonable number of points can be used to define the profile Points 1 and 4 are at the intersections of the water surface with the channel boundar profile.

The cross-sectional area may be calculated by subtracting the series of trapezoid from the rectangle $ADda$. That is

$$A = ADda - ABba - BCcb - CDdc$$

or, in terms of the coordinates,

$$A = \left(\frac{y_1 + y_4}{2}\right)(x_4 - x_1) - \left(\frac{y_1 + y_2}{2}\right)(x_2 - x_1)$$
$$- \left(\frac{y_2 + y_3}{2}\right)(x_3 - x_2) - \left(\frac{y_3 + y_4}{2}\right)(y_4 - y_3$$

When the various terms are multiplied together and rearranged, we get

$$A = \frac{1}{2}[(x_1y_2 + x_2y_3 + x_3y_4 + x_4y_1) - (y_1x_2 + y_2x_3 + y_3x_4 + y_4x_1)] \quad \text{(7–45}$$

The multiplication process of Eq. 7–45 is shown schematically in Fig. 7–35b, where the arrows denote the multiplication operations with signs as shown.

The wetted perimeter is calculated from coordinate point to coordinate point,

$$P = [(x_1 - x_2)^2 + (y_1 - y_2)^2]^{1/2} + [(x_2 - x_3)^2$$
$$+ (y_2 - y_3)^2]^{1/2} + [(x_3 - x_4)^2 + (y_3 - y_4)^2]^{1/2} \quad \text{(7–46}$$

This equation is also easily extended for any number of points. The hydraulic radius as usual is $R = A/P$, and the top width B is simply

$$B = x_4 - x_1 \quad \text{(7–47}$$

Bear in mind that in the actual calculations the intersections of the water surface with the boundary must be located by interpolation. These two points will be labeled (Lx,Ly) and (Rx,Ry) at the left and right banks, respectively. All boundary points with elevations above Ly and Ry must be excluded, and the two new points become the first and last points in Eqs. 7–45 through 7–47. For example, Eq. 7–47 becomes

$$B = Rx - Lx$$

Water Surface Profiles

The calculation of the water surface profile for a subcritical flow will commence at the downstream section with a known or estimated water surface elevation WS. A moderate error in the initial water surface will disappear within a few sections, particularly if the error is on the low side. The flow is assumed steady with a known discharge. The above equations can be used along with the water surface elevation WS, and the discharge to determine the hydraulic properties V and $V^2/2g$, and the

Note that Q, n, x, A, and P are all values that normally would be obtained or calculated from actual data. Here they are picked arbitrarily.

Correction to Section 2

$$\text{Fr} = \frac{4.35}{\sqrt{(32.2)(2.92)}} = 0.449$$

$$\Delta WS = \frac{5.93 - 6.10}{1 - (0.449)^2 + \dfrac{(3)(1.276 \times 10^{-3})(1000)}{(2)(2.92)}} = -0.12 \text{ ft}$$

Correction to Section 3

$$\text{Fr} = \frac{5.14}{\sqrt{(32.2)(2.78)}} = 0.543$$

$$\Delta WS = \frac{8.51 - 8.53}{1 - (0.543)^2 + \dfrac{(3)(1.667 \times 10^{-3})(1500)}{(2)(2.78)}} = -0.01 \text{ ft}$$

This procedure is not accurate if overbank flow is present and must be modified to account for the greatly different velocities in the channel and floodplain sections. More details may be found in the open channel references at the end of the chapter. The expansion and contraction losses given by Eq. 7–38 may be readily included in Table 7–6. At section 2, for example, the area is greater than that at section 1, therefore, the flow from 2 to 1 is contracting. The velocity heads for the two sections are already available and $C_L = 0.1$ would be appropriate. The head loss, which in this case is about 0.01 ft, would be added to H_{2b} before calculating the correction to WS.

A computer program to calculate water surface profiles in natural channels is included as computer program 4 in Appendix E. The following example (Example 7–12) is based on this program. As written, there is no provision for overbank flow, only a single value of Manning n can be used, and expansion and contraction losses are not included.

EXAMPLE 7–12

Calculate the water surface profiles for discharges of 20,000 and 30,000 cfs in a river for which the six cross sections are provided. Use downstream surface elevations of 22.0 and 24.3 ft for the respective flow rates and assume Manning $n = 0.031$. In the data set that follows, the L dimension is the distance to the next downstream section and all coordinate values are in feet.

	Section 1			Section 2 $L = 1500$ ft			Section 3 $L = 2100$ ft	
Point	x	y	Point	x	y	Point	x	y
1	0	40.5	1	0	43.6	1	0	46.1
2	40	36.3	2	60	19.7	2	40	45.1
3	95	19.7	3	120	18.8	3	140	12.5
4	140	9.8	4	195	16.2	4	240	1.7
5	200	5.3	5	210	7.1	5	275	15.1
6	299	1.7	6	245	4.1	6	290	15.3
7	360	21.8	7	345	20.5	7	305	15.9
8	421	42.9	8	450	25.7	8	310	19.9
			9	480	44.4	9	400	48.1

	Section 4 $L = 2000$ ft			Section 5 $L = 3150$ ft			Section 6 $L = 1855$ ft	
Point	x	y	Point	x	y	Point	x	y
1	0	48.2	1	0	61.1	1	0	64.4
2	100	19.1	2	50	51.2	2	19	57.0
3	113	15.1	3	101	14.3	3	49	29.1
4	230	17.3	4	201	1.1	4	149	19.0
5	340	5.9	5	259	19.6	5	207	14.4
6	395	6.4	6	289	21.1	6	217	13.1
7	461	41.4	7	356	60.9	7	293	18.8
8	503	48.4				8	361	36.1
						9	413	47.0
						10	471	63.3

Solution

The results for the two discharges are obtained by executing the program at the end of the chapter. The final profiles as well as selected geometric and hydraulic characteristics are tabulated below.

DISCHARGE = 20,000 cts

Section	WS (ft)	A (ft)	R (ft)	B (ft)	V (ft/s)	S	H (ft)
1	22.00	3659	13.16	273.20	5.47	0.000419	22.46
2	23.18	2625	7.45	347.83	7.62	0.001737	24.08
3	25.86	3171	13.35	230.01	6.31	0.000546	26.48
4	27.01	4742	12.87	361.02	4.22	0.000257	27.28
5	27.79	3408	14.95	217.91	5.87	0.000407	28.33
6	29.07	2576	9.00	284.02	7.76	0.001402	30.00

DISCHARGE = 30,000 cfs						
Section	WS (ft)	A (ft³)	R (ft)	B (ft)	V (ft/s)	S	H (ft)
1	24.30	4304	14.68	287.47	6.97	0.000588	25.06
2	25.69	3568	8.70	404.72	8.41	0.001718	26.78
3	28.40	3775	14.85	245.89	7.95	0.000753	29.38
4	30.03	5858	15.19	377.11	5.12	0.000304	30.44
5	30.91	4103	17.14	227.48	7.31	0.000526	31.74
6	32.23	3501	11.51	300.16	8.57	0.001229	33.37

The generalized computer program HEC–2 *Water Surface Profiles*,[5] originally developed by Eichert for the Corps of Engineers Hydrologic Engineering Center, is similar to computer program 4 in Appendix E, although it is far more comprehensive. The program calculates and plots water surface profiles, either subcritical or super-critical, in fixed channels. Hydraulic structures such as bridges, culverts, weirs, embankments, and dams can be included in the simulation. With bridges, for instance, once the water surface strikes the bridge deck, orifice flow is assumed, and if the water surface exceeds the top of the bridge (or the approaches) the flow is divided between orifice and weir flow.

Levees can be included in the geometric specification, and flow behind the levee can be excluded until the levee is topped. To indicate some of the inherent flexibility, the Manning n may be specified in different ways, or an estimate of the best Manning n may be determined based on observed high water marks.

7–9 FLOW MEASUREMENT

The measurement of the discharge in a river is generally obtained by measuring the stage and consulting a stage-discharge relationship that has been previously calibrated using a current meter. The entire process is described in Chapter 3. The measurement of the flow rate in artificial channels and even small streams is most frequently made using weirs, gates, or measuring flumes. Each of these three methods will be discussed below.

Weirs

A *weir* is a device placed in the channel that backs up the flow so that in passing over (or occasionally through) it the water drops through critical depth in the process. The discharge may then be related to the upstream depth. If a high tailwater prevails,

---

[5]HEC–2 is one of a series of comprehensive hydrologic engineering computer programs developed by the Hydrologic Engineering Center at Davis, California and available for general use.

the weir will operate in a submerged condition. It may still be possible to estimate the discharge using the tailwater as an additional parameter, but the results usually have a lower accuracy. Only the unsubmerged state will be discussed here.

The weir (see Figs. 7–36 through 7–38) should be positioned carefully in a straight section of channel with its upstream face vertical and perpendicular to the flow direction. Measurement of the upstream depth (or the water surface elevation, as the case may be) must be obtained with considerable care. The ideal location is just upstream of the nonuniform region created by the weir. If measured too close to the weir, the drawdown will influence the reading. If measured too far upstream, the increased water elevation due to channel friction will result in an overestimation of the discharge. If waves are present, the depth measurement is best obtained from a stilling well connected to the main channel at the desired point of measurement.

The discharge equations for various weirs may be derived by integration of the velocity distribution over the flow area, the velocity itself being obtained from the energy equation. A large correction must usually be applied because of the assumptions involved. This theoretical approach may be found in most fluid mechanics texts, and only the actual weir equations will be considered here.

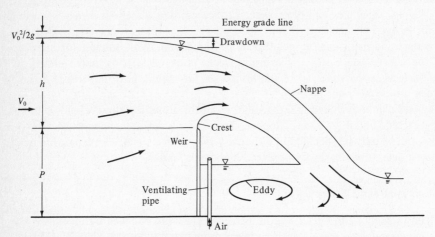

Figure 7–36 Rectangular sharp-crested weir.

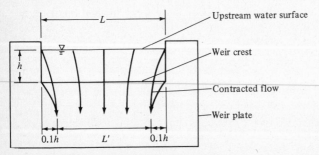

Figure 7–37 Contracted rectangular weir.

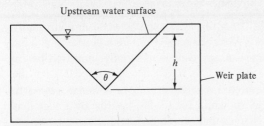

Figure 7–38 Triangular weir.

A *rectangular weir* is shown in Fig. 7–36. A machined sharp-edged crest is best to ensure that the nappe springs free. It is also essential that the crest is horizontal and the region under the nappe is well ventilated. If adequate ventilation is not provided, the air will be drawn out, the pressure reduced, and the upstream head will be reduced. If the crest extends over the entire width of the channel, the discharge is given by

$$Q = C_d \,(2/3)\sqrt{2g}\, L\left[\left(h + \frac{V_0^2}{2g}\right)^{3/2} - \left(\frac{V_0^2}{2g}\right)^{3/2}\right] \tag{7–50}$$

where V_0 is the approach velocity, h is the head above the crest, L is the crest length, and C_d is a discharge coefficient. If the approach velocity can be ignored (i.e., $P \gg h$), the equation reduces to the more convenient form,

$$Q = C_d \,(2/3)\sqrt{2g}\, Lh^{3/2} \tag{7–51}$$

The coefficient C_d may be obtained from an experimental equation by Rehbock,

$$C_d = 0.605 + \frac{1}{305h} + 0.08\,\frac{h}{P} \tag{7–52}$$

Because of the dimensional term, Eq. 7–52 is fully valid only in U.S. customary units: however, an approximate value $C_d = 0.62$ is applicable over most of the normal range of weir conditions. Using this numerical value, a weir coefficient $C_w = (0.62)(2/3)(2g)^{1/2}$ may be introduced. Equation 7–51 then simplifies to

$$Q = C_w\, Lh^{3/2} \tag{7–53}$$

where $C_w = 3.33$ in customary units and 1.84 in SI units.

A weir plate such as that shown in Fig. 7–37 is often placed in a channel. This results in a contraction of the flow from both sides, and a reduction in discharge relative to the uncontracted case. An effective crest length L' may be defined as

$$L' = L - (0.1)nh \tag{7–54}$$

where n is the number of side contractions (two in Fig. 7–37, but other numbers are possible). This equation assumes that each contraction is equal to 10 percent of the upstream head, a reasonable, but only approximate, assumption. The discharge past a contracted weir is calculated from the foregoing discharge equations with the length L replaced by L'.

In an outside location it is often not practical to use a weir with a sharp crest because of the possibility of damage. Other materials such as wood or concrete may be used and a calibration of Q versus h is recommended if a high degree of accuracy is to be achieved. In fact, the accuracy of any weir is materially improved by calibration.

If low flow rates are to be measured a *triangular weir* or *V-notch weir* is frequently employed. This form of weir, shown in Fig. 7–38, has the advantage that a larger head h will occur at small discharges than would be found above a rectangular weir. For a notch angle θ (in deg) with a machined sharp-edged weir plate, the discharge is

$$Q = C_d\left(\frac{8}{15}\right)\sqrt{2g} \tan\left(\frac{\theta}{2}\right)h^{5/2} \qquad (7\text{–}55)$$

The discharge coefficient C_d may be replaced by introducing a weir coefficient defined by $C_w = C_d(8/15)(2g)^{1/2} \tan(\theta/2)$. The discharge equation then becomes

$$Q = C_w h^{5/2} \qquad (7\text{–}56)$$

The most common notch angle is 90° for which C_d is approximately 0.585. This results in weir coefficients of about $C_w = 2.50$ in U.S. customary units and $C_w = 1.38$ in SI units.

The disadvantages of any type of weir plate under field conditions have lead to the development of the much sturdier *broad-crested weir* shown in Fig. 7–39. The broad-crested weir can have a number of different forms, of which only one example

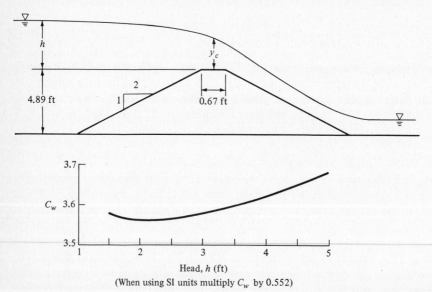

Head, h (ft)
(When using SI units multiply C_w by 0.552)

Figure 7–39 Typical standard broad-crested weir with weir coefficients (plotted from R. E. Horton, "Weir Experiments, Coefficients, and Formulas," U.S.G.S. Water Supply Paper 200, 1907).

is included. The common thread throughout is the absence of a sharp crest. The principle remains the same as that for the rectangular weir since the water surface passes through critical depth (except under submerged conditions), and Eq. 7–53 may again be used. There are a number of standardized shapes for which experimental C_w values are available. Figure 7–39 includes coefficient values for the type of weir shown. In general the broad-crested weirs have weir coefficients that are a little greater than the values for the sharp-crested weir.

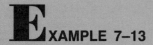XAMPLE 7–13

The measured head above a 1-m-high sharp-crested rectangular weir is 0.5 m. Determine the apparent error in calculating the discharge over the weir if the velocity of approach is ignored. Assume that the weir coefficient $C_w = 1.84$.

Solution

The crest length has no bearing on the solution, provided that end contractions are not present, therefore L will be taken as unity. If V_0 is ignored then

$$Q = (1.84)(0.5)^{3/2} = 0.651 \text{ m}^3/\text{s}$$

Using this value of Q leads to an approach velocity

$$V_0 = \frac{0.651}{1.5} = 0.434 \text{ m/s}$$

With V_0 incorporated into the discharge equation, we get

$$Q = (1.84)\left\{ \left[0.5 + \frac{(0.434)^2}{(2)(9.81)} \right]^{3/2} - \left[\frac{(0.434)^2}{(2)(9.81)} \right]^{3/2} \right\} = 0.668 \text{ m}^3/\text{s}$$

With this improved value for Q, a new approach velocity $V_0 = 0.445$ m/s results. This velocity does not significantly change the above discharge calculation and the best estimate of the discharge is 0.668 m³/s. The error in ignoring V_0 is

$$\text{Error} = \frac{0.668 - 0.651}{0.668} \times 100 = 2.5 \text{ percent}$$

Gates

The primary function of a gate is to regulate the flow. When regulating the flow rate, it is usually necessary to know the discharge, and the gate, especially if properly calibrated, can serve this function as well. A vertical gate or *sluice gate* is shown in Fig. 7–40. The flow passing under the gate experiences a considerable vertical contraction because of the vertical velocity component along the face of the gate. It will be assumed that the gate extends across the entire width of the channel, otherwise there will be side contractions as well.

As with the other measuring devices, the submerged flow condition due to high tailwater will not be considered. Thus, the downstream depth may be related to the opening w by $y_2 = C_c w$, where C_c is a contraction coefficient. Assuming that no energy loss occurs between the upstream and downstream sections, the energy equation states that

$$y_1 + \frac{Q^2}{2gy_1{}^2b^2} = y_2 + \frac{Q^2}{2gy_2{}^2b^2}$$

where b as usual is the channel width. Solving for the discharge,

$$Q = y_1 y_2 b \sqrt{\frac{2g}{y_1 + y_2}} = C_c w b \sqrt{2gy_1} \sqrt{\frac{y_1}{y_1 + y_2}} \qquad (7\text{–}57)$$

Introducing a discharge coefficient

$$C_d = C_c \sqrt{\frac{y_1}{y_1 + C_c w}} = \frac{C_c}{\sqrt{1 + C_c w/y_1}} \qquad (7\text{–}58)$$

leads to a discharge equation that may also be written as

$$Q = C_d w b \sqrt{2gy_1} \qquad (7\text{–}59)$$

The contraction coefficient C_c ranges from about 0.6 upward, depending on the value of w/y_1, with 0.61 a typical value. More information on the coefficients may be obtained from the references at the end of the chapter. These references also cover the case of submerged flow.

A second commonly used gate is the *radial gate* or *Tainter gate* shown in Fig. 7–41. The Tainter gate has an advantage over the vertical gate in that the force of the water is normal to the surface and is therefore transmitted directly to the hinge.

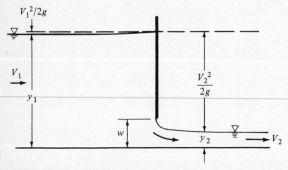

Figure 7–40 Sluice gate.

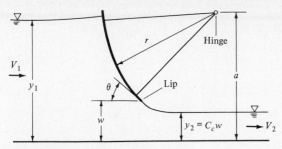

Figure 7–41 Tainter gate.

Hence, the large force due to the water pressure does not have to be overcome when operating the gate. In addition, the trailing edge of the gate gives the water a component in the downstream direction, and less contraction occurs than with the sluice gate. This makes the Tainter gate more hydraulically efficient than the vertical gate.

The analysis of the radial gate is identical to that of the sluice gate and Eqs. 7–57 through 7–59, resulting from the elementary energy analysis, may be used for the radial gate. The coefficients, however, depend on more parameters. Not only does the lip angle θ change with w as a given gate is opened, but different values for the radius of curvature r and location of the hinge a change the entire configuration from gate to gate. An approximate equation given by Henderson for C_c is[6]

$$C_c = 1 - 0.75\left(\frac{\theta}{90}\right) + 0.36\left(\frac{\theta}{90}\right)^2 \tag{7–60}$$

where the angle θ is specified in deg.

Venturi Flumes

A *venturi flume* is the open channel counterpart of the venturi tube used to measure the discharge in a pipe. Just as the venturi tube creates a reduced pressure in the contracted throat, the venturi flume has a contracted section in which the depth of water decreases, usually to a depth below y_c. There are two major advantages to a venturi flume. First, they generally have a much lower head loss than a weir. Second, if the water transports a sediment load, a portion of the sediment will deposit upstream of the weir, thereby altering the flow pattern and the weir coefficient. The venturi flume, on the other hand, tends to be self-scouring.

The best-known flume of this type is the *Parshall flume*[7] shown in Fig. 7–42. This meter was designed by R. L. Parshall to measure flow rates in irrigation channels. The Parshall flume can vary in size from a minimum throat width $b = 3$ in. up to $b = 50$ ft. However, the style and dimensions shown in Fig. 7–42 are for widths of 1 ft to 8 ft. The 8-ft size can measure discharges up to 140 cfs.

---

[6]See the Henderson reference listed at the end of the chapter.
[7]R. L. Parshall, "Measuring Water in Irrigation Channels with Parshall Flumes and Small Weirs," U.S. Soil Conservation Service, Circular 843, May, 1950.

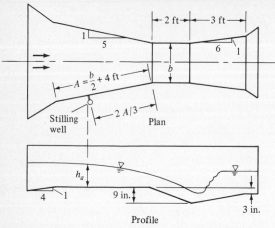

Profile

Figure 7–42 Parshall flume for 1 ft < b < 8 ft.

Only free-flow conditions are shown in Fig. 7–42. The water surface drops through critical depth in the contracted section, followed by a hydraulic jump. The discharge depends on the upstream depth h_a, which is measured in a stilling well. For the 1-ft to 8-ft-width flume shown, the discharge is given by,

$$Q = 4bh_a^{1.522b^{0.026}} \tag{7–61}$$

If the downstream depth drowns out the jump, creating submerged conditions, a second stilling well is required to measure the downstream depth, and the calculated discharge must be reduced to account for the submergence. The accuracy of the measurement can be expected to decrease in this event.

P   PROBLEMS

Section 7–2

7–1. Plot a graph of the ratio of the pressure as given by Eq. 7–1 to the hydrostatic pressure based on the depth versus the slope angle θ for $0° < \theta < 45°$.

7–2. Determine the maximum slope angle for which the assumption of hydrostatic pressure distribution will be accurate within (1) 2 percent and (2) 5 percent.

7–3. Uniform flow with a depth of 6 ft occurs in a rectangular channel that has a width of 15 ft and a slope of 0.0011. Determine the boundary shear stress and the hydraulic radius.

7–4. If the channel of Prob. 7–3 has a discharge of 710 cfs, what is the value of the Manning n?

7–5. The figure shown consists of a main channel and two overbank sections. The slope of all sections is 0.0006, and the Manning n values for the main channel and overbanks are 0.025 and 0.050, respectively. Determine the discharge when the water surface is located as shown. Make whatever assumptions are necessary.

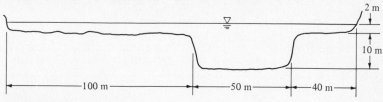

Figure P7–5

7–6. Calculate the normal depth in a smooth concrete, trapezoidal channel with side slopes of 3 on 1, a bed slope of 0.00033, a bottom width of 4.0 m, and a water discharge of 39 m³/s.

7–7. Uniform flow with a depth of 1 m occurs in a trapezoidal brick channel that has side slopes of 2.5 on 1, a bottom width of 2 m, and a channel slope of 0.0003. Determine the boundary shear stress and the discharge.

7–8. Determine the minimum slope necessary to transport a discharge of 100 cfs in a trapezoidal gravel channel that has a bottom width of 8 ft and side slopes of 1.5 on 1, if the depth is not to exceed 1.5 ft.

7–9. Calculate the normal depth in a rectangular timber channel ($n = 0.015$) that has a width of 8 ft, a slope of 0.004, and a discharge of 265 cfs.

7–10. Calculate the discharge in a 30-ft-wide rectangular channel constructed of smooth concrete. The depth is 7 ft and the channel slope is 0.0006.

7–11. Calculate the discharge in a 4-m-wide rectangular flume constructed of planed timber. The depth is 1.2 m and the slope is 0.01.

7–12. Determine the discharge in a trapezoidal channel that has a channel slope of 0.001, side slopes of 2.5 on 1, a bottom width of 12 ft, and a depth of 6 ft. Use $n = 0.020$.

7–13. Determine the discharge in a trapezoidal channel that has a channel slope of 0.0075, side slopes of 2 on 1, a bottom width of 2 m, and a depth of 1.5 m. Use $n = 0.017$.

7–14. Calculate the normal depth in a trapezoidal channel that has side slopes of 1.5 on 1, a channel slope of 0.0025, a bottom width of 6 ft, and a discharge of 120 cfs. Use $n = 0.016$.

7–15. Determine the normal depth in a rectangular channel that has a width of 2 m and a discharge of 18 m³/s. The channel slope is 0.008 and $n = 0.012$.

7–16. The following data are obtained at two gauging sections in a relatively straight reach of river:

Section	Water surface elevation (m)	Cross-sectional area (m²)	Wetted perimeter (m)
1	113.726	360	116
2	113.394	315	109

Estimate the value of Manning n for the reach and determine the average boundary shear stress. The discharge is 480 m³/s and section 1 is 960 m upstream of section 2.

7–17. The following data are obtained at two gauging sections in a relatively straight reach of river:

Section	Water surface elevation (ft)	Cross-sectional area (ft²)	Wetted perimeter (ft)
1	37.68	4317	751
2	36.91	4910	813

Estimate the value of Manning n for the reach and determine the average boundary shear stress. The discharge is 22,500 cfs and section 1 is 960 ft upstream of section 2.

7–18. Write a computer program that will print out a table of discharge versus depth in either a rectangular or trapezoidal channel, given the bottom width, channel slope, side slopes, Manning n, and the range and calculation increment for the depth.

7–19. Write a computer program to calculate the normal depth in rectangular and trapezoidal channels. Make the program applicable for both U.S. customary and SI units.

7–20. Write a computer program to calculate the required bottom width in a rectangular or trapezoidal channel giving the normal depth, side slopes, channel slope, and Manning n.

Section 7–3

The following twenty problems refer to a transition in a rectangular channel with a bottom elevation change Δz, or a change in width from b_1 to b_2, or both. Flow is from section 1 to section 2. A plus Δz is upward. Analyze each problem using a specific energy diagram and determine the downstream depth. If the stated upstream flow is not possible, calculate the adjusted upstream conditions as well. Problems 7–21 to 7–30 are in U.S. customary units and Problems 7–31 to 7–40 are in SI units.

Problem	Q (cfs)	y_1 (ft)	b_1 (ft)	Δz (ft)	b_2 (ft)
7–21	200	5.5	10	0.3	10
7–22	200	1.8	10	−0.3	10
7–23	200	5.5	10	−0.3	10
7–24	200	1.8	10	0.3	10
7–25	300	6.0	12	0	15
7–26	300	1.5	12	0	15
7–27	250	6.0	12	0	10
7–28	250	4.0	12	0	8
7–29	200	2.0	10	0.3	12
7–30	250	4.8	8	−0.5	10

Problem	Q (m^3/s)	y_1 (m)	b_1 (m)	Δz (m)	b_2 (m)
7–31	20	2	5	0.2	5
7–32	20	2	5	−0.2	5
7–33	25	1	5	−0.2	5
7–34	20	1	5	0	4
7–35	20	2	5	0	4
7–36	20	2	5	0.15	4
7–37	15	3	4	0	3
7–38	15	3	4	0	2
7–39	15	3	4	0	1
7–40	30	2.5	6	0.8	4

7–41. Determine the maximum possible increase in channel bottom elevation that can occur in the channel of Prob. 7–29 without causing an upstream flow adjustment.

7–42. Determine the minimum downstream width in the channel of Prob. 7–34 that will not result in an upstream flow adjustment.

7–43. Determine the critical depth in a trapezoidal channel that has a bottom width of 10 ft and side slopes of 2 on 1, if the discharge is 500 cfs.

7–44. Repeat Prob. 7–43 using a discharge of 100 cfs.

7–45. Determine the critical depth in a trapezoidal channel that has a bottom width of 8 m and side slopes of 2.5 on 1, if the discharge is 400 m^3/s.

7–46. Calculate the critical depth in a triangular channel that has side slopes of 2 on 1 and a discharge of 3 m^3/s.

7–47. A water discharge of 400 cfs results in a depth of 6 ft in a trapezoidal channel that has a bottom width of 8 ft and side slopes of 2 on 1. If the channel cross section is lowered 1.0 ft in a transition, determine the downstream depth.

7–48. A discharge of 40 m^3/s results in a depth of 1 m in a trapezoidal channel that has a bottom width of 2 m and side slopes of 2.25 on 1. If a transition section increases the bottom width to 3 m while maintaining the same side slopes and bottom elevation, determine the downstream depth.

7–49. Write and operate a computer program to calculate the critical depth in a trapezoidal channel.

7–50. Write and operate a computer program to analyze rectangular channel transitions. Input should include the discharge, upstream width and depth, downstream width, and the change in bottom elevation. If a given flow is not possible, then the program should determine the adjusted upstream conditions as well. Note that you may want to use separate convergence tests for subcritical and supercritical flows, but each type of transition does not have to be considered separately.

Section 7–4

7–51. A hydraulic jump occurs in a 40-ft-wide rectangular channel. If the upstream and downstream depths are 3.5 ft and 12.5 ft, respectively, determine (1) the flow rate, (2) the head loss, and (3) the power loss.

7–52. What are the upstream and downstream Froude numbers in Prob. 7–51?

7–53. A hydraulic jump occurs in a wide rectangular channel. The upstream depth and velocity are 0.4 m and 13 m/s, respectively. Determine the downstream depth, the head loss, and length of the jump.

7–54. If a discharge of 5000 cfs has a depth of 2.5 ft just ahead of a hydraulic jump in a 45-ft-wide rectangular channel, determine the downstream depth.

7–55. A hydraulic jump occurs in a 20-m-wide rectangular channel. If the upstream and downstream depths are 0.8 m and 3 m, respectively, determine (1) the discharge and (2) the head loss in the jump.

7–56. A hydraulic jump with upstream and downstream depths of 1 ft and 8 ft, respectively, occurs in a wide rectangular channel. Determine the discharge per unit width and the upstream and downstream Froude numbers.

7–57. If the velocity and depth downstream of a hydraulic jump are 5 ft/s and 6 ft, respectively, evaluate the upstream depth and the head loss in the jump.

7–58. A 15-ft-wide rectangular channel transports water with a depth of 2 ft and an average velocity of 15 ft/s. If a downstream gate is suddenly closed, determine the velocity at which the surge will move upstream and the resulting depth.

7–59. A steady flow with a discharge of 30 m^3/s and depth of 1.5 m occurs in a 5-m-wide rectangular channel. If the upstream discharge is abruptly doubled, determine the resulting surge velocity and the subsequent depth and velocity.

7–60. Prepare a computer program that when given the depth, width, and discharge in a rectangular channel will determine whether a hydraulic jump is possible and, if so, will calculate the downstream depth and velocity, the head loss and power loss in the jump, and the length of the jump.

Section 7–5

7–61. The discharge in a long 12-ft-wide rectangular channel is 345 cfs, and the normal depth is 5 ft. At a specific section in the channel the depth is 2 ft. Include all work or reasoning in answering the following questions: (1) Is the channel steep or mild? (2) Is the flow subcritical or supercritical? (3) Does the depth increase, decrease, or remain constant as the flow proceeds downstream? (4) Sketch a possible physical situation depicting the given conditions. (5) If $n = 0.020$, calculate the channel slope. (6) If $n = 0.020$, calculate the friction slope at the specified section.

7–62. The discharge in a 5-m-wide rectangular channel is 10 m^3/s, and the normal depth is 0.4 m. At a specific section in the channel the depth is 2 m. Answer the questions given in Prob. 7–61 for this channel.

7–63 through 7–68. Sketch and label all possible water surface profiles for each of the following transition problems. On mild or steep slopes assume that the sections all have sufficient length so that the transition curves may be completed in the length shown. Flow is from left to right and, as usual, the short dashes represent critical depth while the long dashes identify the normal depth.

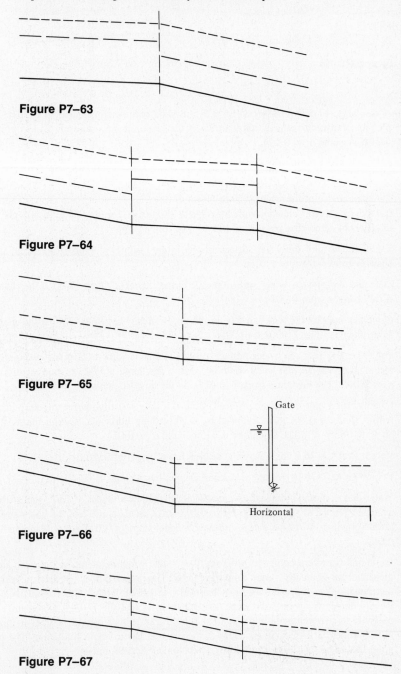

Figure P7–63

Figure P7–64

Figure P7–65

Gate

Horizontal

Figure P7–66

Figure P7–67

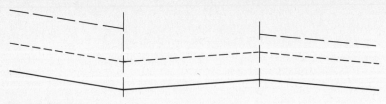

Figure P7–68

7–69. A discharge of 3.7 m³/s occurs in a 3.5-m-wide rectangular channel. Calculate and plot the transition water surface profile if the slope changes abruptly from 0.006 to 0.014. Assume $n = 0.014$.

7–70. Calculate and plot the water surface profile in the channel of Prob. 7–69 if the channel slope changes abruptly from 0.014 to 0.006.

7–71. Calculate and plot the water surface profile in the vicinity of a grade change in a 10-ft-wide rectangular channel with Manning $n = 0.014$. The discharge is 400 cfs and the channel slope changes from 0.016 to 0.0016.

7–72. Calculate and plot the water surface profile in the channel of Prob. 7–71 if the slope changes from 0.0016 to 0.016.

7–73. Calculate and plot the water surface profile in the channel of Prob. 7–71 if the channel slope changes from 0.0016 to 0.0001.

7–74. Calculate and plot the water surface profile in the channel of Prob. 7–71 if the channel slope changes from 0.0001 to 0.0016.

7–75. Calculate and plot the water surface profile in the vicinity of a grade change in a rectangular channel that has a width of 20 ft, a discharge of 900 cfs, and a resistance $n = 0.015$. The upstream slope is 0.015 and the downstream slope is 0.00044.

7–76. Calculate and plot the water surface profile if a long, 10-ft-wide rectangular channel with a slope of 0.00048 is followed by a 100-ft-long horizontal section that ends in an abrupt drop-off. The discharge is 215 cfs and $n = 0.014$.

7–77. Repeat Prob. 7–76 if the horizontal section has a length of 2500 ft.

7–78. Repeat Prob. 7–76 if the long upstream channel has a slope of 0.04.

7–79. Water in a 4-m-wide rectangular channel discharges from under a sluice gate with an average velocity of 9 m/s and a depth of 0.35 m. Calculate and plot the water surface profile downstream of the gate. Assume a channel slope of 0.005 and a Manning $n = 0.012$.

7–80. Write and operate a computer program to calculate water depth versus location along a channel for gradually varied flow transitions. The program does not need to analyze the transition, as the user can input the initial and final depths. Other input includes the discharge, the width, the slope, the Manning n, and the vertical increment. The program should terminate with the final depth regardless of increment size. It should also terminate at a specified location if the channel has a length limitation. The program should be operational in both U.S. customary and SI units.

Section 7–6

7–81. Analyze the gradually varied water surface profile in the vicinity of an abrupt expansion in a rectangular channel. The discharge is 420 cfs, the slope is 0.0005, and the width increase is from 9 to 12 ft. Assume $n = 0.015$ and the loss coefficient is $C_L = 0.95$. Include the effect of the expansion (Section 7–3) as well as the energy loss.

7–82. Repeat Prob. 7–81 for a rectangular channel with $Q = 4$ m³/s, $S = 0.0004$, $b_1 = 3.6$ m, and $b_2 = 4.0$ m. Assume n and C_L are unchanged.

7–83. Repeat Prob. 7–81 if the expansion is replaced by a contraction in width from 12 ft to 9 ft. Assume a contraction coefficient $C_L = 0.21$.

7–84. Repeat Prob. 7–82 for an abrupt contraction in width from 4 m to 3 m. Use a contraction coefficient $C_L = 0.22$.

7–85. A reservoir similar to that in Fig. 7–31 has a head of 5 m above the bed of a 20-m-wide rectangular channel. The channel has a slope of 0.005 and a Manning $n = 0.017$. Determine the discharge and normal depth in the channel.

7–86. Repeat Prob. 7–85 if the channel slope is 0.00045.

7–87. A reservoir similar to that in Fig. 7–31 has a head of 20 ft above the bed of a 100-ft-wide rectangular channel. The channel has a slope of 0.0004 and a Manning $n = 0.020$. Determine the discharge and normal depth in the channel.

7–88. Repeat Prob. 7–87 if the slope is increased to 0.002.

7–89. A discharge of 700 cfs occurs in a 15-ft-wide rectangular channel that has a slope of 0.00012 and $n = 0.013$. Estimate the increase in water surface elevation above normal depth if the channel has a bend with a 150-ft radius.

7–90. A discharge of 19 m³/s occurs in a trapezoidal brick channel that has a bottom width of 2 m, side slopes of 2 on 1, and a slope of 0.00022. Estimate the increase in water surface elevation above normal depth if the channel has a bend with a radius of (1) 20 m and (2) 50 m.

Section 7–7

7–91. Write a computer program to evaluate the normal depth in a pipe, given the diameter, discharge, slope, and Manning n, in either U.S. customary or SI units.

7–92. Write a computer program to evaluate the critical depth in a pipe, given the diameter and discharge in either U.S. customary or SI units.

7–93. Plot the specific energy diagram for a 10-ft-diameter pipe and a discharge of 200 cfs.

7–94. A discharge of 200 cfs flows at a depth of 7 ft in a 10-ft-diameter pipe. Use the specific energy diagram of Prob. 7–93 to determine the downstream depth if a transition section increases the pipe elevation by 0.5 ft. Ignore energy loss.

7–95. Repeat Prob. 7–94 if the transition lowers the pipe by 0.5 ft.

7–96. Determine the discharge in a 1-m-diameter pipe if the depth is 0.4 m and the slope is 0.002. Assume $n = 0.014$.

7–97. Determine the discharge in a 4-ft-diameter concrete pipe ($n = 0.012$) if the depth is 1 ft and the pipe slope is 0.0002.

7–98. Determine the normal depth in a 4-ft-diameter concrete pipe ($n = 0.012$) laid to a slope of 0.0005 if the discharge is (1) 20 cfs and (2) 200 cfs.

7–99. What minimum slope is required to transport a flow rate of 0.1 m³/s in a 35-cm-diameter pipe, if the design depth is 0.8 D. Use $n = 0.013$.

7–100. Write a computer program similar to that of Prob. 7–80 to analyze water surface profiles in a circular pipe.

Section 7–8

7–101. Verify that Eqs. 7–45 and 7–46 are correct for a rectangular channel.

7–102 through 7–105. Use the river data tabulated below to generate the necessary geometric parameters for the next four problems. Although U.S. customary units are shown, the latter two problems use the same numerical values with ft and ft² replaced by m and m², respectively.

Stage	Section 1		Section 2		Section 3	
	P	A	P	A	P	A
(ft)	(ft)	(ft²)	(ft)	(ft²)	(ft)	(ft²)
100	200	1000				
101	203	1090	201	1210		
102	208	1240	209	1420	250	1400
103	216	1470	218	1630	262	1710
104	225	1750	227	1850	273	2020
105	236	2090	237	2080	281	2330
106	248	2400	247	2330	298	2660
107	262	2950	258	2620	313	3000
108			272	3000	328	3370
109					350	3780

All flow is in the main channel, which has a Manning $n = 0.032$. Section 2 is 2500 ft upstream of Section 1 and Section 3 is 2100 ft upstream of Section 2. There are no expansion or contraction losses.

7–102. Determine the water surface elevation at Sections 2 and 3 for a discharge of 6800 cfs. At this discharge the stage at Section 1 is 102.5 ft.

7–103. Repeat Prob. 7–102 for a discharge of 10,000 cfs and a downstream stage of 104 ft.

7–104. Assume that all geometric data is in SI units and calculate the water surface elevation at Sections 2 and 3 for a flow rate of 1700 m³/s. At this discharge the downstream stage is 102.5 m.

7–105. Repeat Prob. 7–104 for a discharge of 3150 m³/s and a downstream stage of 104 m.

7–106. Modify computer program 4 in Appendix E to reflect expansion and contraction losses as calculated by Eq. 7–38. The coefficients should be treated as input data with default values of 0.3 and 0.1, respectively.

Section 7–9

7–107. Calculate the discharge over a 1-m-high sharp-crested rectangular weir with a crest length of 1.5 m, if the upstream head is (1) 0.01 m, (2) 0.1 m, and (3) 1 m. Include the approach velocity and use Eq. 7–50 with $C_d = 0.62$.

7–108. Determine the error in each part of Prob. 7–107 if the approach velocity is ignored.

7–109. Calculate the discharge over a 1-ft-high sharp-crested rectangular weir that has a crest length of 3 ft, if the upstream head is (1) 0.01 ft, (2) 0.1 ft, and (3) 1 ft. Include the approach velocity and use Eq. 7–50 along with Eq. 7–52.

7–110. Determine the error in each part of Prob. 7–109 if you assume $C_d = 0.62$.

7–111. Determine the error in each part of Prob. 7–109 if the approach velocity is ignored.

7–112. Repeat Prob. 7–107 assuming that a weir plate with a crest length of 1 m is available in the 1.5-m-wide channel.

7–113. Repeat Prob. 7–109 assuming that a weir plate with a crest length of 2.25 ft is available in the 3-ft-wide channel.

7–114. Determine the discharge past a triangular weir with a notch angle of 90 deg if the head is (1) 0.1 ft and (2) 1 ft.

7–115. What error in discharge would result if there were a 1 deg error in the notch angle in Prob. 7–114?

7–116. What head would be required above a 90 deg angle triangular weir to pass the same discharge as a sharp-crested rectangular weir with a crest height of 1.5 ft, a crest length of 3.5 ft, and a head of (1) 0.05 ft and (2) 0.5 ft. Ignore the approach velocity. Assume that the rectangular weir extends over the full width of the channel.

7–117. If a V-notch weir has a discharge of 0.89 cfs at a head 0.7 ft, what discharge may be expected at a head of 1.1 ft?

7–118. Determine the maximum positive and negative percent errors that will occur in the measurement of discharge using the broad-crested weir of Fig. 7–39 within the range 1.5 ft $< h <$ 5 ft if a constant coefficient $C_w = 3.6$ is assumed.

7–119. Determine the discharge in a 6-ft-wide rectangular channel if the depth upstream of a sluice gate is 3.4 ft and the opening is (1) 0.3 ft and (2) 0.7 ft. Assume $C_c = 0.61$.

7–120. A discharge of 2.5 cfs occurs in a 2-ft-wide flume. Assuming a constant contraction coefficient $C_c = 0.61$ (a poor assumption as the sluice gate opening is increased), plot a graph of y_1 versus gate opening w. Determine the upstream and downstream Froude numbers for selected points.

7–121. Determine the discharge in a 1-m-wide rectangular channel if the depth upstream of a radial gate is 1.3 m, the opening is 0.05 m, and the lip angle is 25 deg.

7–122. Repeat Prob. 7–121 if the lip angle is 60 deg, with the other values remaining unchanged.

7-123. Determine the discharge passing through a Parshall flume that has a throat width of 2 ft, if the depth h_a is (1) 2 ft and (2) 4 ft.

7-124. What depth h_a may be expected in the Parshall flume of Prob. 7-123, if the discharge is 50 cfs?

References

Brater, E. F. and H. W. King, *Handbook of Hydraulics,* 6th ed., New York: McGraw-Hill, 1976.

Brebbia, C. A. and A. J. Ferrante, *Computational Hydraulics,* London: Butterworths, 1983.

Chow, V. T., "Open Channel Flow," Section 24 in *Handbook of Fluid Dynamics,* V. L. Streeter (ed.). New York: McGraw-Hill, 1961.

Chow, V. T., *Open Channel Hydraulics,* New York: McGraw-Hill, 1959.

French, R. H., *Open-Channel Hydraulics,* New York: McGraw-Hill, 1985.

Henderson, F. M., *Open Channel Flow,* New York: Macmillan, 1966.

Leliavsky, S. L., *River and Channel Hydraulics,* London: Chapman and Hall, 1965.

Mahmood, K. and V. Yevjevich (eds.), *Unsteady Flow in Open Channels* (3 vols.). Fort Collins, CO: Water Resources Publications, 1975.

Posey, C. J., *Fundamentals of Open Channel Hydraulics,* Allenspark, CO: Rocky Mountain Hydraulics Laboratory, 1969.

Rouse, H. (ed.), *Engineering Hydraulics,* New York: Wiley, 1950.

Smith, P. D., *Basic Hydraulics,* London: Butterworths, 1982.

chapter 8

Mechanics of Sediment Transport

Figure 8–0 Debris flow below Mt. St. Helens following eruption (U.S. Geological Survey).

8–1 INTRODUCTION

The previous chapter covered the characteristics of flow in an open channel that had a prescribed and unchanging channel boundary. This could have been either a lined (as with concrete) channel or a natural channel that was stable and therefore not changing significantly with time. The emphasis here will be on channels that are free to interact with the flow and, in the case of a river, even owe their cross-sectional shape to the flow itself. The problems facing the engineer at this point are obviously more complex than those of the previous chapter. In a rigid channel, the shape and slope are fixed and the depth is a dependent variable determined by the shape and slope as well as the discharge and roughness. In the natural channel, the cross section, depth, slope, and even the roughness depend on the water and sediment discharges. However, the sediment discharge itself depends on these other parameters. Consequently, changes in any one of the variables triggers a chain of adjustments that often can be at best only qualitatively evaluated.

The types of sediment-related problems facing the engineer and requiring his or her consideration include the following. From the water-quality point of view, a high sediment load adversely affects the appearance of the water, directly affects the variety of life the water can support, and reduces the quantity of light available for aquatic food supply. In addition to being indicative of serious upstream land or channel erosion, high-sediment loads may cause depositional problems interfering with navigation and shortening the useful life of dams. Projects such as dams, which have a pronounced effect on the sediment load, upset the equilibrium of the river system, often leading to serious downstream scour problems. Structures such as bridge piers, which cause a local acceleration of the flow, may induce local scour at the piers, which can threaten the structure itself. Further, the designer of unlined water-carrying channels must ensure that they are essentially nonscouring and nondepositing.

To quote a couple of examples that illustrate the magnitude of the problem, consider that during the decades following the discovery of gold in California, hydraulic monitors were used to direct jets of water with heads up to 400 ft against the gold bearing Sierra Nevada foothills. Entire hillsides were literally washed through sluice boxes which separated the gold and returned the debris to the streams. By the time hydraulic mining was prohibited in the 1880s over one and one-half billion cubic yards of sediment were deposited in the lower reaches of the Sacramento River. Further, the streambed upstream at Sacramento was raised over 11 feet by channel deposition. As another example, many of our reservoirs are being rapidly filled with sediment deposits; more than one moderate-size reservoir has been completely filled with sediment within a single year of the dam closure!

To understand and attack these problems, we will first examine the relevant sediment properties. This will be followed by a section dealing with incipient motion and the estimation of the quantity of sediment being transported. The design of stable (i.e., nonscouring and nondepositing) channels will be treated next. The final section will include specific problems such as bridge pier scour and the impact of a dam on the river channel and will conclude with the analysis of these problems by mathematical modeling.

8–2 SEDIMENT PROPERTIES

The equations governing sediment behavior depend on a number of sediment properties, just as fluid mechanics relationships are functions of the fluid properties. The more significant of these properties follow.

Sediment Size

Sediment size is usually expressed by the sediment diameter. This is the length dimension felt to best represent a particle or the average size of a group of particles. By its very name this implies a somewhat spherical shape. For most sediment samples, size analysis may be obtained by sieving. This leads directly to a breakdown of the sediment distribution into classes based at least in part on the opening dimensions of successive sieves. A commonly used size classification, proposed by the American Geophysical Union Subcommittee on Sediment Terminology,[1] is given in Table 8–1. This classification, which is based on a geometric series with a ratio of two, will be used throughout the chapter. The analysis of a sediment sample will be expressed as the fraction of the total sample falling into each size class. The representative sediment diameter for each size is usually the geometric mean of the class (i.e., the square root of the product of the upper and lower limits of the class). A number of different diameter definitions are in common usage, but the accuracy of most transport calculations does not justify their further consideration here.

The analysis of a sediment sample can be plotted as an accumulative function from which the percent finer of any size can be determined. The notation "percent finer" can be explained by an example. If 0.2 lb of a 5-lb sediment sample passes through a sieve with a 1 mm opening, then 4 percent of the material is finer than 1 mm in diameter. Thus $d_4 = 1$ mm. This is illustrated in Fig. 8–1. The median size of the sample, indicated as d_{50}, and identifying the diameter of that sediment size for which 50 percent of the sample is finer is so indicated, as are two other selected sizes. Additional parameters such as the standard deviation and the skewness of the sample can also be used to improve the definition of sediment size distribution. Again, the final accuracy usually does not warrant the additional effort.

[1]E. W. Lane, "Report of the Subcommittee on Sediment Terminology," *Transactions* American Geophysical Union, Vol. 28, No. 6, Washington D.C., Dec. 1947, pp. 936–938.

Table 8–1 SEDIMENT SIZE CLASSIFICATION

Class Name	Symbol	Size range				Geometric mean (ft)
		Microns	Millimeters	Inches		
Very large boulders	VLB		4096–2048	160–80		9.502
Large boulders	LB		2048–1024	80–40		4.751
Medium boulders	MB		1024–512	40–20		2.376
Small boulders	SB		512–256	20–10		1.188
Large cobbles	LC		256–128	10–5		0.594
Small cobbles	SC		128–64	5–2.5		0.297
Very coarse gravel	VCG		64–32	2.5–1.3		0.1485
Coarse gravel	CG		32–16	1.3–0.6		0.0742
Medium gravel	MG		16–8	0.6–0.3		0.0371
Fine gravel	FG		8–4	0.3–0.16		0.0186
Very fine gravel	VFG		4–2	0.16–0.08		0.00928
Very coarse sand	VCS	2000–1000	2–1			4.64×10^{-3}
Coarse sand	CS	1000–500	1–0.5			2.32×10^{-3}
Medium sand	MS	500–250	0.5–0.25			1.16×10^{-3}
Fine sand	FS	250–125	0.25–0.125			5.80×10^{-4}
Very fine sand	VFS	125–62	0.125–0.062			2.89×10^{-4}
Coarse silt		62–31	0.062–0.031			
Medium silt		31–16	0.031–0.016			
Fine silt		16–8	0.016–0.008			
Very fine silt		8–4	0.008–0.004			
Coarse clay		4–2	0.004–0.002			
Medium clay		2–1	0.002–0.001			
Fine clay		1–0.5	0.001–0.0005			
Very fine clay		0.5–0.24	0.0005–0.00024			

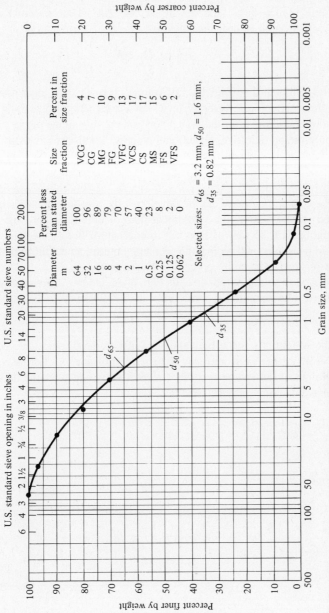

Figure 8–1 Characteristics of a typical sediment distribution (Vanoni, 1975, p. 25).

Sediment Specific Gravity

Although the specific gravity of the sediment can vary over a considerable range from heavy magnetite to very light lava particles, the most common mineral component is quartz, which has a specific gravity of 2.65. Unless knowledge to the contrary is available, this value is usually used.

Fall Velocity

Along with the sediment diameter, the fall velocity is one of the most useful parameters. The fall velocity is the velocity at which a sediment particle falls through a fluid. Its importance stems from the realization that the magnitude of the fall velocity reflects not only the particle size, shape, and weight, but also the relevant characteristics of the fluid through which it falls. Further, if a sediment particle is to be lifted off of the streambed or maintained in suspension, it is this fall velocity that must be overcome.

The concept of fall velocity can be developed by considering a sphere of diameter d released at zero velocity in a quiescent fluid of infinite extent (Fig. 8–2a). Its initial acceleration will be that of free fall in a vacuum, but as the fall velocity w_s increases, fluid resistance will reduce the acceleration until an equilibrium or terminal velocity w_t is reached. The entire process is shown in Fig. 8–2b.

Considering downward as positive in Fig. 8–2a, adding forces on the sphere, and setting the results equal to the mass times the acceleration gives

$$F_W - F_B - F_R = Ma$$

Here F_W is the weight, F_B is the buoyant force due to the displaced fluid, F_R is the resistant force, M is the mass, and a is the acceleration dw_s/dt. Taking γ_s and γ as the specific weight of the sphere and fluid, respectively, and ρ_s as the density of the sphere, the above equation may be written

$$\frac{\pi}{6} d^3 \gamma_s - \frac{\pi}{6} d^3 \gamma - F_R = \rho_s \frac{\pi}{6} d^3 \frac{dw_s}{dt} \qquad \text{(8–1)}$$

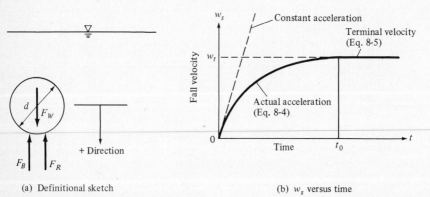

(a) Definitional sketch

(b) w_s versus time

Figure 8–2 Fall velocity of a sphere.

The drag force F_R on a sphere can in general be expressed in terms of a drag coefficient C_D according to

$$F_R = C_D \frac{\pi}{4} d^2 \rho \frac{w_s^2}{2} \tag{8–2}$$

where C_D is given in Fig. 8–3 and ρ is the fluid density. Equation 8–1 cannot be integrated in general because of the complexity of the relationship between C_D and the Reynolds number, $\text{Re} = w_s d / \nu$. However, in the range $\text{Re} < 1$, we have $C_D = 24/\text{Re}$, which may be combined with Eq. 8–2 to yield the Stokes equation

$$F_R = 3\pi \rho \nu w_s d \tag{8–3}$$

If this equation for F_R is substituted into Eq. 8–1, the equation can be integrated from zero to time t with the result

$$w_s = \frac{d^2 g}{18\nu} \left(\frac{\rho_s}{\rho} - 1 \right) \left\{ 1 - \exp \left[\frac{-18\nu t}{(\rho_s/\rho) d^2} \right] \right\} \tag{8–4}$$

This equation is only appropriate in the range $\text{Re} < 1$, and even then it is only approximate because the additional fluid that is accelerated along with the sphere is not considered. Equation 8–4 gives the fall velocity as a function of time, similar to the graph of Fig. 8–2b. Allowing t to approach infinity immediately gives the terminal velocity

$$w_t = \frac{d^2}{18\nu} \left(\frac{\rho_s}{\rho} - 1 \right) g \tag{8–5}$$

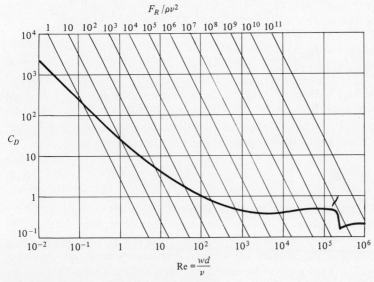

Figure 8–3 Drag coefficient for a sphere.

for the same Reynolds number range. We will generally be interested in the terminal velocity. Equation 8–5 may of course be used to calculate w_t, but suffers from its limited range of validity. We must look for a more universal procedure, but first note that the ratio w_s/w_t can be used to determine the time required for a sphere to reach its terminal velocity. This ratio may be written

$$\frac{w_s}{w_t} = 1 - \exp\left[\frac{-18\nu}{(\rho_s/\rho)d^2}\right] \tag{8–6}$$

Assuming that the terminal velocity is reached when w_s equals 99 percent of w_t, the time required, t_0 may be obtained from

$$0.99 = 1 - \exp\frac{-18\nu}{(\rho_s/\rho)d^2}$$

or

$$t_0 = \frac{(0.256)(\rho_s/\rho)d^2}{\nu} \tag{8–7}$$

This equation can be used as a guide for justifying the use of w_t rather than w_s (see Example 8–1). However, the equation becomes increasingly inaccurate as the sphere size, or specifically Re, increases.

To obtain a general solution for the terminal fall velocity, Eq. 8–2 may be substituted into Eq. 8–1. Although the equation may not be integrated in general, the right-hand side may be set equal to zero and the terminal velocity evaluated from the resulting equilibrium conditions. Thus

$$\frac{\pi}{6}d^3\gamma_s - \frac{\pi}{6}d^3\gamma - C_D\frac{\pi}{4}d^2\rho\frac{w_t^2}{2} = 0$$

or, upon rearranging

$$w_t = \left[\frac{4}{3}\frac{d}{C_D}\left(\frac{\rho_s}{\rho} - 1\right)g\right]^{1/2} \tag{8–8}$$

The two axes of Fig. 8–3 do not permit a direct determination of C_D as a function of the Reynolds number because the Reynolds number itself includes the unknown velocity. The auxiliary $F_R/\rho\nu^2$ scale can be used since this parameter is independent of the velocity. Under equilibrium conditions F_R is the submerged weight

$$F_R = \frac{\pi}{6}d^3(\gamma_s - \gamma) \tag{8–9}$$

In addition, note that both ρ and ν pertain to the fluid.

Assuming quartz spheres (sp. gr. = 2.65) and either air or water as the fluid, the curve of Fig. 8–3 can be reworked into the form given in Fig. 8–4, which allows for the direct reading of the terminal fall velocity. Sediment particles are somewhat less than spherical, and for a given diameter based on a sieve analysis, they usually

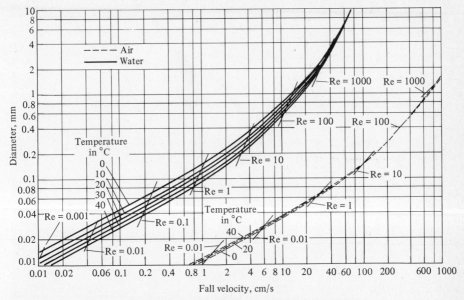

Figure 8–4 Terminal fall velocity of quartz particles in air and water.

have a fall velocity a little less than that of a sphere of the same diameter.[2] Nevertheless, these procedures should serve as at least a first approximation. When other complicating factors such as fluid turbulence, which has a pronounced effect on the fall velocity, are considered, it is apparent that the estimated fall velocity will be an approximation at best.

Although sediment shape has not been considered in this section, computer program 5 in Appendix E incorporates particle shape to give improved values of the fall velocity. The accompanying comments provide some background on particle shape.

EXAMPLE 8–1

Determine the time required for a particle of fine sand to reach 99 percent of its terminal fall velocity. Calculate the fall velocity and determine whether it is

[2]For information on more accurate evaluation of the fall velocity, refer to the references at the end of the chapter or "Some Fundamentals of Particle Size Analysis, A Study of Methods Used in Measurement and Analysis of Sediment Loads in Streams," *Report No. 12*, Subcommittee on Sedimentation, Interagency Committee on Water Resources, St. Anthony Falls Hydraulic Laboratory, Minneapolis, MN, 1957. This report forms the basis of computer program 5 in Appendix E.

in the Stokes range. Assume that the sand particle is quartz and that the water temperature is 50°F.

Solution

The geometric mean diameter of fine sand is 5.80×10^{-4} ft, and the kinematic viscosity of the water is 1.41×10^{-5} ft²/s. From Eq. 8–7,

$$t_0 = \frac{(0.256)(2.65)(5.80 \times 10^{-4})^2}{1.41 \times 10^{-5}} = 0.016 \text{ s}$$

This is such a short time interval that the use of the terminal rather than the time-variable fall velocity is probably justified. However, it is assumed in the above that the Reynolds number is in the Stokes range. This may be verified by first using Eq. 8–5 to obtain the fall velocity. Note that this equation also assumes that the Reynolds number is within the Stokes range. From Eq. 8–5,

$$w_t = \frac{(5.80 \times 10^{-4})^2(2.65 - 1)(32.2)}{(18)(1.41 \times 10^{-5})} = 0.070 \text{ ft/s}$$

and

$$\text{Re} = \frac{(0.070)(5.80 \times 10^{-4})}{1.41 \times 10^{-5}} = 2.90 > 1$$

This Reynolds number is a little above the Stokes range, but close enough that the calculated time is still reasonably correct. The actual terminal velocity can be found from either Fig. 8–3 or Fig. 8–4. Figure 8–4 provides the more rapid answer: For $d = 5.80 \times 10^{-4}$ ft $= 0.177$ mm and a temperature of 50°F or 10°C, $w_t = 1.7$ cm/s $= 0.056$ ft/s.

To illustrate the use of Fig. 8–3, calculate F_R by Eq. 8–9,

$$F_R = \frac{\pi}{6}(5.80 \times 10^{-4})^3(62.4)(2.65 - 1) = 1.05 \times 10^{-8} \text{ lb}$$

and therefore

$$\frac{F_R}{\rho v^2} = \frac{1.05 \times 10^{-8}}{(1.94)(1.41 \times 10^{-5})^2} = 27.3$$

Using this value and reading down from the top scale along the diagonal gives $C_D = 13$. Thus

$$1.05 \times 10^{-8} = (13)\left(\frac{\pi}{4}\right)(5.80 \times 10^{-4})^2 \frac{(1.94)w_t^2}{2}$$

and solving, $w_t = 0.056$ ft/s, which verifies the use of the previous graph.

8–3 SEDIMENT TRANSPORT

Introduction and Definitions

Sediment transport can be separated into two somewhat distinct modes of transport. The portion of the sediment moving in the immediate vicinity of the bed is referred to as the bed load. This is material that hops, rolls, or slides along the bed. Bed-load transport provides the primary mode of transport for coarse material or low transport rates. As the energy of the flow is increased, more and more of the sediment is thrown into suspension. The sediment transported in this fashion is called suspended load. For the finer sediments, this may constitute 90 to 95 percent of the sediment transport.

There is a continual exchange of material between the regions of suspended load and bed load and between the stationary bed and the transported sediment. The transport equations to be introduced in this section all stem from the premise that these exchanges are occurring continuously. As a corollary of this assumption, for a given set of hydraulic parameters, the more of any size class found in the bed, the greater will be the transportation of that particular size. This leads to the frequently used relationship, that if a given size fraction (say the ith size in the bed sample) constitutes a p_i fraction of the total bed sample, then the transport of that size will be the p_i fraction of the transport that would result if only that size were present in the bed. To illustrate, consider Table 8–2 where a bed sample is composed of 20 percent FS, 45 percent MS, and 35 percent CS. The transport capacities given in the third column would be determined by an equation of the type to be considered later in the section. For each size, however, the transport capacities are based on the assumption that the bed is entirely composed of that size fraction. The actual transport of each size is given in the fourth column as the product of columns two and three. The total transport is the final sum. Not all transport equations are based directly on this procedure, and at best, this is only an approximation. Left out in this type of calculation is the effect that any one size found in the bed might have on another size.[3]

Table 8–2 USE OF p_i

Size	Fraction in bed sample p_i	Transport capacity (tons/day)	Actual transport (tons/day)
(1)	(2)	(3)	(4)
FS	0.20	120,000	24,000
MS	0.45	27,000	12,100
CS	0.35	6,500	2,300
		Total transport =	38,400

[3]One such effect is the hiding or sheltering of smaller sizes when in the vicinity of larger sizes.

The foregoing concepts are only applicable to certain size fractions, most typically the sand and larger sizes. The finer materials, usually the silts and clays, can be transported with such ease that once they find their way into the channel, they are swept or washed through with only trace amounts found in the bed. For this reason, the transport equations do not apply and these sizes have to be handled separately. This material is called the wash load,[4] and the transport of the larger sizes—those found in the bed—is called the bed material load. The wash load and the bed material load together compose the total load. Figure 8–5 illustrates the relationship among the various transport terms. When wash load is not present, the terms bed material load and total load are often used interchangeably.

The bed material load will generally be expressed as a weight of sediment per unit time, that is, tons/day or N/s. The notation used for the transport will be either G_s or g_s, depending on whether the transport rate is with respect to the entire cross section or on a unit-width basis. In the latter case, typical units will be tons/day/ft or N/s/m. If wash load is not present, then the above notation for bed material load applies to the total load as well. In a similar fashion, the bed load will be expressed by G_{sb} and g_{sb}, and the suspended load by G_{ss} and g_{ss}. When the transport refers to the ith size class, the subscript i will also be appended to the term.

As changes in hydraulic parameters lead to increasing sediment transport rates, there is a progressive change in the form of the stream bed itself. These characteristic patterns, called bed forms, are shown in Fig. 8–6 in the order of increasing transport. Under very low transport conditions a ripple bed will often occur. Sediment transport is restricted to particles rolling up the relatively flat face, sliding down the steeper face, and remaining buried for extended periods of time as the ripple itself moves downstream. Ripples have a maximum height and length of 0.1 ft and 1.0 ft, respectively, and they rarely form in sediment greater than 0.6 mm in diameter. The bed pattern may be nearly two dimensional. Typical Manning n values associated with a ripple bed range between 0.016 and 0.022.

Increased transport leads to combined ripples and dunes and finally dunes. Although dunes have cross section profiles that are similar to ripples, they are rarely two dimensional (except in laboratory flumes) and there is no apparent upper bound

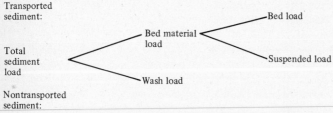

Figure 8–5 Relationships among sediment definitions.

[4]In steep, primarily gravel bed streams, many of the sand sizes may become part of the wash load as well.

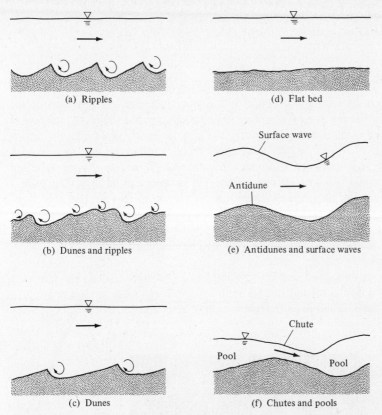

(a) Ripples

(d) Flat bed

Surface wave

Antidune →

(b) Dunes and ripples

(e) Antidunes and surface waves

Chute

Pool

Pool

(c) Dunes

(f) Chutes and pools

Figure 8–6 Bed forms in an alluvial channel (Vanoni, 1975, p. 160).

on their size. In the Mississippi and Missouri rivers, dunes with heights of 20 to 30 ft and lengths as long as 3000 ft have been observed. The Manning n values may vary over the wide range from 0.018 up to 0.045.

The bed forms tend to wash out with further increase in the sediment transport, leaving a flat bed with a Manning n as low as 0.009 to 0.016. By this point, there is very heavy sediment transport near the bed and usually a significant suspended load. Continuing, the next bed form, called an antidune, is characterized by a symmetrical profile as shown in Fig. 8–6, in which water surface waves form and are in phase with the sand wave. Because the greatest erosion rate is on the downstream faces of the antidunes and deposition follows on the upstream faces, the bed forms frequently appear to move upstream. Resistance remains low, with n in the 0.010 to 0.018 range.

Antidunes are often unstable and tend to grow in amplitude until the water wave breaks, forming a surge or hydraulic jump that in the process washes out much of the bed form. The antidunes will again grow on the nearly flat bed and the cycle will repeat. Chutes and pools are the extreme case in which the supercritical chute is followed by a hydraulic jump and a subcritical pool. Because the violent breaking wave or jump only endures for a short period of time, the Manning *n* remains relatively low at 0.013 to 0.025.

Measurement of Sediment Transport

The measurement of sediment transport is a regular but relatively new part of the U.S. Geological Survey stream sampling program. Suspended sediment transport is usually measured using a *depth-integrating sediment sampler,* an example of which is shown in Fig. 8–7. The sampler is designed to cause minimum interference with the flow. Water and suspended sediment enters through the nozzle and are collected in a pint milk bottle. As with current meters, the sampler can be supported from a bridge or cableway, or alternatively carried by a wader in smaller streams. The sampler is lowered (and raised) through the water at a constant vertical velocity from the surface to about 0.5 ft from the stream bed. Since the sample collected at each point is proportional to the respective water velocity, the collected sample represents an integrated sample.

The sample is then dried and analyzed by sieving or other means. The transport can be expressed by one of the various sediment discharge units or as a concentration. Typical concentration units include weight of sediment per unit weight of sam-

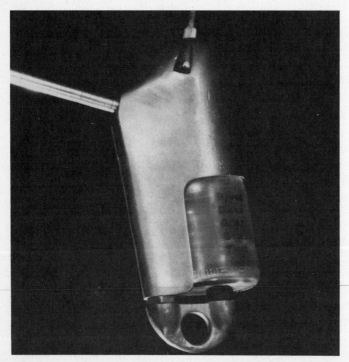

Figure 8–7 Depth-integrating suspended sediment sampler (U.S. Geological Survey).

ple (sediment plus water). If this ratio is multiplied by 10^6, the concentration is in *parts per million* (ppm).

The unmeasured load (bed load plus suspended load in the bottom 0.5 ft) is frequently estimated by using a bed load equation such as the Meyer-Peter and Muller equation or the modified Einstein equation discussed later. Pits and various other devices have been proposed to trap and measure the bed load. The *Helley-Smith bed load sampler* shown in Fig. 8–8 has been extensively studied and is coming into increased usage. The sample collected from the lower portion of the stream is analyzed similarly to the suspended sediment sample.

Incipient Motion

Incipient motion of bed particles represents the critical condition between transport and no transport. Therefore, it is of interest in the consideration of sediment transport. In fact, many of the transport equations base the sediment transport capacity on the amount by which a selected hydraulic parameter, such as the shear stress or discharge, exceeds the value of that parameter at incipient motion. Alternately, this same condition of incipient motion represents the desired target in the design of nonscouring unlined channels.

Figure 8–8 Helley-Smith bed load sampler (U.S. Geological Survey).

The concept of incipient motion can be introduced by the following very approximate model based on a more rigorous approach by White.[5] With reference to Fig. 8–9, we wish to evaluate the conditions leading to the incipient motion of particle A. With motion impending, the particle will be just on the verge of lifting off other particles such as B, and the entire force will be taken by the point of support shown.

The hydraulic force F_H acting on the particle, although actually a more complicated mechanism, will be treated as due entirely to the shear stress τ_0, which acts over a portion of the surface area of the particle. For a particle diameter d_s, the effective surface area will be proportional to d_s^2. Calling the constant of proportionality c_1, this force becomes

$$F_H = \tau_0 \, c_1 \, d_s^2$$

Assuming that the distance a_1 is also proportional to d_s, say $a_1 = c_2 d_s$, then the overturning moment due to the flow is

$$\tau_0 c_1 d_s^2 \, (c_2 d_s)$$

This is balanced by the submerged weight of the particle, which is proportional to $(\gamma_s - \gamma) d_s^3$. Finally, assuming that a_2 is proportional to d_s, the righting moment is

$$c_3 \, (\gamma_s - \gamma) \, d_s^3 \, (c_4 d_s)$$

where c_3 and c_4 are additional constants of proportionality. At the point of incipient motion, the overturning moment will just equal the righting moment, and the shear stress associated with this condition will be denoted as the critical shear stress τ_c. Thus,

$$\tau_c \, c_1 c_2 d_s^3 = (\gamma_s - \gamma) c_3 c_4 d_s^4$$

or

$$\tau_c = \frac{c_3 c_4}{c_1 c_2} (\gamma_s - \gamma) d_s \qquad\qquad \text{(8–10)}$$

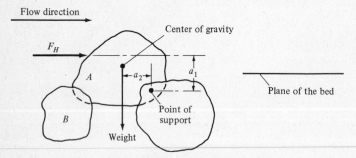

Figure 8–9 Definitional sketch for initiation of motion.

[5]C. M. White, "The Equilibrium of Grains on the Bed of a Stream," *Proceedings* Royal Society of London, Series A, No. 958, Vol. 174, Feb. 1940, pp. 322–338.

Combining the four constants into a single constant C, we have

$$\frac{\tau_c}{(\gamma_s - \gamma)d_s} = C \tag{8–11}$$

White's experiments covered a variety of conditions and fluids, but those results obtained under turbulent conditions frequently had values of C quite close to 0.1.

A more complete development of the critical shear stress can be obtained using dimensional analysis. Assuming that the critical shear stress is given by

$$\tau_c = f(\gamma_s - \gamma, d_s, \rho, \mu) \tag{8–12}$$

and taking τ_c, d_s, and ρ as repeating variables leads to

$$\frac{\tau_c}{(\gamma_s - \gamma)d_s} = \phi\left(\frac{\tau_c^{1/2}\rho^{1/2}d_s}{\mu}\right)$$

The second term can be rearranged as follows:

$$\frac{\tau_c^{1/2}\rho^{1/2}d_s}{\mu}\frac{1/\rho}{1/\rho} = \frac{\sqrt{\tau_c/\rho}\, d_s}{\nu} = \frac{u_{*c}d_s}{\nu}$$

Thus,

$$\frac{\tau_c}{(\gamma_s - \gamma)d_s} = \phi\left(\frac{u_{*c}d_s}{\nu}\right) \tag{8–13}$$

The first dimensionless term is the same parameter associated with the analysis by White, whereas the second term, consisting of the critical shear velocity (see Eq. 7–4), a length, and the kinematic viscosity, is called the critical shear velocity Reynolds number (or sometimes the critical boundary Reynolds number) Re_{*c}. This approach was originally presented by Shields[6] in conjunction with what is now known as the Shields diagram (Fig. 8–10). The parameters τ_* and Re_* in Fig. 8–10 are noncritical values of the dimensionless shear stress and the shear velocity Reynolds number, respectively. The curve on the Shields diagram represents the experimental relationship implied by Eq. 8–13 and can be used to evaluate the critical shear stress, that is, the shear stress at incipient motion. The additional parameter shown within the diagram is the result of combining τ_* and Re_* so as to eliminate the shear stress. It accordingly may be used to obtain the critical shear stress directly. Note also that selected data from White are also included on the graph.

Differences in the flume size, flow structure,[7] and method of determining the critical conditions all contribute to the variation in results. One method of determining the critical shear stress in a flume is to plot a graph of the shear stress as a

[6]A. Shields, "Anwendung der Aenlichkeitsmechanik und der Turbulenzforschung auf die Geschiebebewegung," *Mitteilungen der Preussischen Versuchsanstalt fur Wasserbau und Schiffbau,* Berlin, Germany, translated into English by W. P. Ott and J. C. van Uchelen, California Institute of Technology, Pasadena, Calif., 1936.
[7]The White tests were usually conducted in a boundary layer, whereas the other results were probably obtained in fully developed turbulent flows.

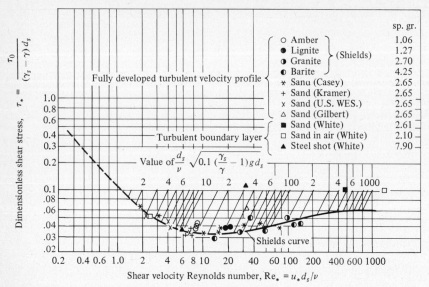

Figure 8–10 Shields diagram (Vanoni, 1975, p. 96).

function of sediment transport. Such a plot can be extrapolated to conditions of zero transport, yielding τ_c.

For quartz sediments in water, the Shields diagram can be rearranged into the more convenient form of Fig. 8–11. The solid lines represent the Shields critical

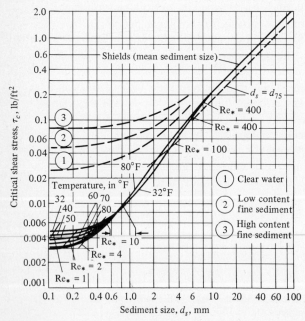

Figure 8–11 Critical shear stress for quartz particles in water (Vanoni, 1975, p. 99).

shear stress curve for different water temperatures. The dashed lines will be discussed later in conjunction with the design of stable channels.

EXAMPLE 8–2

Determine the critical shear stress for the fine sand particle of Example 8–1 if this sediment size forms the bed of a wide ($R = y$) channel with a depth of 1 m and a slope of 0.0008. Will this critical shear stress be exceeded by the flow?

Solution

The solution will first be obtained using Fig. 8–10. At 10°C $\nu = 1.308 \times 10^{-6}$ m²/s. The parameter needed to enter Fig. 8–10 is

$$\frac{d_s}{\nu}\sqrt{0.1\left(\frac{\gamma_s}{\gamma} - 1\right)gd_s}$$

$$= \frac{0.177 \times 10^{-3}}{1.308 \times 10^{-6}}\sqrt{(0.1)(2.65 - 1)(9.81)(0.177 \times 10^{-3})} = 2.29$$

whereupon $\tau_{*c} = 0.065$. Solving for the critical shear stress,

$$\tau_c = (0.065)(2.65 - 1)(9810)(0.177 \times 10^{-3}) = 0.186 \text{ N/m}^2$$

This value can be checked by comparison with Fig. 8–11. For $d_s = 0.177$ mm and $T = 50°F$, the shear stress may be read directly as $= 0.004$ lb/ft² $= 0.19$ N/m², which is the same answer obtained previously.

The actual shear stress is

$$\tau_0 = \gamma RS = (9810)(1)(0.0008) = 7.85 \text{ N/m}^2$$

This greatly exceeds the critical shear stress for the particle and transport may be expected. Where does the flow plot on the Shields diagram?

Bed Load

Equations that are specifically intended to calculate the rate of sediment transport as bed load will be considered here. Equations that were developed to calculate both the bed load and the suspended load will be considered in the subsection on total load. There are many equations of each type, and no attempt will be made to even mention all of the equations. In fact, of the three bed load equations to be included here, the first is chosen because of its historic interest and the second for its ease of application.

The Duboys-Straub Equation

This bed load equation was first presented by Duboys in 1879.[8] It is based on a rather inaccurate model, but it represents one of the first significant developments in the estimation of sediment transport and is still in use today. Duboys assumed that the bed load moved as a series of layers such as the n layers shown in Fig. 8–12. In this sketch, the water flow is from left to right, the top layer of the moving bed is the nth layer, and the first layer is just at the point of incipient motion. As implied in Fig. 8–12, the velocity of the various layers is assumed to increase linearly, with the first layer stationary (but at incipient motion), the second layer having a velocity Δv, the third layer having a velocity $2\Delta v$, and so on up to the nth layer, which has a velocity $(n - 1)\Delta v$. Each layer has a thickness d', probably proportional to, but not necessarily equal to, the sediment size.

The average velocity of the moving sediment is $(n - 1)\Delta v/2$, and the submerged weight of the sediment in motion on a per-unit-area basis is $(\gamma_s - \gamma)nd'$. The product of the weight of sediment per unit area and its average velocity will be the weight discharge of sediment per unit width. Thus,

$$g_{sb} = (n - 1)\frac{\Delta v}{2}(\gamma_s - \gamma)nd' \tag{8–14}$$

Further, the driving force required to move this material must overcome the friction between the n layers of the movable bed and the stationary material below it. This force may be expressed as the product of a coefficient of friction $\mathscr{F}$ and the submerged weight, or $\mathscr{F}(\gamma_s - \gamma)nd'$. This is the force per unit area that must be overcome by the shear force of the overlying flow. On a per-unit-area basis, the latter is simply the shear stress τ_0 and

$$\tau_0 = \mathscr{F}(\gamma_s - \gamma)nd' \tag{8–15}$$

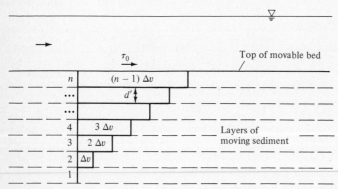

Figure 8–12 Duboys' layers of moving sediment.

[8]The Duboys equation, Schoklitsch equation, and Meyer-Peter and Muller equation as well as many other procedures may be found in *Sedimentation Engineering*, an ASCE manual listed in the references at the end of the chapter. In many cases, the original reference is not readily accessible, but the serious researcher will find it listed in the ASCE manual.

At the point of incipient motion of the bed particles, the top layer will be just on the verge of motion and the shear stress associated with this state is τ_c. Therefore, incipient motion of the bed corresponds to $n = 1$, and Eq. 8–15 becomes

$$\tau_c = \mathscr{F}(\gamma_s - \gamma)d'$$

Combining this equation with Eq. 8–15 gives

$$n = \frac{\tau_0}{\tau_c} \tag{8–16}$$

Substitution of the above equation for n into Eq. 8–14 yields the transport equation

$$g_{sb} = \frac{1}{2}(\gamma_s - \gamma)d'\Delta v\left(\frac{\tau_0}{\tau_c}\right)\left(\frac{\tau_0}{\tau_c} - 1\right)$$

or

$$g_{sb} = \psi\tau_0(\tau_0 - \tau_c) \tag{8–17a}$$

where

$$\psi = \frac{(\gamma_s - \gamma)\,d'\Delta v}{2\tau_c^2} \tag{8–17b}$$

Neither τ_c nor ψ can be calculated, although τ_c could conceivably be obtained from the Shields diagram (Figs. 8–10 or 8–11). Straub ran a series of flume tests to evaluate τ_c and ψ and his results are included in Fig. 8–13. For the units associated with Fig. 8–13, the transport rate per unit width g_{sb} must be in lb/s/ft. Multiplying

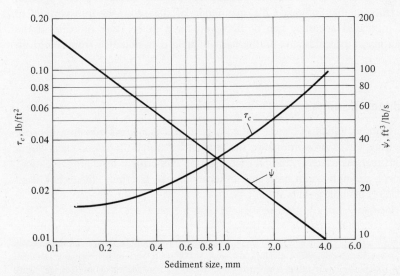

Figure 8–13 Experimental values of τ_c and π for the Duboys equation (Vanoni, 1975, p. 191).

the unit transport by the channel width b and incorporating the conversion factor, 1 lb/s = 43.2 tons/day, gives the transport rate in tons/day,

$$G_{sb} = 43.2b\psi\tau_0 (\tau_0 - \tau_c) \tag{8–18}$$

To apply this equation, the shear stress would be calculated by $\tau_0 = \gamma RS$ (Eq. 7–3).[9] Further, the d_{50} size is recommended for use with Fig. 8–13 as best representing the sediment distribution. The procedure may also be used for the calculation of sediment transport by size fractions. In this case τ_{ci} and ψ_i would be determined for each size fraction by using the geometric mean diameter of each size class found in the bed. The transport of the ith size fraction becomes

$$G_{sbi} = 43.2p_ib\psi_i\tau_0(\tau_0 - \tau_{ci}) \tag{8–19a}$$

and upon calculating the total transport is

$$G_{sb} = \sum_i G_{sbi} \tag{8–19b}$$

where p_i is the fraction of the ith size found in the bed and the subsequent summation is over all of these same sizes.

The Duboys equation, and in fact most of the transport equations that follow, are based on transport per unit width. This means that the cross section must be approximated by a rectangle. This is probably a reasonable assumption for many wide alluvial rivers, but it is certainly subject to question in other rivers.

Although the original model was intended as a means of estimating the bed load, the results from the Duboys equation, rightly or wrongly, are often interpreted as the bed material load and the subscript b dropped from the transport terms. In fact, in the few available evaluations of transport equations, the Duboys-Straub procedure often overestimates the total sediment transport. Part of the problem is undoubtedly rooted in the laboratory data, since a straight rectangular flume cannot duplicate many of the factors that influence the magnitude of the sediment transport in an actual river. Increased accuracy can be achieved for a given river or canal if the Duboys equation can be calibrated on the basis of field data obtained from the given channel.

The Schoklitsch Equation

Whereas the Duboys equation can be called a critical shear stress equation because the sediment transport is proportional to the amount by which the actual shear stress exceeds the critical shear stress, the present equation is a critical discharge equation. Formulated on a per-unit-width basis and in terms of size fractions, the Schoklitsch equation in U.S. customary units is

---

[9]Since uniform flow rarely occurs in natural channels, ideally the slope should be the friction slope. When averaged over a considerable length of channel, the slope of the water surface is essentially identical. This distinction will not be made in this chapter because the accuracy of the equations does not justify it. Therefore, the slope will be denoted without subscripts.

$$g_{sb} = \sum_i p_i \frac{25.03}{\sqrt{d_{si}}} S^{3/2} (q - q_{ci}) \qquad \text{(8–20a)}$$

where the transport g_{sb} (lb/s/ft) is the sum of the individual transport rates, g_{sbi}. The transport of the ith size is a function of the fraction of the ith size found in the bed p_i the geometric mean of the size fraction d_{si} (ft), the slope S, the discharge per unit width q, and the critical discharge per unit width of the ith size q_{ci}. The critical discharge is given by

$$q_{ci} = 0.0638 \frac{d_{si}}{S^{4/3}} \qquad \text{(8–20b)}$$

The total bed load transport in tons/day becomes

$$G_{sb} = 43.2 b g_{sb} \qquad \text{(8–21)}$$

In SI units the corresponding Schoklitsch equation is

$$g_{sb} = \sum_i p_i \frac{2194}{\sqrt{d_{si}}} S^{3/2} (q - q_{ci}) \qquad \text{(8–22a)}$$

and the critical discharge is

$$q_{ci} = \frac{0.0194 d_{si}}{S^{4/3}} \qquad \text{(8–22b)}$$

When using SI units, d_s is expressed in m, q is in m³/s/m and the resulting transport becomes N/s/m.

The Schoklitsch equation is only a fair bed-load equation, probably best suited for gravel-bed streams. The calculated results must be interpreted as bed-load transport. If material is carried in suspension, it must be calculated separately. If sediment transport has been measured with a depth-integrating sampler, the bed load may be added to the measured load to provide a reasonable estimate of the bed material load.

EXAMPLE 8–3

Using the Schoklitsch procedure, calculate the gravel transport in a 100-ft-wide river if the discharge is 10,000 cfs and the slope is 0.002. The bed consists of 10 percent VFG, 55 percent FG, and 35 percent MG.

Solution

The discharge per unit width $q = 100$ cfs/ft. The transport calculations based on Eqs. 8–20 are set up in tabular form on the following page:

Size	d_s (ft)	p_i	q_c	g_s (lb/s/ft)	G_s (tons/day)
VFG	0.00928	0.10	2.35	0.227	980
FG	0.0186	0.55	4.71	0.860	3720
MG	0.0371	0.35	9.39	0.369	1590
				Total transport $=$	6290 tons/day

In the above tabulation

$$q_{ci} = \frac{(0.0638)d_{si}}{(0.002)^{4/3}}$$

In addition, the total transport of each size given in the final column is obtained from the respective value of transport per unit width using Eq. 8–21.

The Meyer-Peter and Muller Equation

The final bed-load equation, the Meyer-Peter and Muller equation, is another critical shear stress equation. This procedure is somewhat more difficult to apply, but it is possibly the best bed-load equation. In more or less original form, the equation may be written,

$$\left(\frac{k}{k'}\right)^{3/2} \gamma RS = 0.047(\gamma_s - \gamma)d_m + 0.25\left(\frac{\gamma}{g}\right)^{1/3} \left(\frac{\gamma_s - \gamma}{\gamma_s}\right)^{2/3} g_{sb}^{2/3} \qquad \text{(8–23a)}$$

where the effective diameter d_m is given by

$$d_m = \sum_i p_i d_{si} \qquad \text{(8–23b)}$$

and

$$\frac{k}{k'} = \sqrt{\frac{f'}{8}} \frac{V}{\sqrt{gRS}} \qquad \text{(8–23c)}$$

All other terms are as defined previously. Since the equation is dimensionally homogeneous, it may be used with any consistent set of units.

The Meyer-Peter and Muller equation is frequently applied to sediment size fractions, whereupon the d_m is replaced by $p_i d_{si}$ and the resulting bed load transport of the ith size is g_{sbi}. Total bed load transport can be obtained by summing the individual size fractions as before.

The ratio, k/k', in Eqs. 8–23a and 8–23c requires some explanation. The slope S, representing the friction slope, is a measure of the rate of energy dissipation or loss. Based on the assumption that the bed-load transport is related to the energy loss associated with the grain roughness, as opposed to the bed-form roughness, the quantity $(k/k')^{3/2}S$ represents that part of the total slope. In Eq. 8–23c, the Darcy-Weisbach resistance coefficient associated with the grain roughness f' is taken from the Moody diagram. As discussed previously, the noncircular cross section requires that the pipe characteristic length D be replaced by $4R$. Consequently, $\text{Re} = 4VR/\nu$. In addition, d_{90} is recommended for the roughness height, leading to a relative roughness, $d_{90}/4R$. Entering the Moody diagram with these values of Reynolds number and relative roughness will provide the required f'.

Although the procedure was based on flume studies, the experimental data contained broad ranges of slope, specific gravity, and sediment sizes. Although not included here, a number of simplifications have been made, particularly to make the procedure more compatible with computer computations.

Suspension of Sediment

To calculate the suspended load, it is first necessary to determine how the sediment concentration varies throughout a vertical section of the flow. As usual, we will assume a steady, uniform, two-dimensional, turbulent flow. Figure 8–14 will serve as a definitional sketch wherein a typical velocity profile and an assumed sediment concentration curve are shown.

A number of factors contribute to the suspension of sediment, including secondary currents, impact between particles, and obstructions such as bridge piers that locally alter the flow pattern. However, it is usually assumed that the suspension of sediment is primarily due to the turbulence of the flow. In addition to the velocity distribution shown in Fig. 8–14, there are turbulent velocity components in all three coordinate directions.

The question arises as to how the vertical turbulent velocity component, which has equal intensity in both the upward and downward directions, can counteract the effect of gravity as manifested through the fall velocity, which always acts in the downward direction. To understand this point, consider first the product of velocity

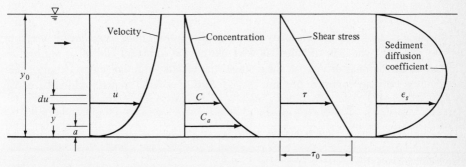

Figure 8–14 Definitional sketch for suspension of sediment.

and concentration. This product, with the concentration in any convenient units, will represent a discharge of sediment per unit area.[10]

Thus, the product of concentration and fall velocity Cw_s is the rate of downward movement of sediment due to gravity, whereas the product of concentration and a vertical turbulent velocity component is the rate of vertical movement of sediment due to the turbulence. The latter will be in the direction of the turbulent component and hence both upward and downward. However, the upward component moves sediment from a region of higher concentration to a region of lower concentration, whereas the downward component moves sediment from a region of lower concentration to that of a higher concentration. Assuming that all sediment was initially near the bed, this type of diffusion process would in itself ultimately lead to a completely uniform distribution across the vertical section. However, gravity is continually returning sediment to a lower level and under equilibrium conditions the downward movement by gravity must equal the net upward diffusion of sediment by the turbulence. This may be expressed by the diffusion equation,[11]

$$Cw_s = -\epsilon_s \frac{dC}{dy} \tag{8–24}$$

where ϵ_s is a sediment diffusion coefficient or mixing coefficient. In Fig. 8–14 and Eq. 8–24, the variables u, C, τ (discussed shortly), and ϵ_s are all assumed to be functions of y only, while the fall velocity, w_s, is assumed dependent only on time.

Equation 8–24 may be integrated to get the concentration distribution. To do so, it will be further assumed that the actual fall velocity w_s may be replaced by the terminal fall velocity w_t, and that the sediment diffusion coefficient may be replaced with ϵ, the momentum diffusion coefficient of the flow itself. This latter quantity is also called the kinematic eddy viscosity, because its introduction in fluid mechanics through

$$\tau = \rho\epsilon \frac{du}{dy}$$

parallels the laminar flow equation

$$\tau = \rho v \frac{du}{dy}$$

wherein v is the kinematic viscosity. The hypothesis from fluid mechanics is that the shear stress τ, which in turbulent flow is due to the turbulent eddies, is similar in nature to the laminar or viscous shear stress that results from molecular activity. The parallel is only approximate because the viscosity is a fluid property while the eddy

---

[10]Typical units for concentration include lb(sed)/ft^3, N(sed)/m^3, parts (of sediment) per million parts of water (ppm), and ft^3(sed)/ft^3(water). If the concentration is in N(sed)/m^3 and the velocity is in m/s, the product becomes N(sed)/s/m^2.

[11]A derivation and discussion of the diffusion equation may be found in *Sedimentation Engineering* (see footnote 8).

viscosity is flow dependent, varying throughout the flow. The equating of ϵ_s and ϵ implies that the sediment mixes at the same rate as the fluid itself. While this may be reasonably valid for the finer material, it should become less and less accurate for larger sediment.

Proceeding on this basis, we assume

$$\epsilon_s = \epsilon = \frac{\tau_0/\rho}{du/dy} \tag{8-25}$$

The shear stress can be expected to vary linearly as shown in Fig 8–14. Thus, from Eq. 7–2,

$$\tau_0 = \gamma RS \approx \gamma y_0 S$$

at the stream bed for the assumed two-dimensional flow, while at any intermediate depth

$$\tau = \gamma(y_0 - y)S$$

Upon combining

$$\tau = \tau_0 \frac{y_0 - y}{y_0} = \tau_0\left(1 - \frac{y}{y_0}\right) \tag{8-26}$$

The von Karman-Prandtl logarithmic equation[12] may be used to describe the velocity distribution shown in Fig. 8–14,

$$\frac{u}{u_*} = \frac{2.303}{\kappa} \log_{10} y + A \tag{8-27}$$

where u_* is the shear velocity ($\sqrt{\tau_0/\rho}$), κ is the von Karman universal constant, and A is a constant dependent on whether the boundary is smooth or rough. Consequently, the denominator in Eq. 8–25 can be expressed by

$$\frac{du}{dy} = \frac{u_*}{\kappa y} \tag{8-28}$$

When this result is substituted into Eq. 8–25 we get

$$\epsilon_s = \frac{(\tau_0/\rho)\kappa y}{u_*}\left(1 - \frac{y}{y_0}\right) = u_*\kappa y\left(1 - \frac{y}{y_0}\right) \tag{8-29}$$

This vertical distribution of ϵ_s is shown in Fig. 8–14.

Upon returning to the diffusion equation (Eq. 8–24), making the foregoing substitutions and separating variables gives

$$\frac{dC}{C} = -\frac{w_t y_0}{u_* \kappa} \frac{dy}{y(y_0 - y)}$$

---

[12]See any text on fluid mechanics.

This equation may now be integrated from some reference height a (shown in Fig. 8–14) to any arbitrary height y. Letting

$$z = \frac{w_t}{u_* \kappa}$$

(8–30)

and integrating, we get

$$\frac{C}{C_a} = \left[\left(\frac{y_0 - y}{y_0 - a} \right) \left(\frac{a}{y} \right) \right]^z$$

(8–31)

where C is the concentration at height y and C_a is the concentration at the reference height a.

In the application of Eq. 8–31, the fall velocity may be determined from either Fig. 8–3 or Fig. 8–4, the shear velocity from

$$u_* = \sqrt{\tau_0/\rho} = \sqrt{g y_0 S}$$

as previously derived, and κ may be assumed equal to 0.40. This value of the von Karman universal constant is generally accepted for most clear water flows. However, for relatively large bed forms κ may be quite unpredictable, while for high sediment transport rates κ generally drops below 0.40 to values as low as 0.20. The most serious limitation to the equation, however, is that initial knowledge of the concentration C_a is required. This point will be discussed further under the subject of total load.

The data plotted on Fig. 8–15 is often offered as a verification of Eq. 8–31.

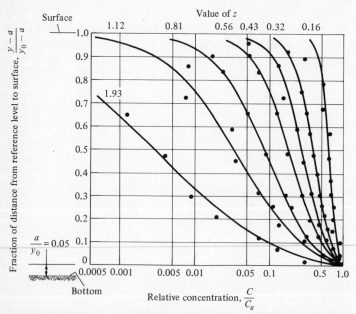

Figure 8–15 Verification graph for Eq. 8–31 (Vanoni, 1975, p. 80).

This figure provides a reasonable comparison of flume and Missouri River data with the theoretical distribution equation. However, it is a verification in form only, because the data points are plotted using best-fit values of z rather than z values calculated using Eq. 8–30. In fact, it has been well established that Eq. 8–30 significantly overestimates the value of z. Nevertheless, this procedure at least provides a starting point. In addition to describing the distribution of suspended sediment, it provides a means of evaluating the suspended load, as will be discussed shortly, and it will play a role in the subsequent subsection on total load as well.

Suspended Load

The calculation of suspended load cannot stand alone since it is tied to the bed load below it. In this subsection, the concept will be discussed, but a usable procedure will not be developed until the following subsection on total load. Referring once again to Fig. 8–14, the sediment discharge per unit width through the element of height dy is

$$g_{ss} = \int_a^{y_0} Cu \, dy \tag{8–32}$$

Integration of this expression, from $y = a$ up to $y = y_0$ at the surface, would yield the suspended load that is transported above the elevation a. One possibility would be to substitute Eq. 8–31 for C and a logarithmic equation such as Eq. 8–27 for u. This would obviously be difficult to integrate, but the integration has been carried out by Einstein[13] and the integral expressed graphically as a function of a and z.

Because of its complexity, the details of this procedure will not be pursued further. In addition to the integrational difficulties, the problem of the unknown concentration C_a must be resolved before the suspended load can be expressed in an usable form. This problem has been resolved in different ways, as will be discussed next.

Total Load

After a brief reference to the classic theoretical procedure developed by Einstein, two total load procedures will be considered in detail. The first, developed by Toffaleti for the U.S. Corps of Engineers, breaks down the calculations into bed load and suspended load. All calculations are then made by size fractions. The second procedure, which is based on the English experience, was presented by Ackers and White. This procedure recognizes the fundamental differences in the suspended load and bed load modes of transport but does not actually separate the two types of transport. In addition, the calculations are all based on a representative sediment size. Note that the term total load as used here usually refers to only the bed material load.

[13]H. A. Einstein, "The Bed Load Function for Sediment Transport in Open Channels, *Technical Bulletin 1026,* U.S. Dept. of Agriculture, Soil Conservation Service, Washington, D.C., 1950.

The Einstein Method

The classic method for the calculation of total load was developed by Einstein. In Einstein's procedure, the bed load is based on the probability that a given particle will move and the resulting transport is dependent on the ratio of the lift force on the particle to the submerged weight of the particle. As previously mentioned, the suspended load is obtained by Eq. 8–32. To relate suspended load, and particularly C_a, to the bed load, Einstein first assumed that the bed-load transport of each size is restricted to a region above the bed with a thickness of two grain diameters ($2d_{si}$). Observed bed-load transport in flumes supports this premise, especially when the bed is relatively flat. Based on the calculated bed-load rate, he evaluated the bed-load concentration and then assumed that since the layer of bed load is so thin, the aforementioned bed-load concentration could be used as the concentration at a, namely C_a. Thus, the otherwise unknown lower limit in the sediment suspension equation (Eq. 8–31) is determined from the bed load, and the total load evaluated. The procedure is lengthy and only moderately accurate. It is mentioned here primarily for its historic significance and because it served, at least conceptually, as a model for the procedure that follows.

The Einstein procedure has been modified to incorporate measured transport based on the depth-integrating sediment sampler. The suspended load concentration curves of the modified Einstein procedure[14] are matched against the sediment concentration in the measured zone and Einstein's concepts are then used to calculate the remainder of the suspended load and the bed load. This procedure, based as it is on measured transport data, includes the wash load as well as the bed material load. Consequently, the total load is calculated.

The Toffaleti Method

A few introductory comments will preface the development of this rather complex total-load procedure. The Toffaleti method[15] is not as theoretically sound as Einstein's procedure, which to a degree it emulates. However, the empirical portions are based on extensive field measurements. The procedure also makes use of Eq. 8–32, albeit in a much simpler form than Einstein used. Whereas Einstein obtained the reference concentration C_a from the bed load, Toffaleti proceeded in the opposite direction and calculated the bed load on the basis of the suspended sediment concentration curve.

As in most transport calculations, the channel cross section is assumed rectangular. Toffaleti suggests that the actual hydraulic radius R be taken as the depth.

---

[14]The original reference is B. R. Colby, and C. H. Hembree, "Computations of Total Sediment Discharge, Niobrara River Near Cody, Nebr.," *Water Supply Paper 1357*, U.S. Geological Survey, Washington, D. C., 1955. The procedure has subsequently been repeated modified. For a good discussion with references, the reader is referred to *Sedimentation Engineering*, ASCE Manual 54.

[15]The procedure and notation that follow are based on *Sedimentation Engineering* rather than the original paper: F. B. Toffaleti, "Definitive Computations of Sand Discharge in Rivers," *Journal of the Hydraulics Division*, ASCE, Jan. 1969, pp 225–248.

Additional input requirements are the width b, the average velocity V, the temperature in °F, the slope S, and the bed material composition. The calculations proceed by size fractions, and in addition, the d_{65} size is required from the bed distribution. The procedure was developed in U.S. customary units and no attempt is made to include alternate equations in SI units.

The first key point in the Toffaleti method will be developed with respect to the definitional sketch of Fig. 8–16. Note that velocity and concentration distributions have been assumed to be similar to those of Fig. 8–14. Rather than use the more complicated form of Eq. 8–31, Toffaleti found that a simpler exponential expression could be used for the concentration distribution, provided that the exponent was allowed to vary with depth. In particular, he divided the suspended sediment portion of the depth into the three regions shown. Like Einstein, he established the thickness of the bed-load zone as equal to two grain diameters.

After introducing a basic exponent z_i, similar to that of Eq. 8–30, but here defined by

$$z_i = \frac{w_{ti}V}{c_z RS} \qquad (8\text{–}33)$$

Toffaleti found that the concentration curve in the three regions could be fitted to the following three equations:

$$C_i = C_{Li}\left(\frac{y}{R}\right)^{-0.756z_i} \qquad \text{(lower zone)} \qquad (8\text{–}34a)$$

$$C_i = C_{Mi}\left(\frac{y}{R}\right)^{-z_i} \qquad \text{(middle zone)} \qquad (8\text{–}34b)$$

and

$$C_i = C_{Ui}\left(\frac{y}{R}\right)^{-1.5z_i} \qquad \text{(upper zone)} \qquad (8\text{–}34c)$$

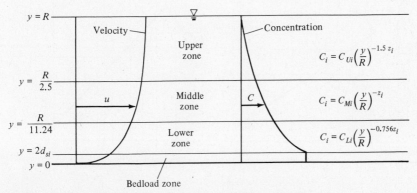

Figure 8–16 Definitional sketch for the Toffaleti method.

In Eq. 8–33, c_z is a temperature-dependent coefficient given by

$$c_z = 260.67 - 0.667T \tag{8–35}$$

where T is the temperature in °F. The concentration coefficients C_{Li}, C_{Mi} and C_{Ui} in Eqs. 8–34 are evaluated using the second key point of the Toffaleti method as follows.

Rather than attempt a fully theoretical solution, Toffaleti tried to correlate various parameters with river data. The best results were obtained with an empirical equation for the lower zone transport (see Fig. 8–16) given by

$$g_{ssLi} = \frac{0.600 p_i}{\left(\dfrac{T_T A k_4}{V^2}\right)^{5/3} \left(\dfrac{d_{si}}{0.00058}\right)^{5/3}} \tag{8–36}$$

Here the subscripts on the transport term indicate that it is the suspended load in the lower zone for the ith size. The resulting transport is in tons/day/ft of width. The grain diameter d_{si} is expressed in ft and, as usual, p_i is the fraction of the ith size found in the bed. The temperature coefficient T_T is

$$T_T = 1.10(0.051 + 0.00009T) \tag{8–37}$$

and A is a function of $(10^5 \nu)^{1/3}/10u'_*$ as defined in Fig. 8–17a. The quantity u'_* pertains to that portion of the shear velocity associated with the grain roughness (as

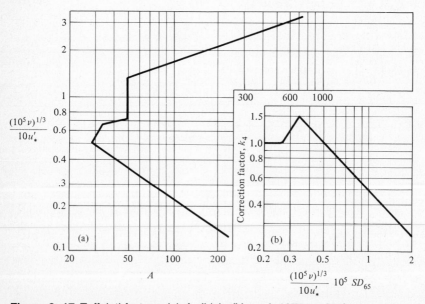

Figure 8–17 Toffaleti factors: (a) *A, (b) k_4* (Vanoni, 1975, p. 211).

opposed to that portion associated with the bed-form roughness). The concept of evaluating u'_* from u'_* is too lengthy to pursue in detail. For our purposes u'_* may be determined from Fig. 8–18 by entering the appropriate graph using the dimensionless parameters $V/\sqrt{gd_{65}S}$ and V^3/gvS, and reading the value of V/u'_* from the left-hand scale. The final variable k_4 (see Fig. 8–17b) is a correction factor, usually needed only for flume data, based on $(10^5 v)^{1/3} 10^5 Sd_{65}/10u'_*$. This parameter is normally less that 0.25 and thus, $k_4 = 1.0$. The product of Ak_4 must not fall below 16.0. If it does, the product is arbitrarily set at this value.

The lower zone transport g_{ssLi} can also be expressed in terms of Eq. 8–32, with the concentration given by Eq. 8–34a and the velocity distribution given by a power relationship rather than the difficult-to-integrate logarithmic equation used by Einstein. In fluid mechanics, a one-seventh-power law in the form of

$$u \sim y^{1/7}$$

is frequently assumed to define the velocity distribution. The Toffaleti method assumes a generalization of this equation as follows:

$$u = (1 + n_v)V\left(\frac{y}{R}\right)^{n_v} \tag{8–38}$$

where

$$n_v = 0.1198 + 0.00048T \tag{8–39}$$

or approximately one-seventh. Thus, Eq. 8–32 may be written for the lower zone as

$$g_{ssLi} = 43.2p_i \int_{2d_{si}}^{R/11.64} C_{Li}\left(\frac{y}{R}\right)^{-0.756z_i} (1 + n_v)V\left(\frac{y}{R}\right)^{n_v} dy$$

$$= \frac{43.2p_i(1 + n_v)VC_{Li}}{R^{n_v - 0.756z_i}} \int_{2d_{si}}^{R/11.64} y^{n_v - 0.756z_i} dy$$

or

$$g_{ssLi} = M_i\left[\frac{y^{1 + n_v - 0.756z_i}}{1 + n_v - 0.756z_i}\right]_{2d_{si}}^{R/11.64} \tag{8–40}$$

where

$$M_i = \frac{43.2p_i(1 + n_v)VC_{Li}}{R^{n_v - 0.756z_i}} \tag{8–41}$$

Finally,

$$g_{ssLi} = M_i\frac{\left(\dfrac{R}{11.24}\right)^{1 + n_v - 0.756z_i} - \left(2d_{si}\right)^{1 + n_v - 0.756z_i}}{1 + n_v - 0.756z_i} \tag{8–42}$$

By equating this expression to g_{ssLi} from Eq. 8–36, the coefficients M_i and C_{Li} can be determined.

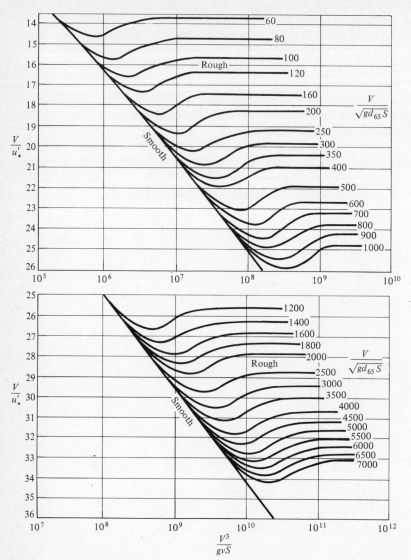

Figure 8–18 Graphical solution of u'_* (ASCE J. of Hyd. Div., Toffaleti, Jan., 1969, p. 236).

Evaluating both Eq. 8–34a and Eq. 8–34b at $y = R/11.24$, the boundary between the lower and middle zones, yields

$$C_i = C_{Li}\left(\frac{1}{11.24}\right)^{-0.756z_i} = C_{Mi}\left(\frac{1}{11.24}\right)^{-z_i}$$

or

$$C_{Mi} = C_{Li}\left(\frac{1}{11.24}\right)^{0.244z_i}$$

With the value of C_{Mi} determined, the transport in the middle zone can be calculated from

$$g_{ssMi} = \frac{43.2p_i(1 + n_v)VC_{Mi}}{R^{n_v - z_i}}\int_{R/11.24}^{R/2.5} y^{n_v - z_i}\, dy$$

which, after integration becomes

$$g_{ssMi} = M_i\frac{\left(\dfrac{R}{11.24}\right)^{0.244z_i}\left[\left(\dfrac{R}{2.5}\right)^{1 + n_v - z_i} - \left(\dfrac{R}{11.24}\right)^{1 + n_v - z_i}\right]}{1 + n_v - z_i} \tag{8–43}$$

Continuing in the same fashion, first evaluating Eqs. 8–34b and 8–34c at the boundary between the middle and upper zones, and then integrating the result from $y = R/2.5$ to $y = R$, gives the upper zone transport equation,

$$g_{ssUi} = M_i\frac{\left(\dfrac{R}{11.24}\right)^{0.244z_i}\left(\dfrac{R}{2.5}\right)^{0.5z_i}\left[R^{1+n_v-1.5z_i} - \left(\dfrac{R}{2.5}\right)^{1+n_v-1.5z_i}\right]}{1 + n_v - 1.5z_i} \tag{8–44}$$

The suspended load of size d_{si} can now be obtained by summing the results from Eqs. 8–36, 8–43, and 8–44,

$$g_{ssi} = g_{ssLi} + g_{ssMi} + g_{ssUi} \tag{8–45}$$

The bed load is determined for each sediment size by taking the product of C_i and u (both evaluated at $y = 2d_{si}$) and the thickness of the bed layer $2d_{si}$. Thus

$$g_{sbi} = 43.2p_iC_{Li}\left(\frac{2d_{si}}{R}\right)^{-0.756z_i}(1 + n_v)V\left(\frac{2d_{si}}{R}\right)^{n_v}(2d_{si})$$

or

$$g_{sbi} = M_i(2d_{si})^{1+n_v-0.756z_i} \tag{8–46}$$

Before continuing, the concentration C_i of each size must be evaluated at $y = 2d_{si}$. If any value of C_i exceeds 100 lb/ft^3, then M_i must be multiplied by the ratio of 100 lb/ft^3 to the value of C_i at $y = 2d_{si}$. The value of 100 lb/ft^3 was chosen as the maximum reasonable concentration, and this adjustment of the entire concentration curve is made to ensure that the bed-load concentration does not exceed that value.[16] Finally, the bed material load per foot of width is obtained from

$$g_{si} + g_{ssi} + g_{sbi} \qquad (8\text{--}47)$$

Summing over the range of sizes and multiplying by the width gives the total bed material load in tons/day.

The Toffaleti method is rather lengthy and is best handled by a computer program. The procedure has been found to give reasonable estimates of sand transport, particularly in large rivers. However, the procedure is not recommended for the calculation of gravel transport. The entire procedure is illustrated in the step-by-step example that follows.

EXAMPLE 8–4

Use the Toffaleti method to calculate the suspended load and bed load by size fractions, then sum to get an estimate of the bed material load for a large river based on the following hydraulic data: $V = 5.08$ ft/s; $R = 36.1$ ft; $T = 45°F$ (7°C); $b = 1320$ ft; $S = 0.0000336$.

Bed size distribution

Size	p_i
VFS	0.07
FS	0.92
MS	0.01

with $d_{65} = 0.000561$ ft

Solution

Because of the length and complexity of the procedure, each step is clearly indicated. The procedure as set up can be used for any number of sediment sizes. The numerical substitutions for the various concentration and transport calculations are not included, however, and the reader should verify each step for at least one size.

Step 1. Calculate the velocity component n_v using Eq. 8–39.

---

[16]This adjustment is indicative of a weakness in the procedure. The matching of the lower zone transport to Eq. 8–36 on occasion leads to unrealistically high concentrations near the bottom of the channel.

$$n_v = 0.1198 + (0.00048)(45) = 0.1414$$

Step 2. Determine the fall velocity w_{ti} in ft/s, for each size fraction using Fig. 8–4 (or computer program 5 in Appendix E).

Size	d_{si} (ft)	d_{si} (mm)	w_{ti} (cm/s)	w_{ti} (ft/s)
VFS	0.00029	0.0880	0.45	0.015
FS	0.00058	0.177	1.6	0.052
MS	0.00116	0.354	4.4	0.144

Step 3. Calculate the concentration coefficient c_z using Eq. 8–35.

$$c_z = 260.67 - (0.667)(45) = 230.7$$

Step 4. Calculate the concentration coefficient z_i for each size fraction using Eq. 8–33.

$$z_i = \frac{w_{ti}(5.08)}{(230.7)(36.1)(0.0000336)}$$

Size	z_i
VFS	0.272
FS	0.944
MS	2.61

Step 5. Determine the grain roughness shear velocity u_*' from Fig. 8–18 ($v = 1.53 \times 10^{-5}$ ft²/s).

$$\frac{V}{\sqrt{gd_{65}S}} = \frac{5.08}{\sqrt{(32.2)(0.000561)(0.0000336)}} = 6.53 \times 10^3$$

$$\frac{V^3}{gvS} = \frac{(5.08)^3}{(32.2)(1.53 \times 10^{-5})(0.0000336)} = 7.92 \times 10^9$$

From Fig. 8–18, $V/u_*' = 33.3$ and

$$u_*' = \frac{5.08}{33.3} = 0.153 \text{ ft/s}$$

Step 6. Determine the factor A from Fig. 8–17a.

$$\frac{(10^5 v)^{1/3}}{10 u'_*} = \frac{(1.53)^{1/3}}{(10)(0.153)} = 0.753$$

Therefore $A = 49$

Step 7. Calculate the factor T_T using Eq. 8–37.

$$T_T = 1.10[0.051 + (0.00009)(45)] = 0.0606$$

Step 8. Determine k_4 from Fig. 8–17b.

$$\frac{(10^5 v)^{1/3} 10^5 S d_{65}}{10 u'_*} = \frac{(1.53)^{1/3}(10^5)(0.0000336)(0.000561)}{(10)(0.153)} = 1.42 \times 10^{-3}$$

and $k_4 = 1.0$ as expected. In addition $A k_4$ is greater than 16.0.

Step 9. Calculate g_{ssLi} using Eq. 8–36.

$$\left(\frac{T_T A k_4}{V^2}\right)^{5/3} = \left[\frac{(0.0606)(49)(1)}{(5.08)^2}\right]^{5/3} = 0.0272$$

$$g_{ssLi} = \frac{0.600 P_i}{(0.0272)\left(\dfrac{d_{si}}{0.00058}\right)^{5/3}}$$

Size	d_{si} (ft)	p_i	g_{ssLi} (tons/day/ft)
VFS	0.00029	0.07	4.90
FS	0.00058	0.92	20.29
MS	0.00116	0.01	0.069

Step 10. Calculate the various exponents for the concentration-velocity (Cu) products for each size fraction. This includes the following:

Size	z_i	$0.5 z_i$	$0.244 z_i$	$0.756 z_i$
VFS	0.272	0.136	0.0664	0.206
FS	0.944	0.472	0.230	0.714
MS	2.61	1.305	0.637	1.973

Size	$n_v - 0.756 z_i$	$1 + n_v - 0.756 z_i$	$1 + n_v - z_i$	$1 + n_v - 1.5 z_i$
VFS	−0.0646	0.936	0.869	0.733
FS	−0.573	0.428	0.197	−0.275
MS	−1.832	−0.832	−1.469	−2.774

Step 11. Calculate the transport coefficient M_i for each size fraction using Eq. 8–42 with g_{ssLi} from step 9.

Size	d_{si} (ft)	g_{ssLi} (tons/day/ft)	M_i
VFS	0.00029	4.90	1.539
FS	0.00058	20.29	5.454
MS	0.00116	0.069	0.000370

Step 12. Calculate the lower zone concentration coefficient C_{Li} for each size fraction using Eq. 8–41. (This is necessary only for the check on C_i at $y = 2d_{si}$ in Step 13.)

Size	d_{si} (ft)	M_i	C_{Li}
VFS	0.00029	1.539	0.0696
FS	0.00058	5.454	0.00303
MS	0.00116	0.000370	2.07×10^{-7}

Step 13. Calculate the concentration at $y = 2d_{si}$ for each size fraction using Eq. 8–34a. Note that this calculation treats the bed as if it were composed of 100 percent size d_{si}. If any value of C_i exceeds 100 lb/ft^3, then C_{Li} (and hence M_i) must be adjusted so that C_i at $y = 2d_{si}$ just equals 100 lb/ft^3.

Size	d_{si} (ft)	C_{Li}	C_i (lb/ft^3)
VFS	0.00029	0.0696	0.59
FS	0.00058	0.00303	4.89
MS	0.00116	2.07×10^{-7}	38.62

All C_i values are below 100 lb/ft^3 and therefore are satisfactory.

Step 14. Calculate the middle zone and upper zone suspended loads g_{ssMi} and g_{ssUi} for each size fraction using Eqs. 8–43 and 8–44.

Size	M_i	g_{ssMi} (tons/day/ft)	g_{ssUi} (tons/day/ft)
VFS	1.539	14.16	22.11
FS	5.454	15.70	9.78
MS	0.000370	8.49×10^{-5}	5.11×10^{-6}

Step 15. Calculate the bed load g_{sbi} for each size fraction using Eq. 8–46.

Size	d_{si} (ft)	M_i	g_{sbi} (tons/day/ft)
VFS	0.00029	1.539	1.44×10^{-3}
FS	0.00058	5.454	0.302
MS	0.00116	0.000370	0.0576

Step 16. Calculate the suspended load and the bed material load using Eqs. 8–45 and 8–47. Multiply each value by the width to get totals for the cross section. Finally, sum the various size fractions to get the total transport by each mode.

Size	Bed load (tons/day)	Suspended load (tons/day)	Bed material load (tons/day)
VFS	1.9	53,344	53,346
FS	398.6	60,416	60,815
MS	76.0	91	167
Total	477	114.851	115,328

The total estimated transport of the three sediment sizes is 115,000 tons/day with about 500 tons/day of the total transport as bed load.

The Ackers and White Method

The final transport procedure to be introduced is a relative newcomer to the list, but one that shows considerable promise. Similar to other transport calculations, the Ackers and White method[17] consists of a transport equation that is calibrated against one set of data and then verified by testing it with additional data. The procedure is dimensionally consistent and therefore usable with either U.S. customary or SI units. A disadvantage is that the calculations cannot be made by size fractions and a representative diameter must be used.

The Ackers and White method represents a significant departure from the concepts used by Einstein and Toffaleti as no attempt is made to involve the suspended sediment concentration. However, the procedure recognizes the inherent difference between the transport of large material as bed load and the transport of fine material primarily as suspended load. To this end, the transport of the coarser material is assumed dependent on an effective shear stress that is proportional to the mean ve-

[17]P. Ackers, and W. R. White, "Sediment Transport: New Approach and Analysis," *Journal of the Hydraulics Division* ASCE, Nov. 1973, pp. 2041–2060.

locity and evaluated by using the logarithmic equation for flow over a rough boundary. This is expressed by

$$\sqrt{\frac{\tau_0}{\rho}} = \frac{V}{\sqrt{32}\,\log_{10}\left(\alpha\dfrac{y_0}{d_s}\right)} \qquad \text{(coarse sediment)} \qquad\qquad \text{(8–48)}$$

where α is a coefficient found within the context of this procedure to equal 10, but in fluid mechanics is assumed to equal to 12.3. The representative diameter d_s is chosen as the d_{35} size, and all other variables are as defined previously. The finer material, on the other hand, is supported by the flow turbulence, the intensity of which is assumed to depend on the total energy degradation. For a uniform flow this is proportional to the slope. Thus, the shear stress is not divided as discussed in conjunction with the Toffaleti method, and

$$\sqrt{\frac{\tau_0}{\rho}} = u_* = \sqrt{gy_0 S} \qquad \text{(fine sediment)} \qquad\qquad \text{(8–49)}$$

These two concepts are combined dimensionlessly in the following equation, which defines the mobility number F_{gr},

$$F_{gr} = \frac{u_*^n}{\sqrt{gd_s(s-1)}} \left[\frac{V}{\sqrt{32}\,\log_{10}\left(\alpha\dfrac{y_0}{d_s}\right)} \right]^{1-n} \qquad\qquad \text{(8–50)}$$

Here s is the specific gravity of the sediment and n is a factor that reflects the sediment size. As limits, n ranges from a value of 0 for coarse material ($d_{gr} \geqslant 60$) to a value of 1 for fine material. Intermediate values of n were determined experimentally and found to fit the equation

$$n = 1.000 - 0.56 \log_{10} d_{gr} \qquad\qquad \text{(8–51)}$$

where d_{gr} is the dimensionless grain size expressed by

$$d_{gr} = d_s \left[\frac{g(s-1)}{\nu^2} \right]^{1/3} \qquad\qquad \text{(8–52)}$$

The minimum value of d_{gr} is 1.0, which roughly corresponds to the lower limit of the sand sizes. The sediment was found to be coarse when $d_{gr} \geqslant 60$. The transition range therefore becomes $1.0 \leqslant d_{gr} < 60$.

The dimensionless transport G_{gr} (defined below in Eq. 8–55) was experimentally determined to be a function of the particle mobility. The resulting equation is

$$G_{gr} = C\left(\frac{F_{gr}}{A} - 1\right)^m \qquad\qquad \text{(8–53)}$$

and the best-fit relationships for the coefficients and exponents are as follows:

Transition range ($1 \leqslant d_{gr} < 60$):

$$\log_{10} C = 2.86 \log_{10} d_{gr} - (\log_{10} d_{gr})^2 - 3.53 \qquad\qquad \text{(8–54a)}$$

$$M = \frac{9.66}{d_{gr}} + 1.34 \tag{8-54b}$$

$$A = \frac{0.23}{\sqrt{d_{gr}}} + 0.14 \tag{8-54c}$$

Coarse range ($d_{gr} \geq 60$):

$$C = 0.025 \tag{8-54d}$$

$$m = 1.50 \tag{8-54e}$$

$$A = 0.17 \tag{8-54f}$$

It should be noted in Eq. 8–53 that A is the value of F_{gr} corresponding to conditions of zero transport.

The resulting dimensionless transport is related to the sediment flux X (in lb sediment/lb water, or N sediment/N water) by

$$G_{gr} = \frac{X}{s} \frac{y_0}{d_s} \left(\frac{u_*}{V}\right)^n \tag{8-55}$$

Once the sediment flux is evaluated, the actual transport may be obtained in U.S. customary units using

$$G_s = (43.2)Q\gamma X \quad \text{(tons/day)} \tag{8-56a}$$

while the SI counterpart is

$$G_s = Q\gamma X \quad \text{(N/s)} \tag{8-56b}$$

As stated previously, the Ackers and White method uses a representative diameter that was chosen as the d_{35} size. Computer modeling of sediment problems[18] usually requires a breakdown by size fractions to allow for selective scour and deposition. If necessary, the transport of each size G_{si} should be determined by using the geometric mean of the respective size class and the appropriate p_i value. Adding the various sizes would then yield an estimate of the total transport as in the previous transport equations.

EXAMPLE 8–5

Repeat Example 8–4 using the Ackers and White method. Use $d_{35} = 4.92 \times 10^{-4}$ ft.

---

[18]See Section 8–5.

Solution

The dimensionless grain size is obtained from Eq. 8–52,

$$d_{gr} = 4.92 \times 10^{-4} \left[\frac{(32.2)(2.65 - 1.0)}{(1.53 \times 10^{-5})^2} \right]^{1/3} = 3.00$$

The variables C, m, and A are evaluated from Eqs. 8–54, and n is determined from Eq. 8–51,

$$\log_{10} C = 2.86 \log_{10} (3.00) - [\log_{10} (3.00)]^2 - 3.53$$

$$C = 0.00405$$

$$m = \frac{9.66}{3.00} + 1.34 = 4.56$$

$$A = \frac{0.23}{\sqrt{3.00}} + 0.14 = 0.273$$

and

$$n = 1.00 - 0.56 \log_{10} (3.00) = 0.733$$

The shear velocity is

$$u_* = \sqrt{gy_0 S} = \sqrt{(32.2)(36.1)(0.0000336)} = 0.198 \text{ ft/s}$$

and the particle mobility F_{gr} determined from Eq. 8–50 is

$$F_{gr} = \frac{(0.198)^{0.733}}{\sqrt{(32.2)(4.92 \times 10^{-4})(2.65 - 1)}}$$

$$\left\{ \frac{5.08}{\sqrt{32} \log_{10} \left[\frac{(10)(36.1)}{4.92 \times 10^{-4}} \right]} \right\}^{1 - 0.733}$$

$$= 1.143$$

Equation 8–53 provides the dimensionless transport G_{gr},

$$G_{gr} = (0.00405) \left(\frac{1.143}{0.273} - 1 \right)^{4.56} = 0.799$$

From Eq. 8–55

$$X = \frac{(0.799)(2.65)(4.92 \times 10^{-4})}{(36.1) \left(\frac{0.198}{5.08} \right)^{0.733}} = 3.113 \times 10^{-4} \text{ lb sediment/lb water}$$

In addition, the water discharge is

$$Q = (5.08)(36.1)(1320) = 242,072 \text{ cfs}$$

Finally, the sediment transport can be calculated using Eq. 8–56a,

$$G_s = (43.2)(242,072)(62.4)(3.113 \times 10^{-4}) = 203,100 \text{ tons/day}$$

The comparison of the Ackers and White method with that of Toffaleti in Example 8–4 is reasonably close for sediment calculations.

Conclusion

A number of sediment transport equations have been considered in this section. These as well as the many others that were not considered often give a range of estimates that are disconcerting to say the least. In fact, the engineer must frequently be satisfied with a relatively rough estimate of the sediment transport. Fortunately, sediment discharge measurements are being obtained in more and more streams as part of the regular United States Geological Survey stream gauging program. Whenever they are available, actual transport measurements are preferred to estimates based solely on transport equations. Limited measurements are, of course, useful complements to the calculated transport and increase the overall confidence in the numerical values.

One of the increasing uses of the predictive equations is in conjunction with mathematical modeling of sediment problems. The transport equations are coupled with the hydraulic equations in computer models to predict various river adjustments, such as changes in the river profile due to scour or deposition. This application of the sediment transport equations is somewhat forgiving in that the different transport functions usually have more impact on the rate of change of the river bed than on the ultimate magnitude of the changes.

8–4 DESIGN OF UNLINED CHANNELS

The economic considerations associated with the construction of water-carrying canals frequently dictates that an unlined channel be used rather than a completely rigid, lined channel such as was generally assumed in the previous chapter. Whereas providing adequate capacity economically was of primary concern in the lined channel, a number of additional factors become equally important in the present case. If the water to be conveyed has a relatively high sediment load, then the channel design must ensure that the velocity remains adequate to avoid deposition. At the same time, the velocity cannot be so high that it initiates scour of the channel banks or bed. If the water entering the channel is relatively clear, then even more care is required to avoid scour problems.

The trapezoidal cross section, by far the most common shape for unlined channels, will be the only type of channel addressed here. There are several approaches to the design of nonscouring, nondepositing channels, each of which has certain advantages and disadvantages. The method that follows, called the *tractive force method,* was primarily pioneered by the U.S. Bureau of Reclamation.[19] Other methods are described in the references at the end of the chapter.

The tractive force method is based on the tractive force or shear stress acting on the channel boundary. A critical shear stress such as that given by Figs. 8–10 and 8–11 may be used as a design value. The channel must then be designed so that the critical shear stress is not exceeded anywhere along the channel perimeter. Previously, we developed and subsequently used the equation $\tau_0 = \gamma RS$. Unfortunately, this relationship provides only the average shear stress around the perimeter, a single numerical value that is not adequate for the present purposes. Only when the channel is very wide and $R = y_0$, so that $\tau_0 = \gamma y_0 S$, does τ_0 equal the actual shear stress along the boundary . The Bureau of Reclamation performed an extensive analysis of the actual shear-stress distribution in trapezoidal channels. Their results are generalized in Fig. 8–19. The shear stress distribution along the bottom is defined by τ_b. The maximum value of this stress occurs at the center where

$$\tau_b = k_b \gamma y_0 S \tag{8–57a}$$

In a wide channel, k_b approaches unity, but drops off for narrower channels. On the side slopes, the distribution is defined by τ_s, and the maximum value, which occurs at $y/y_0 = 0.1$ to 0.2, is

$$\tau_s = k_s \gamma y_0 S \tag{8–57b}$$

The values of k_b and k_s depend on the angle of the side slopes (or z) and the width to depth ratio b/y_0. Typical values are given in Fig. 8–19 for the particular side slope $z = 1.5$. Although there are some variations in k_b and k_s, it is usually satisfactory for design purposes to simply use $k_b = 1.0$ and $k_s = 0.75$. This provides the designer with relationships for the maximum shear stresses on both the bottom and the side slopes in terms of the relevant hydraulic parameters of slope and depth.

The relative stability of a particle on the side slope versus one on the channel bottom must now be investigated. This can be done by first examining the stability of a particle reposing on a side slope with no flow and then extending the analysis to flow conditions. Assume still water and a particle resting on a side slope that has

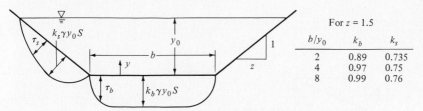

Figure 8–19 Boundary shear stress distribution in a trapezoidal channel.

[19]E. W. Lane, "Design of Stable Channels," *Transactions* ASCE, Vol. 120, 1955, pp. 1234–1279.

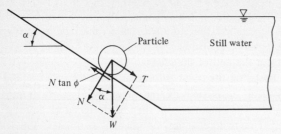

Figure 8–20 Particle stability in still water.

a slope angle α, as shown in Fig. 8–20. Note that z, usually used to define the side slope, is related to α by cot $\alpha = z$. Analyzing the stability of the particle in the plane of the cross section by the algebraic summation of forces along the line of the slope leads to the following. For a particle of weight W, the normal component of the weight becomes

$$N = W \cos \alpha$$

while the tangential component is

$$T = W \sin \alpha$$

To be in equilibrium T must equal the frictional force $N\mathscr{F}$ developed between the particle and the side slope. The coefficient of friction $\mathscr{F}$ may be alternatively expressed by the tangent of the angle of repose of the particles composing the side slope ϕ. Thus,

$$W \sin \alpha = N \tan \phi = W \cos \alpha \tan \phi$$

or simply

$$\tan \alpha = \tan \phi$$

which demonstrates that in still water the maximum angle of the side slope cannot exceed the angle of repose of the particles.

With flow in the channel, the analysis includes the forces previously involved in Fig. 8–20 and an additional force in the direction of the flow. This is shown in Fig. 8–21, which is a sketch that looks directly at the side slope. In this projection, we

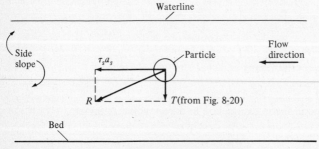

Figure 8–21 Particle stability in flowing water.

see the tangential force down the slope and the force due to the flow, the latter expressed by the produce of τ_s and the effective area over which it acts a_s. The resultant force on the given particle is

$$R = \sqrt{T^2 + (\tau_s a_s)^2} = \sqrt{(W \sin \alpha)^2 + (\tau_s a_s)^2}$$

Equating this to the frictional force

$$N\mathcal{F} = N \tan \phi = W \cos \alpha \tan \phi$$

and squaring gives

$$(W \sin \alpha)^2 + (\tau_s a_s)^2 = W^2 \cos^2 \alpha \tan^2 \phi$$

whereupon

$$\tau_s = \frac{W}{a_s} \sqrt{cos^2 \alpha \tan^2 \phi - \sin^2 \alpha}$$

or

$$\tau_s = \frac{W}{a_s} \cos \alpha \tan \phi \sqrt{1 - \frac{\tan^2 \alpha}{\tan^2 \phi}} \tag{8–58}$$

Applying this equation to the bed where $\alpha = 0$ gives

$$\tau_b = \frac{W}{a_s} \tan \phi \tag{8–59}$$

so that the shear stress ratio K becomes

$$K = \frac{\tau_s}{\tau_b} = \cos \alpha \sqrt{1 - \frac{\tan^2 \alpha}{\tan^2 \phi}} \tag{8–60}$$

which after some arranging leads to

$$K = \frac{\tau_s}{\tau_b} = \sqrt{1 - \frac{\sin^2 \alpha}{\sin^2 \phi}} \tag{8–61}$$

Equation 8–61 enables the ratio of the side slope shear stress to the bed shear stress to be evaluated in terms of the channel shape and the angle of repose of the material. This equation, along with Eqs. 8–57, provides a rational design procedure for unlined channels constructed from noncohesive material.

The allowable bed shear stress τ_b is obtained from Fig. 8–11. If SI units are desired, the allowable shear stress in U.S. customary units may be multiplied by 47.88 to convert to N/m^2. According to the Bureau of Reclamation procedure, the median or d_{50} size is used for material less than 5 mm in diameter. The critical shear stress, here designated as τ_b, is read using the appropriate curve depending on whether the water is clear or has a low or high content of fine sediment. Water with a high content of fine sediment has much of its transport capacity satisfied initially and is therefore less likely to scour the channel. Hence, a higher shear stress is permitted. By the same token, all three curves yield shear stress values considerably in excess of that from the Shields curve. This reflects the difference between essentially zero transport using the Shields curve and a small allowable bed material transport in the less conservative, but more economical, Bureau of Reclamation design procedure. For larger material, the d_{75} rather than the d_{50} size is entered in Fig. 8–11.

The angle of repose for noncohesive material may be estimated from either Fig. 8–22 as proposed by Lane,[20] or from Fig. 8–23, a relationship presented by Simons and Albertson.[21] The former covers only the gravel sizes, while the latter extends over a much broader range of sediment sizes. Some care is necessary as the two figures are not completely consistent.

Based on the angle of repose, a side slope can be selected (with $z > \cot \phi$), and Eq. 8–61 may then be used to evaluate τ_s. Next, the allowable depth can be calculated using first Eq. 8–57a with $k_b = 1.0$ and second Eq. 8–57b with $k_s = 0.75$. The controlling depth will be the lesser of the two values. Finally, the width may be determined from the Manning equation. The entire design process is illustrated in Example 8–6.

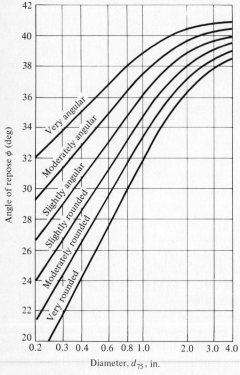

Figure 8–22 Angle of repose (Lane, 1955, p. 1247).

[20]See footnote 19.

[21]D. B. Simons, and M. L. Albertson, "Uniform Conveyance Channels in Alluvial Material," *Journal of the Hydraulics Division* ASCE, May, 1960, pp. 33–71.

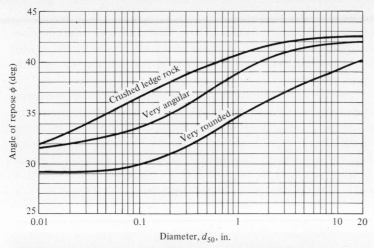

Figure 8–23 Angle of repose (Simon and Albertson, 1960).

This procedure provides the basis for the design of an unlined trapezoidal channel. If the incoming water carries much of a sediment load, it is necessary to check that the resulting velocity is above the minimum value that could result in deposition problems. One approach is to use the equation developed by R. C. Kennedy in 1895 based on his experiences with the irrigation canals in India and Pakistan. For silty water, the Kennedy equation is (in U.S. customary units)

$$V_0 = 0.84 \, y_0^{0.64} \tag{8–62a}$$

This equation may be converted to SI units, whereupon,

$$V_0 = 0.55 \, y_0^{0.64} \tag{8–62b}$$

Entering either of these equations with the previously determined design depth gives an estimate of the nonsilting velocity V_0. The average velocity based on the tractive force method should exceed this value.

If the incoming water has a very heavy sediment load, the engineer should consider using settling basins to reduce the sediment content. If topography or other design factors require a sinuous channel, then an uneven shear stress will result in the bends. The design can proceed as before, but the allowable bed shear stress τ_b must be reduced according to the following factor:

Channel plan	Reduction in τ_b (%)
Slightly sinuous	10
Moderately sinuous	25
Very sinuous	40

Long channels through dry, windy regions may lead to additional complications. Wind-driven sediments entering the channel in large quantities may result in unexpected depositional problems. If the wind blows along a long, relatively straight reach, the resulting waves not only complicate the transport relationship, but also lead to increased bank erosion.

In conclusion, the design of unlined channels cannot be completely separated from factors such as the costs of the right of way and construction or from other complications such as the cross-sectional changes required because of different soil types. In addition, the designer must be willing to accept the seepage losses associated with an unlined channel. Finally, whether the channel is lined or not, evaporation losses may become important in long channels. In this case, the possible economic advantage of a narrower channel should be considered.

EXAMPLE 8–6

Design a trapezoidal channel to carry a discharge of 1600 cfs on a slope of 0.0005. The boundary material is moderately angular gravel with $d_{75} = 1$ in.

Solution

On the basis of the d_{75} size, Fig. 8–11 may be used to determine the allowable bed stress. With $d_{75} = 1$ in. $= 25.4$ mm, we get $\tau_b = 0.42$ lb/ft^2.

From Fig. 8–22, the angle of repose is $\phi = 37.4$ deg, thus cot $\phi = 1.31$. With reference to Fig. 8–19, the side slope must be picked so that $z > $ cot ϕ. A reasonable choice is $z = 1.75$, whereupon $\alpha = \cot^{-1}z = \cot^{-1} 1.75 = 29.7$ deg. From Eq. 8–61,

$$K = \frac{\tau_s}{\tau_b} = \sqrt{1 - \frac{\sin^2 29.7 \text{ deg}}{\sin^2 37.4 \text{ deg}}} = 0.578$$

or

$$\tau_s = 0.578 \ \tau_b = (0.578)(0.42) = 0.24 \text{ lb/ft}^2$$

Taking $k_b = 1.0$ and $k_s = 0.75$, the depth is determined from Eq. 8–57a as

$$y_0 = \frac{\tau_b}{\gamma S} = \frac{0.42}{(62.4)(0.0005)} = 13.46 \text{ ft}$$

while the side slope equation (Eq. 8–57b) gives

$$y_0 = \frac{\tau_s}{0.75 \gamma S} = \frac{0.24}{(0.75)(62.4)(0.0005)} = 10.26 \text{ ft}$$

The lower value controls and the design depth is 10.26 ft.

The Manning equation is required at this point to determine the necessary bottom width. The Manning n may be estimated from Table 7–1 using the d_{75} size,

$$n = 0.031 d_{75}^{1/6} = (0.031)(\frac{1}{12})^{1/6} = 0.020$$

The Manning equation becomes

$$1600 = \frac{1.49}{0.020} \frac{A^{5/3}}{P^{2/3}} (0.0005)^{1/2}$$

where

$$A = b(10.26) + (1.75)(10.26)^2$$

$$= 184.2 + 10.26b$$

and

$$P = b + (2)(10.26)\sqrt{1 + (1.75)^2}$$

$$= 41.36 + b$$

Therefore,

$$\frac{(184.2 + 10.26b)^{5/3}}{(41.36 + b)^{2/3}} = \frac{(1600)(0.020)}{(0.0005)^{1/2}} = 960.5$$

Solving by iteration, $b = 11.4$ ft. The resulting dimensions for the trapezoidal channel include a bottom width of 11.4 ft, side slopes of 1.75 on 1, and a design depth of 10.26 ft. Consulting Fig. 7–6, an additional freeboard of 3.8 ft should be added to this depth. In addition, Eq. 8–62 should be checked to ensure that channel deposition will not be a problem. For a depth of 10.26 ft, this becomes

$$V_0 = (0.84)(10.26)^{0.64} = 3.73 \text{ ft/s}$$

The cross-sectional area is 300.86 ft^2, and the design velocity is accordingly,

$$V = \frac{Q}{A} = \frac{1600}{300.86} = 5.32 \text{ ft/s}$$

This is well in excess of the Kennedy nonsilting velocity and should be satisfactory.

8–5 ANALYSIS OF SEDIMENT PROBLEMS

Selected topics are included in this section to illustrate typical engineering solutions to sediment problems. Although the section does not do justice to the broad spectrum of important sediment problems that must be solved, the two major subjects are examples of repeated concern to the engineer. Scour at bridges will be considered first. This is closely related to the subject of bridge hydraulics treated in Section 11–

4. Understanding bridge scour is extremely important, because the undermining of one or more bridge piers can lead to the failure of the entire structure. The second problem is the impact of a dam on the sediment equilibrium of a river. Both the upstream reservoir sedimentation and the possible downstream scour will be discussed. The final portion of this section will deal with the more general topic of computer modeling of sediment problems.

Bridge Scour

The presence of a bridge usually results in a reduction in the cross-sectional area available for the flow. This leads to the various head loss and backwater problems considered in Section 11–4. Along with the other effects of the bridge, there is often an increased velocity at the bridge section. This contraction of the area and increase in the velocity produces a greater transport capacity at the bridge than exists along the rest of the channel. Thus, there is a tendency for the bed to scour at the bridge, enlarging the opening and reducing the transport capacity until it is more in line with that of the rest of the river. It should be mentioned that this scour may have a beneficial result in that the larger bridge opening will tend to reduce the backwater upstream of the bridge. However, this effect should not usually be encouraged because the damage due to scour may far outweigh the possible benefit.

Contraction of scour can be estimated by combining a sediment transport equation with the Manning equation. A number of restrictive assumptions are required and the result can only be used as a guide. To evaluate the contraction scour, refer to the sketch in Fig. 8–24 where section 1 is upstream of the bridge and section 2 is the bridge section itself. Assuming two-dimensional flow with the hydraulic radius equal to the depth, the Manning equation may be written as

$$Q = \frac{1.49}{n} byR^{2/3}S^{1/2}$$

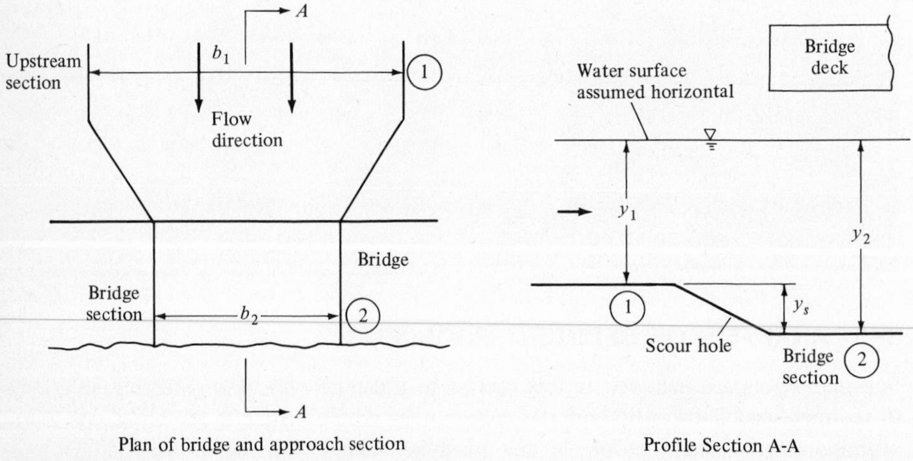

Plan of bridge and approach section

Profile Section A-A

Figure 8–24 Contraction scour at a bridge.

or

$$q = \frac{1.49}{n} y^{5/3} S^{1/2} \tag{8–63}$$

Selecting the Schoklitsch equation (Eqs. 8–20) as the transport equation with $q \gg q_c$ gives

$$g_s = \frac{25.03 \ S^{3/2}}{d_s^{1/2}} (q - q_c)$$

or

$$g_s = \frac{25.03 \ S^{3/2} q}{d_s^{1/2}} \tag{8–64}$$

The slope may be eliminated between Eqs. 8–63 and 8–64 to yield

$$\frac{g_s}{q} \frac{d_s^{1/2}}{25.03} = \frac{n^3 q^3}{(1.49)^3 y^5}$$

which rearranges to

$$\frac{g_s}{q} = 7.57 \frac{n^3}{d_s^{1/2}} \frac{Q^3}{b^3 y^5} \tag{8–65}$$

As flow passes through the contraction, both Q and g_s/q remain constant. If it is further assumed that n and d_s are constant, Eq. 8–65 may be written between the two sections of Fig. 8–24 to give

$$y_1^5 b_1^3 = y_2^5 b_2^3$$

or

$$\frac{y_2}{y_1} = \left(\frac{b_1}{b_2}\right)^{3/5} \tag{8–66}$$

If the change in water surface elevation is small relative to both the upstream depth and the scour hole, then

$$y_2 = y_1 + y_s$$

where y_1 and y_2 are the respective depths and y_s is the depth of scour. Substituting for y_2 in Eq. 8–66, the depth of scour is

$$\frac{y_s}{y_1} = \left(\frac{b_1}{b_2}\right)^{3/5} - 1 \tag{8–67}$$

Depending on which transport function is used, the exponent in Eq. 8–67 will vary somewhat. However, the other assumptions that were made probably have more effect on the accuracy of Eq. 8–67 than does the choice of transport equation. Note also that the numerical constants dropped out of the equations during the derivation, and consequently Eqs. 8–66 and 8–67 are appropriate with either U.S. customary or SI units.

The second type of bridge scour is due to the local acceleration of the water in the immediate vicinity of a bridge pier. As the water encounters the pier, a vortex, known as a horseshoe vortex because of its shape, forms just above the bed and around the pier, as shown in Fig. 8–25. The accompanying high velocity in such close proximity to the bed readily removes the bed material, leaving a scour hole around the base of the pier. One way to reduce this problem is to place riprap on that portion of the bed near the pier.

Recent research has contributed significantly to the understanding of pier scour. However, a final consensus has yet to be reached, and the standard remains the work by Laursen.[22] The Laursen procedure, which is based on a laboratory study conducted at the University of Iowa, differentiates between a river that is transporting a bed load and one that is not. The latter situation, which will not be considered here, might occur at a bridge crossing the floodplain of a river. The so-called clear-water scour hole will usually be deeper than that predicted by Laursen, since less of the transport capacity will be satisfied by the upstream supply of sediment.

The Laursen procedure is based on rectangular piers aligned with the flow, with coefficients to account for other conditions. The equation that best fits his results is

$$\frac{y_p}{b} = 1.5 \, K_1 K_2 \left(\frac{y_1}{b}\right)^{0.3} \tag{8–68}$$

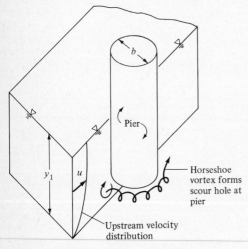

Figure 8–25 River section at bridge pier.

[22] E. M. Laursen, "Scour at Bridge Crossings," *Transactions* ASCE, Vol. 127, Part 1, 1962, pp. 166–209.

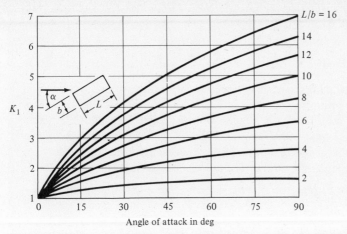

Figure 8–26 Laursen bridge scour shape factor for Eq. 8–68 (Laursen, 1962).

Here b is the pier width, y_1 is the depth of the approach flow, and the depth of the scour hole is y_p. If the pier is not aligned with the flow, increased scour will result. This is reflected by the coefficient K_1, for which experimentally determined values can be obtained from Fig. 8–26. The rectangular shape represents the worst geometric case. A second coefficient, K_2 (Table 8–3), provides an estimate of the reduction in scour with various rounded pier shapes. However, if the pier is not aligned with the flow, then Table 8–3 should be ignored and K_2 should always be taken as unity.

Many additional scour equations are also available. A Froude number effect shows up in many, but not all, of them. The Laursen equation, which ignores the Froude number, is most applicable in the low to moderate (but less than unity) range. Fortunately, this includes the majority of rivers.

The total scour may be estimated by adding the contraction and local scour values. This is probably conservative since it does not allow for any interaction between the two processes.

Table 8–3 Shape Coefficient K_2 for Pier Nose Forms*

Nose form	Length/width ratio	Sketch	K_2
Rectangular			1.00
Semicircular			0.90
Elliptic	2:1		0.80
	3:1		0.75
Lenticular	2:1		0.80
	3:1		0.70

*(To be used only for piers aligned with the flow.)

EXAMPLE 8–7

Calculate the maximum scour depth expected at a bridge section as shown in Fig. 8–24, if the upstream channel width is 200 ft and the width at the bridge is 150 ft. There are three piers aligned with the flow, each of which has a width of 2.5 ft, a length/width ratio of 10, and a semicircular nose. Note that the total pier width must be subtracted from the channel width at the bridge to get the clear opening at the bridge. The upstream depth is 8 ft.

Solution

The clear opening at the bridge is $150 - (3)(2.5) = 142.5$ ft. The contraction scour is obtained from Eq. 8–67:

$$\frac{y_s}{8} = \left(\frac{200}{142.5}\right)^{0.6} - 1$$

Solving, $y_s = 1.8$ ft. The local scour is given by Eq. 8–68 with $K_1 = 1$ and $K_2 = 0.90$:

$$\frac{y_p}{2.5} = (1.5)(1)(0.90)\left(\frac{8}{2.5}\right)^{0.3}$$

whereupon $y_p = 4.8$ ft. The estimated total scour at the bridge is the sum of the contraction and local scour or 6.6 ft.

Dam Deposition and Scour

Most rivers are more or less in equilibrium. Although a flood may cause some scour and possibly deposition, the river geometry is likely to change very little on an annual basis. The construction of a dam is almost certain to upset this equilibrium in both the upstream and downstream directions. In fact, the effects of the dam may extend for considerable distances and disturb tributary equilibrium as well. The reservoir created by a dam reduces the stream velocity. This in turn reduces the sediment transport capacity, resulting in deposition. The tendency for scour to occur below the dam is perhaps less obvious. The relatively clear water released from the reservoir frequently scours the stream bed in an attempt to satisfy the transport capacity of the stream. This subsection will consider both of these problems.

As the river enters a reservoir, the transport capacity immediately begins to drop. The largest material will settle out first, with the finer material carried farther into the reservoir. Much of the wash load may be carried completely through the reservoir. Thus, a segregated delta forms in the reservoir, as shown in Fig. 8–27. This deposition reduces the storage capacity of the reservoir. It is rarely practical to remove the deposits, and the ultimate fate of most reservoirs is to fill with sediment.

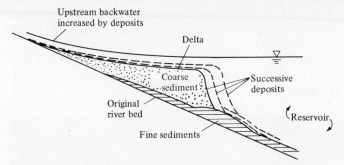

Figure 8–27 Typical reservoir delta (Vanoni, 1975, p. 349).

This factor must be recognized in the design stage, and the engineer must determine the useful life of the structure. A backwater (or M–1) profile also forms above the dam, as shown in Fig. 8–27. This initiates deposition above the level of the reservoir pool, and this deposition forces the backwater profile even farther upstream. Consequently, the planner must be alert to possible flooding problems for some distance upstream of the reservoir itself. A computer model, such as that described in the following subsection, is useful to study this kind of a problem.

The delta formation within a reservoir may also be modeled. However, a generalized trap efficiency curve is useful for estimating the time required for the reservoir to fill. This is more economical to use and may suffice during the planning stages. Figure 8–28 is based on a study by Brune[23] of 44 reservoirs. Here the *trap efficiency,* or percent trapped, is given as a function of the capacity/inflow ratio. The inflow is the annual inflow, and both the inflow and reservoir capacity must be in the same units, for example, acre-feet, cubic meters, or second-foot-days.[24] The upper of the envelope curves would be used for coarse sediment and the lower used

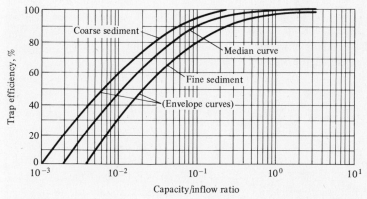

Figure 8–28 Reservoir trap efficiency (after Brune) (Vanoni, 1975, p. 590).

[23]G. M. Brune, "Trap Efficiency of Reservoirs," *Transactions* American Geophysical Union, Vol. 34, June, 1953, pp. 407–418.
[24]Refer to Section 3–5 for an explanation of the various units.

when the incoming load is composed primarily of fine material. A number of relevant parameters are not included in this graph, and at best, it gives an estimate for an average reservoir. The use of this figure to calculate the functional life of a reservoir is illustrated in Example 8–8. A specific weight of the deposits is required. A reasonable estimate is 70 lb/ft^3 or 11 kN/m^3; however, the specific weight usually increases with time due to compaction.

EXAMPLE 8–8

Use the Brune graph to estimate the useful life of a reservoir that has an initial capacity of 5.5×10^8 m^3, an average annual inflow of 6.5×10^9 m^3, and an average sediment inflow of 3.1×10^{10} N. Assume that the useful life of the reservoir is reached when its capacity is reduced to 20 percent of the original capacity. Use the median curve, and assume that the deposits have a specific weight of 11,000 N/m^3.

Solution

The calculations are tabulated below. The capacity has been divided into even increments ranging from the initial to the final values. The second column is the capacity/inflow ratio for each capacity. The third column is the trap efficiency for the respective capacity/inflow ratio (obtained from Fig. 8–28), and the fourth column is the average trap efficiency over the increment. For example, as the capacity drops from 5.5×10^8 to 4.4×10^8 m^3 the average trap efficiency is 85.5 percent. The respective trap efficiency yields the annual weight and volume of sediment deposited, the volume being based on the specific weight of the deposits. The incremental volume of 1.1×10^8 m^3 is divided by the annual volume deposited to get the years required to fill each increment. These values are tabulated in the final column and then added to get 203 years as the useful life of the reservoir.

Capacity (10^8 × m^3)	Capacity/ inflow ratio	Trap efficiency (%)	Average for increment (%)	Annual sediment trapped weight 10^{10}N	Volume 10^{6}m^3	Years required to fill
(1)	(2)	(3)	(4)	(5)	(6)	(7)
5.5	0.085	87	—	—	—	—
4.4	0.068	84	85.5	2.65	2.41	46
3.3	0.051	81	82.5	2.56	2.33	47
2.2	0.034	73	77.0	2.39	2.17	51
1.1	0.017	60	66.5	2.06	1.87	59
					Total =	203 years

The scour problem below the dam is considerably more complex than the reservoir deposition. The storage capability of the reservoir attenuates the hydrograph so that the peak flows downstream of the dam have a lower magnitude (but longer duration) than the reservoir inflow. This reduces the transport capacity downstream of the dam. However, the reservoir has intercepted most of the normal sediment load of the stream, so the water leaving the dam has virtually none of its transport capacity satisfied. Consequently, a potential scour problem may exist. In gravel bed streams, the reduction in transport capacity due to the reduced flows usually outweighs the tendency for scour due to the clear water. As a result, there will often be little or no scour below a dam on a gravel stream.

In sand bed channels, on the other hand, the reduction in the peak flow will reduce the maximum transport capacity, but a significant amount of downstream scour may still be anticipated. As the bed scours, an *armor layer* may form on the surface, which limits the ultimate scour depth. This armoring occurs because the lower peak flows and/or the increase in depth that accompanies the scour rapidly reduces the transport capacity of the larger sizes to such an extent that they accumulate on the surface of the bed as the overall bed level decreases. Once these larger sizes cover most or all of the surface, the transport of the finer sizes must cease as well. The result is a one-layer-thick armor surface with the protected finer material below. This armor layer will remain stable until a discharge capable of moving this large material occurs. It is possible to estimate the scour depth that will precede the formation of an armored bed, but this type of analysis is more ideally suited to the mathematical model described next.

Computer Modeling of Sediment Problems

As with other types of hydraulic problems, the computer model has become a useful tool in evaluating sediment-related problems. Unfortunately, the low accuracy associated with many of the sediment transport calculations may be expected to carry over to the computer model as well. Some use has been made of two- and three-dimensional models, but the mathematical complexity and additional computer costs often limit the engineer's choice to the one-dimensional model. This becomes particularly true when long-term simulations are required.

Although many sediment models exist, the only model to be discussed here will be HEC–6, developed by Thomas for the Corps of Engineers Hydrologic Engineering Center. This is a one-dimensional model that sequentially adjusts the sediment transport and channel geometry to a changing hydrograph. It is suitable for subcritical flow in rivers, channels, and reservoirs, but not estuaries and tidal channels. The hydrograph is simulated by breaking it up into a series of discrete discharges, each occurring over a specified period of time. Each discharge is treated as a steady flow over the time interval, and the water surface profile is determined in much the same way as was discussed in Section 7–8. Overbank flow may be included, and levees may be specified.

The sediment transport capacity in the main channel is calculated section by section using any of a wide choice of transport functions. The actual transport rate is based on the transport capacity and the availability of material in the bed. The channel geometry is adjusted using the sediment continuity equation:

$$\frac{\partial G_s}{\partial x} + B \frac{\partial y}{\partial t} = 0 \tag{8-69}$$

Here, G_s is the sediment transport, x is the distance along the channel, B is the width of the movable bed (the channel width subject to scour or deposition), y is the elevation of the bed, and t is time. This equation is solved in finite difference form to obtain Δy, the change in bed elevation during the time interval Δt, as a function of ΔG_s. This latter quantity is the difference between the actual transport rate at a given section and the sediment load entering the section. If the incoming load is greater than the actual transport rate, deposition will result. Scour will occur when the actual transport rate is greater than the incoming load. As the bed change takes place, an iterative process provides for the adjustment between the transport rate and the bed distribution. In the case of scour, armoring is simulated by allowing nontransportable sizes to accumulate on the bed and thereby reduce the supply of available material. The calculated value of Δy at each section is used to adjust vertically all channel coordinates within the movable bed width.

The water surface profile is evaluated section by section, starting with the downstream section and working upstream. The subsequent sediment calculations proceed in the opposite direction. Once the sediment is routed through the model and the channel geometry adjusted, the entire process is repeated for the next discharge. Thus, changes in the bed elevation, water surface elevation, and bed graduation are all modeled as functions of both location and time. Weirs, dams, and tributary inflow can all be accommodated. A variety of special options is also available in the model. These include features such as channel dredging and gravel mining.

Data requirements include geometric data, hydraulic data, and sediment data. The geometric data are essentially the same as the requirements for HEC–2 discussed in the previous chapter. That is, the cross sections are specified by coordinate points (stations and elevations) and the distance between them. Manning n values must be known or estimated for main channel and overbank regions. The banks of the main channel and the limits of the movable bed must usually be identified. In addition to the hydrograph, the hydraulic requirements include the water temperature and a stage/discharge relationship at the downstream end of the model. The sediment data includes the bed distribution at each section and the incoming sediment load by size fractions.

It is usually very difficult to get a fully satisfactory verification of a sediment model because of the number of parameters involved and the limited data that are normally available. In the verification process, the model should be run in an attempt to duplicate historic conditions. As with the fixed-bed mathematical models, the water surface profiles should compare with past flood events. If a record of scour or deposition is available, the model should be tested against this. If a river has had a period of stability, then the model should also behave in a stable fashion.

HEC–6 has been applied to a large number of different projects with reasonable success. Figure 8–29 illustrates the modeling of the scour and armor formation below

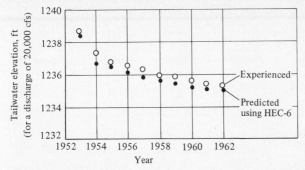

Figure 8–29 Modeling scour below Fort Randall Dam (Thomas and Prasuhn, 1977).

Fort Randall Dam on the Missouri River following closure.[25] As the downstream bed scours, there is an accompanying drop in the water surface. Figure 8–29 compares the experienced shift or drop in the water level below the dam for a discharge of 20,000 cfs with the tailwater level as predicted using HEC–6. The depth of scour cannot be assumed equal to the drop in tailwater elevation, since the bed slope decreases in the process. In fact, the scour depth will probably exceed the drop in the water level.

The model was used to evaluate the impact of two large upstream dams on a gravel bed stream, particularly with respect to gravel supply for fish spawning.[26] The proposed dams will be located on different branches of the river, thereby requiring a special version of HEC–6. The model was run using over 30 years of stream flow records to establish preproject conditions. The model was then rerun using the operating schedules for the reservoirs to provide a regulated discharge hydrograph in each branch of the river. The reservoir sites served as the upstream ends of the model. In the preproject run, the incoming sediment load was based on gauging station records. In the postproject run, it was assumed that the reservoirs would trap all of the sands and gravels and the incoming sediment load was reset to zero. Tributaries downstream of the dams were left unchanged with respect to both discharge and sediment load. Table 8–4 gives the results of the long-term simulations with respect to gravel supply and balance. Sand discharge, although present, is not included in this table.

All transport values are in tons/year. Preproject gravel (and small cobbles) entering the study reach are given in columns 2, 3, and 4. Column 5 is the total incoming gravel by size fractions. Gravel leaving the system under preproject conditions is found in column 6. A comparison of columns 5 and 6 shows that the incoming load for each size fraction is in relatively good balance with the movement out of the system of that size. Project conditions assume that no gravel enters the study at the dam sites, but that the tributary contribution remains unchanged. Consequently, project outflow of gravel in column 7 can be compared with column 4,

[25]W. A. Thomas and A. L. Prasuhn, "Mathematical Modeling of Scour and Deposition," *Journal of the Hydraulics Division* ASCE, Aug. 1977, pp. 851–863.

[26]A. L. Prasuhn, and E. F. Sing, "Modeling of Sediment Transport in Cottonwood Creek," *Proceedings* ASCE Specialty Conference, Aug. 1980, pp. 209–220.

Table 8–4 AVERAGE ANNUAL TRANSPORT OF GRAVEL IN TONS (34-YEAR HYDROGRAPH)

	Preproject					Project
Sediment Size	South branch dam site	North branch dam site	Supply from tributaries	Total supply	Outflow	Outflow
(1)	(2)	(3)	(4)	(5)	(6)	(7)
VFG	410	4,300	2,675	7,385	4,340	2,320
FG	420	3,200	1,750	5,370	3,420	1,500
MG	420	2,000	700	3,120	2,950	960
CG	580	860	460	1,900	2,970	1,050
VCG	400	120	175	695	900	265
SC	20	0	70	90	60	0
Total	2,250	10,480	5,830	18,560	14,640	6,095

where it is seen that the gravel balance again remains in equilibrium, albeit at a lower level.

A final example illustrates the use of HEC–6 to model the contraction scour at a bridge under extreme flood conditions. The results are shown in Fig. 8–30.[27] In this case, as the flood hydrograph passes, the water surface hits the bridge deck (alternately specified at elevations of 365 and 367 ft), causing a pressure flow to occur under the bridge. The formation of the scour hole and the partial refilling of the recession limb are clearly indicated. The lower bridge deck elevation increased

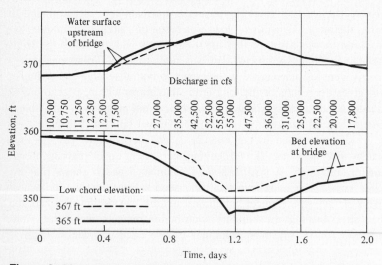

Figure 8–30 Modeling contraction scour at a bridge (Prasuhn, 1981).

[27]A. L. Prasuhn, "Modeling Scour, Backwater and Debris at Bridges" *Proceedings* ASCE Specialty Conference, Aug. 1981, pp. 852–859.

the scour depth by approximately 2 ft. However, because of the increased scour, there was almost no increase in the upstream stage.

 PROBLEMS

Note: For all problems in this chapter, unless it is otherwise stated, assume that the sediment is quartz with a specific gravity of 2.65, the water temperature is 10°C (50°F), and the channel width is approximately rectangular.

Section 8–2

8–1. Plot the following sediment bed material sample on semilog graph paper and determine the d_{35}, d_{50}, and d_{90} sizes (see Fig. 8–1):

Size	Percent of Sample
VFS	8
FS	17
MS	35
CS	25
VCS	15

8–2. Repeat Prob. 8–1 for the following bed sample:

Size	Percent of Sample
VFG	27
FG	36
MG	23
CG	11
VCG	3

8–3. Repeat Prob. 8–1 for the following bed sample:

Size	Percent of Sample
FS	38
MS	39
CS	23

8–4. A given bed sample has a $d_{85} = 5$ mm and a $d_{15} = 0.45$ mm. Assume that the sample plots as a straight line on semilog graph paper, and determine the fraction of each size using the size classification given in Table 8–1.

8–5. Repeat Prob. 8–4 if $d_{95} = 5$ mm and $d_5 = 0.45$ mm.

8–6. Write a computer program based on the size classification of Table 8–1 so that the user can input the size classes by number (e.g., 3 through 6 would be MS through VFG) and the fraction (or percent) of each size in the bed. The p_i values should be placed into an array. The program should then perform a logarithmic interpolation to estimate selected sizes, such as the d_{65} size. In other words, it should be able to determine directly the solutions to the first three problems. Include only the twelve sizes from very fine sand through large cobbles.

8–7. Calculate the time required for a particle of VFS to reach 99 percent of its terminal fall velocity if the particle is quartz and the water temperature is (1) 40° and (2) 70°F. In each case evaluate the fall velocity and determine whether the solution is in the Stokes range.

8–8. Calculate the time required for a quartz particle with a diameter of 0.06 mm to reach 99 percent of its terminal velocity if the water temperature is 10°C.

8–9. Taking the upper limit of the Stokes range at Re = 1, determine the largest diameter quartz sphere that has a terminal fall velocity within the Stokes range if the water temperature is (1) 5°C and (2) 25°C.

8–10. Determine the terminal fall velocity of a grain of coarse sand if the particle has a specific gravity of (1) 1.95, (2) 2.65, and (3) 2.95. Assume a water temperature of 70°F.

8–11. Repeat Prob. 8–10 for a sand grain with a diameter of 0.75 mm.

Section 8–3

8–12. Use Fig. 8–10 to determine the critical shear stress on a bed of very fine sand. The sediment has a specific gravity of 2.65 and the water temperature is 65°F. Verify the answer using Fig. 8–11.

8–13. Use the Shields diagram to determine the critical shear stress on sediment particles that have a diameter of (1) 0.1 mm, (2) 1.0 mm, and (3) 10 mm. Assume a particle specific gravity of 2.45 and a water temperature of 15°C.

8–14. A discharge of 4600 m³/s occurs in an approximately rectangular channel with a depth of 10 m and a width of 70 m. The channel slopes at the rate of ½₀₀₀. The bed consists of FS and MS. Use the Shields diagram to determine whether these sizes will remain stationary or be transported. Assuming that the width and slope do not change with the discharge, determine the discharge that would correspond to incipient motion for each size. Use $n = 0.020$ as needed.

8–15. Repeat Prob. 8–14 if the bed consists of CS and VCS.

8–16. Repeat Prob. 8–14 if the bed consists of VFG and FG.

8–17. A discharge of 15,000 cfs occurs in a 180-ft-wide channel that has a depth of 12 ft and a slope of 0.00022. The streambed consists of equal fractions of FS, MS, and CS. Use the Duboys equation to calculate the sediment transport based on the d_{50} size.

8–18. Repeat Prob. 8–17, calculating the transport by size fractions.

8–19. A discharge of 12,000 cfs occurs in a channel that has a width of 160 ft and a depth of 8 ft. The channel slope is 0.0002. Use the sediment distribution given in Prob. 8–3 and the Duboys equation. Calculate the sediment transport by size fractions and compare with the solution using $d_{50} = 0.31$ mm.

8–20 through 8–24. Calculate the sediment transport using the method indicated for each problem. A river has the bed distribution given in Prob. 8–3. Use $d_{90} = 0.74$ mm, $d_{65} = 0.39$ mm, $d_{50} = 0.31$ mm, and $d_{35} = 0.24$ mm (or your graph of Prob. 8–3) as needed. The discharge is 16,000 cfs, the width is 400 ft, the depth is 10 ft, and the slope is 0.00016. Assume $R = y$.

8–20. The Duboys method using d_{50}.

8–21. The Schoklitsch method (bed load).

8–22. The Meyer-Peter and Muller method (bed load using size fractions).

8–23. The Toffaleti method.

8–24. The Ackers and White method.

8–25 through 8–30. Calculate the sediment transport using the method indicated for each problem. A river has the bed distribution given in Prob. 8–3. Use $d_{90} = 0.74$ mm, $d_{65} = 0.39$ mm, $d_{50} = 0.31$ mm, and $d_{35} = 0.24$ mm (or your graph of Prob. 8–3) as needed. The discharge is 3150 cfs, the width is 100 ft, the depth is 7 ft, and the slope is 0.00023. Assume $R = y$.

8–25. The Duboys method using d_{50}.

8–26. The Schoklitsch method (bed load).

8–27. The Meyer-Peter and Muller method (bed load using d_m).

8–28. Repeat Prob. 8–27 using size fractions.

8–29. The Toffaleti method.

8–30. The Ackers and White method.

8–31 through 8–35. Calculate the sediment transport using the method indicated for each problem. A channel has the bed distribution given in Prob. 8–1. Use $d_{90} = 1.26$ mm, $d_{65} = 0.57$ mm, $d_{50} = 0.41$ mm, and $d_{35} = 0.31$ mm (or your graph of Prob. 8–1) as needed. The discharge is 16,000 cfs, the width is 400 ft, the depth is 10 ft, and the slope is 0.00016. Assume $R = y$.

8–31. The Duboys method using d_{50}.

8–32. The Schoklitsch method (bed load).

8–33. The Meyer-Peter and Muller method (bed load using size fractions).

8–34. The Toffaleti method.

8–35. The Ackers and White method.

8–36 through 8–38. Calculate the sediment transport using the method indicated for each problem. A river has the bed distribution given in Prob. 8–1. Use $d_{90} = 1.26$ mm, $d_{65} = 0.57$ mm, $d_{50} = 0.41$ mm, and $d_{35} = 0.31$ mm (or your graph of Prob. 8–1) as needed. The discharge is 330 m³/s, the width is 100 m, the depth is 3 m, and the slope is 0.00011. Assume $R = y$.

8–36. The Schoklitsch method (bed load).

8–37. The Meyer-Peter and Muller method (bed load using d_m).

8–38. Ackers and White method.

8–39 through 8–42. Calculate the sediment transport using the method indicated for each problem. A river has the bed distribution given in Prob. 8–1. Use $d_{90} = 1.26$ mm, $d_{65} = 0.57$ mm, $d_{50} = 0.41$ mm, and $d_{35} = 0.31$ mm (or your graph of Prob. 8–1) as needed. The discharge is 900 m³/s, the width is 150 m, the depth is 4 m, and the slope is 0.00016. Assume $R = y$.

8–39. The Schoklitsch method (bed load).

8–40. The Meyer-Peter and Muller method (bed load using d_m).

8–41. Repeat Prob. 8–40 using size fractions.

8–42. The Ackers and White method.

8–43 through 8–46. Calculate the gravel transport using the method indicated for each problem. A channel has the bed distribution given in Prob. 8–2. Use $d_{90} = 20.57$ mm and $d_{35} = 4.7$ mm (or your graph of Prob. 8–2) as needed. The discharge is 18,000 cfs, the width is 300 ft, the depth is 7.5 ft, and the slope is 0.0013. Assume $R = y$.

8–43. The Schoklitsch method.

8–44. The Meyer-Peter and Muller method using d_m.

8–45. Repeat Prob. 8–44 using size fractions.

8–46. The Ackers and White method.

8–47 through 8–48. Calculate the gravel transport using the method indicated for each problem. A channel has the bed distribution given in Prob. 8–2. Use $d_{35} = 4.7$ mm (or your graph of Prob. 8–2) as needed. The discharge is 3600 cfs, the width is 110 ft, the depth is 4 ft, and the slope is 0.0033. Assume $R = y$.

8–47. The Schoklitsch method.

8–48. The Ackers and White method.

8–49 through 8–50. Calculate the gravel transport using the method indicated for each problem. A channel has the bed distribution given in Prob. 8–2. Use $d_{35} = 4.7$ mm (or your graph of Prob. 8–2) as needed. The discharge is 75 m³/s, the width is 30 m, the depth is 1 m, and the slope is 0.0053. Assume $R = y$.

8–49. The Schoklitsch method.

8–50. The Ackers and White method.

8–51 through 8–53. Calculate the gravel transport using the method indicated for each problem. A river has the bed distribution given in Prob. 8–2. Use $d_{90} = 20.57$ mm and $d_{35} = 4.7$ mm (or your graph of Prob. 8–2) as needed. The discharge is 315 m³/s, the width is 70 m, the depth is 1.5 m, and the slope is 0.0040. Assume $R = y$.

8–51. The Schoklitsch method.

8–52. The Meyer-Peter and Muller method using size fractions.

8–53. The Ackers and White method.

8–54. Assuming that the width and slope do not change with discharge, calculate the discharge associated with incipient transport for each sediment size in Prob. 8–21.

8–55. Assuming that the width and slope do not change with discharge, calculate the discharge associated with incipient transport for each sediment size in Prob. 8–51.

8–56. Assuming that the width and slope do not change with the discharge, determine the maximum discharge for which there will be no transport in Prob. 8–18. Hint: use the Manning equation as needed.

8–57. Assuming that the width and slope do not change with the discharge, determine the maximum discharge for which there will be no transport in Prob. 8–49.

8–58. Assuming that the width and slope do not change with the discharge, determine the effect on the sediment transport of doubling the discharge in Prob. 8–25. Hint: use the Manning equation to get the depth.

8–59. Assuming that the width and slope do not change with the discharge, determine the effect on the gravel transport of doubling the discharge in the channel of Prob. 8–47.

8–60. Calibrate (i.e., find best values for τ_c and ψ) the Duboys equation for a river with a width of 300 ft and a slope of 0.00015. At a depth of 11 ft, the sediment transport rate is 8310 tons/day, while a depth of 15 ft results in a transport rate of 16,300 tons/day. Assume that the width and slope do not change with the discharge.

8–61. A flow rate of 12,000 cfs occurs in a 160-ft-wide river that has a slope of 0.002. The bed consists of 10 percent VCS, 65 percent VFG, and 25 percent FG. Calculate the sediment transport using the Schoklitsch method.

8–62. Repeat Prob. 8–61 if the slope is 0.0002.

8–63. A streambed consists of VFS, FS, and MS. The channel has a width of 250 ft, a depth of 8 ft, and a slope of 0.00022. Plot the relative concentration (C/C_a) of each size from a reference height 1 ft above the bed up to the surface. Assume that the temperature is 20°C, $R = y$, and $\kappa = 0.40$.

8–64. A streambed consists of MS, CS, and VCS. The channel width is 80 m, the depth is 1.8 m, and the slope is 0.00085. Assume $R = y$. Plot the relative concentration (C/C_a) of each size from a reference height 0.2 m above the bed up to the surface. Assume that the water temperature is 20°C and $\kappa = 0.40$.

8–65. Repeat Prob. 8–64 if $\kappa = 0.20$.

8–66. If the sediment concentrations 1 ft and 2 ft above the streambed are 2000 ppm and 500 ppm, respectively, calculate the concentrations 4 ft above the bed if the depth is (1) 10 ft and (2) 100 ft.

8–67. If the sediment concentrations 10 cm and 30 cm above the streambed are 4000 ppm and 1000 ppm, respectively, plot the concentration curve of C versus y from the 10-cm elevation up to the water surface. The depth is 2 m.

8–68. The velocity distribution in a river is found to vary in accordance with the seventh-power law,

$$\frac{u}{U_0} = \left(\frac{y}{y_0}\right)^{1/7}$$

where u is the velocity at height y, and U_0 is the velocity at the surface. The river is 1.5 m deep, 100 m wide, and a suspended sediment sample obtained at 75 mm above the bed is 10 N/m^3. If the surface velocity is 2 m/s, what is the transport rate of suspended sediment? Use Eq. 8–31 for the sediment concentration and assume that $z = 1$.

8–69. Write a computer program to calculate the sediment transport using the Duboys method. The hydraulic input may include the width, depth, and slope. The sediment input should be by size fractions for use with the computer program prepared for Prob. 8–6, or the d_{50} size can be entered directly.

8–70. Redo Prob. 8–69 so that the sediment transport calculations will be by size fractions.

8–71. Prepare a computer program to calculate the sediment transport using the Schoklitsch procedure. Transport should be by size fractions with transport expressed in tons/day for U.S. customary units or N/s in SI units. Hydraulic requirements should include the discharge, average width, and slope. The sediment input should be the size classifications and p_i values of the streambed. It is suggested that the computer program of Prob. 8–6 be incorporated to provide the sediment input.

8–72. Write a computer program to calculate the sediment transport in either U.S. customary or SI units using the Ackers and White procedure. Input of hydraulic parameters should include the discharge, width, depth, slope, and viscosity. The sediment input should make use of the computer program required for Prob. 8–6, or the d_{35} size may be entered directly.

Section 8–4

8–73 through 8–80. In each of the following problems use the tractive force method to design a trapezoidal channel to transport the discharge shown. Use Eq. 8–62 to verify that the velocity is not so low that silting problems will occur. Add the appropriate freeboard to each channel.

8–73. Design a channel to carry 1100 cfs on a slope of 0.00045. The boundary material will be moderately angular gravel with $d_{75} = 0.5$ in. Assume the Manning $n = 0.020$.

8–74. Design a channel to carry 3000 cfs on a slope of 0.0015. The boundary material will be very angular gravel with $d_{75} = 2$ in. Use a Manning $n = 0.025$.

8–75. Design a channel to deliver 10 m^3/s on a slope of 0.0009. The boundary material is moderately rounded with a $d_{75} = 1.5$ cm. Use a Manning $n = 0.020$.

8–76. Design a channel to deliver 40 m^3/s on a slope of 0.00085. The boundary material is slightly angular with a $d_{75} = 4$ cm. Use a Manning $n = 0.024$.

8–77. Design a channel to deliver 100 cfs on a slope of 0.00018. The boundary material is very angular sand with a median diameter of 2 mm. The Manning n is 0.018. Assume clear water.

8–78. Design a channel to carry 500 cfs with a slope of 0.00015. The boundary material is very rounded sand with a median diameter of 0.5 mm. The Manning n is 0.018. Assume a low content of fine sediment.

8–79. Design a channel to transport 12 m^3/s on a slope of 0.00021. The boundary material is moderately rounded very coarse sand. The Manning $n = 0.020$. Assume a high content of fine sediment.

8–80. Design a channel to transport 5 m^3/s on a slope of 0.00015. The boundary material is moderately rounded coarse sand. The Manning $n = 0.019$. Assume a low content of fine sediment.

Section 8–5

8–81. Determine the depth of scour in a contracted bridge section if the upstream depth is 10 ft and the width change is from 300 ft to 200 ft.

8–82. Determine the depth of scour in a contracted bridge section if the upstream depth is 2 m and the width change is from 50 m to 25 m.

8–83. Estimate the depth of the scour hole at a bridge pier that has a 2:1 elliptic nose and an upstream depth of 15 ft. The pier has a length of 30 ft, a width of 3 ft, and (1) is aligned with the flow, (2) is skewed at 15 deg to the flow, and (3) is skewed at 30 deg to the flow.

8–84. Estimate the depth of the scour hole at a bridge pier that has a 3:1 lenticular nose and an upstream depth of 4 m. The pier has a length of 8 m, a width of 1 m, and (1) is aligned with the flow, (2) is skewed at 30 deg to the flow, and (3) is skewed at 50 deg to the flow.

8–85. Plot a graph of pier scour versus upstream depth over the range 1 ft $< y_1 <$ 20 ft for a bridge pier that has a semicircular nose, a length of 40 ft, a width of 3 ft, and is aligned with the flow.

8–86. Plot a graph of pier scour versus alignment angle over the range of 0 deg $< \alpha <$ 60 deg for a bridge pier that has a semicircular nose, a length of 10 m, and a width of 0.7 m. The upstream depth is 3 m.

8–87. Use Section 8–4 on channel design to estimate the material size (d_{75}) required to prevent contraction scour in the channel bed of Prob. 8–81 if the channel slope is 0.00027. Assume that the water surface elevation remains constant through the contraction.

8–88. Use Section 8–4 on channel design to estimate the material size (d_{75}) required to prevent contraction scour in the channel bed of Prob. 8–82 if the channel slope is 0.0005. Assume that the water surface remains constant through the contraction.

8–89. Determine the useful life of the reservoir in Example 8–8 if the size of the sediment load is unusually fine.

8–90. Repeat Prob. 8–89 if the sediment size is unusually coarse.

8–91. Estimate the useful life (down to 20 percent of original capacity) of a reservoir that has an initial capacity of 60,000 ac-ft, an average inflow of 400 cfs, and an average annual sediment inflow of 300,000 tons. Use the Brune median curve and assume that the deposits have a density of 70 lb/ft^3.

8–92. Repeat Prob. 8–91 for a very coarse sediment load.

8–93. Repeat Prob. 8–91 if the average annual inflow is 600 cfs.

References

Graf, W. H., *Hydraulics of Sediment Transport,* New York: McGraw-Hill, 1971.

Leliavsky, S. L., *An Introduction to Fluvial Hydraulics,* New York: Dover, 1966.

Raudkivi, A. J., *Loose Boundary Hydraulics,* 2nd ed., New York: Pergamon, 1976.

Richards, K., *Rivers,* New York: Methuen, 1982.

Shen, H. W. (ed.), *River Mechanics,* 2 vol., Fort Collins, CO: Water Resources Publications, 1971.

Simons, D. B. and F. Senturk, *Sediment Transport Technology,* Fort Collins, CO: Water Resources Publications, 1977.

Vanoni, V. A. (ed.), *Sedimentation Engineering,* ASCE Manual No. 54, New York: ASCE, 1975.

Yalin, M. S., *Mechanics of Sediment Transport,* 2nd ed., New York: Pergamon, 1977.

chapter
9

Physical Modeling

Figure 9–0 Spillway model for Pine Flat Dam, Kings River, California. Tests were conducted to investigate the hydraulic performance of the spillway, spillway bucket and conduits (courtesy of the U.S. Army Engineer Waterways Experiment Station, Vicksburg, Mississippi).

9–1 INTRODUCTION

We have periodically discussed the concept of mathematical modeling. Although more and more types of hydraulic engineering problems are being solved by mathematical or computer modeling, there will always be the need for physical models as well. The physical model is generally expensive; however, there are many occasions when it is even more expensive to develop a computer model. In addition, there are many hydraulic structure and channel conditions in which the complex flow has yet to be duplicated satisfactorily by a mathematical model. In other cases, usually involving major projects, both physical and mathematical models have been used to increase confidence in the results. The Thames Barrier (Fig. 2.0), which protects the city of London from extreme tides, was carefully analyzed by both types of models in the design stages.

The concept of similitude will be introduced, and this concept, along with the dimensional analysis procedure illustrated in Section 2–3, will provide the basis for the laws of hydraulic modeling that follow. These laws will be applied to a variety of hydraulic engineering situations, culminating in their application to rivers with both fixed and movable beds.

9–2 SIMILITUDE AND THE MODELING LAWS

The key to proper modeling of hydraulic phenomena is the establishment of *similitude* between model and prototype. Three types of similitude are generally required; geometric, kinematic, and dynamic. Once similitude is achieved, there remains the task of correctly interpreting the model results as they apply to the prototype. The subject of similitude will be developed first, and after we have considered how to perform a model study, we will return to the equally important subject of applying the results to the prototype.

Geometric similitude is the requirement that the model be geometrically similar to the prototype. Strictly speaking, this means that each dimension in the model bears exactly the same relationship to the corresponding dimension in the prototype. We can define a length ratio, $L_r = L_m/L_p$, which must have the identical value for each of the physical dimensions involved. When the model is so constructed, it resembles the prototype in at least all the important aspects. Usually, the model will be at a reduced scale and the length ratio will be less than one. Once L_r is chosen, the model may be constructed. Note that the area ratio $A_r = A_m/A_p$ will vary as L_r^2, and the volume ratio will be proportional to L_r^3. We will see later in the chapter that there are conditions that require a relaxation of geometric similitude and a *distorted model* is necessary.

Kinematic similitude is the requirement that the model flow field must be similar to that of the prototype. This means that the velocities and accelerations are all scaled correctly. Finally, *dynamic similitude* requires that all forces and other dynamic quantities are properly scaled. We will return to the concept of kinematic similitude, but only after a discussion of dynamic similitude. Greater emphasis will be placed on the establishment of the principle of dynamic similitude, since once it is achieved and the forces are correctly scaled, kinematic similitude must also be satisfied.

In most hydraulic engineering problems requiring a model study, the significant forces can be divided into four groups: pressure forces, inertia forces, gravity forces, and viscous forces. Surface tension and elastic forces constitute other types of forces. The inertia force F_i can be considered the response of the system to the forces that are attempting to accelerate the flow. Through Newton's second law this may be expressed by

$$F_i = Ma \sim (\rho L^3)\left(\frac{L}{T^2}\right) = \rho\left(\frac{L}{T}\right)^2 L^2 \sim \rho V^2 L^2 \tag{9–1a}$$

Assuming that only the four types of forces are important, the remaining three forces may be written,

Pressure force: $\quad F_p \sim \Delta p A \sim \Delta p L^2$ $\qquad\qquad\qquad\qquad$ **(9–1b)**

Gravity force: $\quad F_g \sim \gamma \forall \sim \gamma L^3$ $\qquad\qquad\qquad\qquad\qquad$ **(9–1c)**

Viscous force: $\quad F_v \sim \tau A \sim \mu \dfrac{dV}{dy} A \sim \mu V L$ $\qquad\qquad\quad$ **(9–1d)**

The pressure force results from an unbalanced pressure distribution, the gravity force is proportional to the weight of the fluid, and the viscous force is a resistance force that attempts to decelerate the flow and dissipate energy. The dynamic viscosity is introduced in the last expression for force using Eq. 2–2.

We will assume for the present that all of these, and only these, types of forces are important to a given model study. Dynamic similitude requires that all of the force ratios be identical in the same way that the length ratios had to be constant for geometric similitude. That is

$$F_{pr} = F_{gr} = F_{vr} = F_{ir} \tag{9–2a}$$

Equation 9–2a when expanded out becomes

$$\frac{F_{pm}}{F_{pp}} = \frac{F_{gm}}{F_{gp}} = \frac{F_{vm}}{F_{vp}} = \frac{F_{im}}{F_{ip}} \tag{9–2b}$$

which may be rearranged and combined with Eqs. 9–1 to yield three independent equations. From the first and fourth terms of Eqs. 9–2,

$$\frac{F_{pm}}{F_{im}} = \frac{F_{pp}}{F_{ip}}$$

or

$$\left(\frac{\Delta p}{\rho V^2}\right)_m = \left(\frac{\Delta p}{\rho V^2}\right)_p \tag{9–3}$$

In a similar manner, the second and fourth terms become,

$$\left(\frac{V^2}{L\gamma/\rho}\right)_m = \left(\frac{V^2}{L\gamma/\rho}\right)_p$$

which leads to

$$\mathrm{Fr}_m = \mathrm{Fr}_p \tag{9-4}$$

Finally, the third and fourth terms combine to

$$\left(\frac{VL\rho}{\mu}\right)_m = \left(\frac{VL\rho}{\mu}\right)_p$$

or simply

$$\mathrm{Re}_m = \mathrm{Re}_p \tag{9-5}$$

The implication is that all four force ratios will be equal and dynamic similitude satisfied provided that Eqs. 9–3, 9–4, and 9–5 are satisfied. The consequences of dimensional analysis are useful at this point. Rather than rederive a general equation to suit our purposes, we can use a result from Example 2–3 in Chapter 2. In that example the Buckingham pi method led to the equation,

$$\frac{F_p}{\rho D^2 V^2} = \phi\left(\frac{y}{L}, \frac{\gamma D}{\rho V^2}, \frac{\mu}{VD\rho}\right)$$

Replacing the diameter D with the more general length dimension L, and introducing the Froude and Reynolds numbers, as was done in Example 2–3, leads immediately to

$$\frac{F_p}{\rho L^2 V^2} = \phi\left(\frac{y}{L}, \mathrm{Fr}, \mathrm{Re}\right)$$

In the more general modeling problem, many length dimensions $(x, y, z, \ldots)$ may be of importance. In the dimensional analysis process, each new independent length will result in an additional ratio of two lengths, so that a more general statement of the above equation is

$$\frac{F}{\rho L^2 V^2} = \phi\left(\frac{x}{L}, \frac{y}{L}, \frac{z}{L}, \ldots, \mathrm{Fr}, \mathrm{Re}\right)$$

The subscript on F has been dropped, reflecting a more general force than that of the original example. Since geometric similitude is assumed to be satisfied, the ratio of each of the various lengths to that of L will automatically assume a constant value between model and prototype. This is demonstrated as follows. From geometric similitude,

$$L_r = \frac{x_m}{x_p} = \frac{y_m}{y_p} = \frac{z_m}{z_p} = \frac{L_m}{L_p} = \text{constant}$$

Thus,

$$\left(\frac{x}{L}\right)_m = \left(\frac{x}{L}\right)_p, \qquad \left(\frac{y}{L}\right)_m = \left(\frac{y}{L}\right)_p$$

and

$$\left(\frac{z}{L}\right)_m = \left(\frac{z}{L}\right)_p$$

At this point, the length ratios will be replaced by the word geometry, implying that geometric similitude is satisfied,

$$\frac{F}{\rho L^2 V^2} = \phi(\text{Geometry, Fr, Re}) \qquad (9\text{–}6)$$

The term in the left-hand side of Eq. 9–6 is essentially identical to those in Eq. 9–3 since $F \sim \Delta p L^2$. Thus, while we have found that dynamic similitude required that Eqs. 9–3, 9–4, and 9–5 each be satisfied, dimensional analysis shows that only the Froude and Reynolds numbers must be held constant to achieve the desired result.

The dimensionless terms can be interpreted in different ways. Although obtained on the basis of force ratios, constant values of Fr and Re state how the model velocity must relate to the prototype velocity to ensure kinematic similitude. They may accordingly be recognized as the necessary conditions for kinematic similitude. Regardless of interpretation, if the model is operated so that both $\text{Fr}_m = \text{Fr}_p$ and $\text{Re}_m = \text{Re}_p$, the left side of Eq. 9–6 must be constant as well. This may be expressed by

$$\left(\frac{F}{\rho L^2 V^2}\right)_m = \left(\frac{F}{\rho L^2 V^2}\right)_p$$

which may be written as either

$$\frac{F_m}{F_p} = \left(\frac{\rho_m}{\rho_p}\right)\left(\frac{L_m}{L_p}\right)^2\left(\frac{V_m}{V_p}\right)^2 \qquad (9\text{–}7a)$$

or

$$F_r = \rho_r L_r^2 V_r^2 \qquad (9\text{–}7b)$$

The application of these equations will be considered subsequently. Note, however, that the procedure whereby F and Δp were related can be used to introduce other dynamic and kinematic terms as needed. For instance, the power P is related to F by $p \sim FV$ and therefore,

$$P_r = F_r V_r \qquad (9\text{–}8)$$

Likewise, we can obtain the acceleration ratio

$$a_r = L_r/t_r^2 \qquad (9\text{–}9)$$

the time ratio

$$t_r = L_r/V_r \qquad (9\text{–}10)$$

and the discharge ratio

$$Q_r = V_r L_r^2 \tag{9–11}$$

Either the gravity or the viscous forces will predominate in many hydraulic model studies. In such situations, only the one dimensionless parameter must be held constant rather than both. This is extremely fortuitous because, and this will be examined later, rarely is it physically possible to design and operate a model so that both Fr and Re are constant.

The modeling procedure can now be summarized as follows. The model is constructed to a convenient scale. This involves a number of constraints that will be examined in the next sections. The model is then operated so that either Fr or Re is held constant as appropriate for the type of forces involved. The various modeling laws are then used to calculate the prototype results.

EXAMPLE 9–1

Assume that geometric similitude is satisfied and that the length ratio is L_r. Determine (1) how the force ratio F_r must depend on the L_r, and (2) how the power ratio P_r must depend on the discharge ratio Q_r. In part (1), evaluate the relationship assuming that gravity forces predominate and hence the Froude law applies, and then repeat assuming viscous forces and the Reynolds law are appropriate.

Solution

(1) From Eq. 9–7b the force ratio is

$$F_r = \rho_r L_r^2 V_r^2$$

This equation is true in general, independent of whether gravity, viscous or other forces predominate. If the Froude law applies, then

$$\left(\frac{V}{\sqrt{Lg}}\right)_m = \left(\frac{V}{\sqrt{Lg}}\right)_p$$

or for $g_r = 1$,

$$V_r = L_r^{1/2}$$

Substituting this equation into that for F_r yields,

$$F_r = \rho_r L_r^3$$

Note that this equation is no longer general, but is only applicable when gravity forces predominate. In the same way, if viscous forces are more important, then the Reynolds number requires that

$$\left(\frac{VL}{\nu}\right)_m = \left(\frac{VL}{\nu}\right)_p$$

or

$$V_r = \frac{\nu_r}{L_r}$$

which when substituted into the general force equation results in

$$F_r = \frac{\rho_r L_r^2 \nu_r^2}{L_r^2} = \rho_r \nu_r^2 = \mu_r \nu_r$$

(2) The power ratio is $P_r \sim F_r V_r$. If this is introduced into the general equation for the force ratio, a general equation for power results,

$$P_r = \rho_r L_r^2 V_r^3$$

From Eq. 9–11,

$$V_r = \frac{Q_r}{L_r^2}$$

Substituting this equation into the equation for P_r gives

$$P_r = \frac{\rho_r Q_r^3}{L_r^4}$$

This is the required answer for P_r. This equation is also general, but in applying this equation Q_r would differ, depending on whether the Froude law or Reynolds law were used.

9–3   APPLICATIONS OF THE MODELING LAWS

The fundamental modeling laws were set forth in the previous section. At this point the application of these various laws to hydraulic engineering problems will be demonstrated with a number of illustrative examples. Froude number modeling will be considered first. This will be followed by applications of the Reynolds number law, and the section will conclude with a consideration of those cases where both Fr and Re are significant. The additional complications associated with river modeling will be reserved for the following section.

Froude Number Modeling

Froude number modeling is necessary when gravitational forces are the predominant forces. This will frequently be the case when a free surface is present or when gravity is the driving force. The general modeling laws were developed in Section 9–2, and specific equations invoking the additional constraint of the Froude number were introduced in Example 9–1. The two examples that follow will illustrate different aspects of the problem.

EXAMPLE 9-2

A tidal basin with a scale of 1:100, that is, $L_r = 0.01$, has been constructed to study various tidal phenomena. If a velocity of 0.15 m/s is observed in the model, what is the magnitude of the corresponding prototype velocity? What is the time period in the model operation that is equivalent to one day in the prototype? If a high tide results in a force of 17 N on a model breakwater, determine the corresponding prototype force.

Solution

Assuming a constant gravitational acceleration, the Froude law may be written

$$\left(\frac{V}{\sqrt{L}}\right)_m = \left(\frac{V}{\sqrt{L}}\right)_p$$

Rearranging and substituting the given velocity and length ratio gives the prototype velocity

$$V_p = V_m \left(\frac{L_p}{L_m}\right)^{1/2} = (0.15)(100)^{1/2} = 1.5 \text{ m/s}$$

Time can be entered into the Froude law through Eq. 9–10 in the following manner. The above Froude law may first be written as

$$V_r = L_r^{1/2}$$

but from Eq. 9–10,

$$t_r = \frac{L_r}{V_r}$$

Combining,

$$t_r = L_r^{1/2}$$

or

$$t_m = t_p \left(\frac{L_m}{L_p}\right)^{1/2}$$

Substituting the numerical values

$$t_m = (1)\left(\frac{1}{100}\right)^{1/2} = 0.1 \text{ days}$$

and thus, 0.1 days or 2.4 hours in the model is equivalent to one day in the prototype.

The general form of the force ratio was given previously in Eqs. 9–7. Upon substituting the Froude law, as in Example 9–1, this becomes,

$$F_r = \rho_r L_r^3$$

The density ratio will equal the salt water specific gravity of 1.03, and the prototype force is calculated to be

$$F_p = F_m \left(\frac{\rho_p}{\rho_m}\right)\left(\frac{L_p}{L_m}\right)^3 = (17)(1.03)(100)^3 = 1.75 \times 10^7 \text{ N} = 17.5 \text{ MN}$$

This example, which deals with tidal phenomena, requires only kinematic similitude to obtain the first two required answers. However, since the solution is based on a constant Froude number, dynamic similitude will be satisfied as well. Thus, the third part of the example can be completed without any additional concern about similitude. Every step was included to ensure that the modeling process is understood. The remaining examples in this section will be presented with somewhat less detail.

EXAMPLE 9–3

A model spillway is built to a scale of 1:64. If the prototype spillway has a design discharge of 10,000 cfs, determine the model discharge required for dynamic similitude. Under conditions of similitude, a hydraulic jump at the base of the model spillway is found to dissipate 0.011 hp. What is the rate of energy dissipation below the prototype spillway?

Solution

Substitution of Eq. 9–11 for the discharge ratio leads to

$$Q_m = Q_p \left(\frac{L_m}{L_p}\right)^{5/2} = (10,000)\left(\frac{1}{64}\right)^{5/2} = 0.305 \text{ cfs}$$

The power ratio was expressed in terms of the discharge ratio in Example 9–1. Using this result

$$P_p = P_m \left(\frac{\rho_p}{\rho_m}\right)\left(\frac{Q_p}{Q_m}\right)^3 \left(\frac{L_m}{L_p}\right)^4$$

Since the density ratio is unity, this leads to

$$P_p = (0.011)\left(\frac{10{,}000}{0.305}\right)^3 \left(\frac{1}{64}\right)^4 = 23{,}100 \text{ hp}$$

Note that if we start with Eq. 9–8 for the power ratio and substitute

$$V_r = L_r^{1/2}$$

from the Froude law, we get an alternative equation

$$P_r = \rho_r L_r^{7/2}$$

and this gives

$$P_p = (0.011)(64)^{7/2} = 23{,}100 \text{ hp}$$

which is the same result as before.

Reynolds Number Modeling

Reynolds number modeling becomes important when viscous forces predominate. This will frequently be the case in confined flows and when resistance plays an important role in determining the shape of the flow field. In flows in which the effects of gravity are relatively insignificant, operation of the geometrically similar model so that the Reynolds number has the same value as in the prototype ensures dynamic similitude.

A difficulty often occurs with Reynolds number modeling. Writing $\text{Re}_m = \text{Re}_p$ as was done in Example 9–1 leads to

$$V_r = \frac{\nu_r}{L_r}$$

If the same fluid is used for both model and prototype, then

$$\frac{V_m}{V_p} = \frac{L_p}{L_m}$$

and the smaller the size of the model, the higher must be the model velocity. This often creates insurmountable operational difficulties that can only be overcome by using a different fluid for the model. If this is done,

$$\frac{V_m}{V_p} = \frac{L_p}{L_m} \frac{\nu_m}{\nu_p}$$

The model velocity can be reduced by using a fluid with a low kinematic viscosity, but the constraints of the preceding equations and the difficulty of finding a fluid with suitable viscosity often mean that a small-scale model will not be satisfactory.

EXAMPLE 9–4

The submerged operation of an underwater vehicle is to be modeled with a 1:10 scale model. If the design velocity is 0.9 m/s, determine the proper speed at which the model is to be tested. Under test conditions, the drag force on the model is measured to be 400 N. Determine the drag force on the prototype and the power requirement for both the model and prototype vehicles. Assume that seawater is used for both the test and actual operating conditions.

Solution

Since the viscosity ratio is unity, the model velocity is

$$V_m = V_p \left(\frac{L_p}{L_m} \right) = (0.9)(10) = 9 \text{ m/s}$$

Substitution of the Reynolds number law into Eq. 9–7a gives

$$F_p = F_m \left(\frac{L_p}{L_m} \right)^2 \left(\frac{V_p}{V_m} \right)^2 = F_m \left(\frac{L_p}{L_m} \right)^2 \left(\frac{L_m}{L_p} \right)^2$$

and $F_p = F_m = 400$ N.
The power requirement for the model is

$$P_m = F_m V_m = (400)(9) = 3600 \text{ W}$$

while for the prototype it is only

$$P_p = F_p V_p = (400)(0.9) = 360 \text{ W}$$

which demonstrates some of the difficulty of Reynolds number modeling at reduced scale, particularly if the same fluid is used throughout.

If the underwater device were to operate close to the surface so that surface waves were generated, the Froude number would have to be considered as well. With adequate submergence, however, gravity has no effect on the flow pattern or on the resulting resistance to the motion.

EXAMPLE 9–5

In the interest of safety and convenience, a new valve design to be used in the pumping of gasoline is to be tested using water at 25°C. The test is to be conducted using a one-fourth-size model. If the design discharge for the gasoline is 0.02 m³/s, determine the water discharge for dynamic similitude. If the measured pressure drop through the model valve at design conditions is 20 kN/m², what is the corresponding pressure drop in the actual valve? Assume ρ = 680 kg/m³ and ν = 4.6 × 10⁻⁷ m²/s for gasoline.

Solution

At 25°C, ν = 8.97 × 10⁻⁷ for water. Combining Eq. 9–11 with $\text{Re}_m = \text{Re}_p$ gives the following equation for the model discharge,

$$Q_m = Q_p\left(\frac{L_m}{L_p}\right)\left(\frac{\nu_m}{\nu_p}\right) = (0.02)\left(\frac{1}{4}\right)\left(\frac{8.97 \times 10^{-7}}{4.6 \times 10^{-7}}\right) = 0.00975 \text{ m}^3/\text{s}$$

The pressure difference, which has not been considered previously, may be substituted into Eq. 9–7a by first noting that $\Delta p \sim F/L^2$, whereupon

$$\Delta p_p = \Delta p_m\left(\frac{\rho_p}{\rho_m}\right)\left(\frac{V_p}{V_m}\right)^2$$

Replacing the velocities with Eq. 9–11 gives,

$$\Delta p_p = \Delta p_m\left(\frac{\rho_p}{\rho_m}\right)\left(\frac{Q_p}{Q_m}\right)^2\left(\frac{L_m}{L_p}\right)^4$$

Upon substituting the numerical values we get

$$\Delta p_p = (20)\left(\frac{680}{1000}\right)\left(\frac{0.02}{0.00975}\right)^2\left(\frac{1}{4}\right)^4 = 0.224 \text{ kN/m}^2$$

Combined Froude Number and Reynolds Number Modeling

If both gravitational and viscous forces are important, then dynamic similitude requires that both Fr and Re be held constant. If one parameter is considerably more important than the other, the one of less significance can be ignored. The extent that the ignored parameter affects the results is sometimes called a scale effect. For example, if Fr is held constant, Re will usually be larger in the prototype than in the model, and this will usually have some influence on the flow. When interpreting the model results, the engineer must at least be aware of this possibility.

We will first examine the general impasse that results when both parameters are held constant. This will be followed by an example that circumvents the difficulty in the important area of modeling ship hydrodynamics.

If we assume that $\text{Fr}_m = \text{Fr}_p$ and $\text{Re}_m = \text{Re}_p$, we may solve for the velocity ratio in each case. This gives the two equations,

$$V_r = L_r^{1/2}$$

and

$$V_r = \frac{v_r}{L_r}$$

Equating the velocity ratios,

$$v_r = L_r^{3/2} \qquad\qquad\qquad\qquad \textbf{(9–12a)}$$

or

$$\frac{v_m}{v_p} = \left(\frac{L_m}{L_p}\right)^{3/2} \qquad\qquad\qquad\qquad \textbf{(9–12b)}$$

Because of the added constraint we have imposed on the modeling process, we have lost one degree of freedom and the length ratio or scale is no longer arbitrary. Rather, it now depends on the choice of fluids as reflected by the ratio of viscosities. In hydraulic engineering studies, water will frequently be the fluid associated with the prototype. As a practical and economical matter, water may be the only choice for the model as well. But Eq. 9–12 reveals that if both model and prototype use the same fluid, then L_r must equal unity, or in other words, the model must be full size. On the other hand, if we select a length ratio, the viscosity ratio must vary as the three-halves power of the length ratio and rarely can any fluid, economic or otherwise, be found with a low enough viscosity to satisfy Eq. 9–12.

Only in very unusual circumstances can Fr and Re be held constant simultaneously. The engineer must find ways around this problem. The example that follows offers one such approach.

EXAMPLE 9–6

It is desired to model a 400-ft-long ship that has a surface area of 27,000 ft^2 below the water line. The design velocity is 30 ft/s and a 16-ft-model is proposed. Both wave forces (gravity) and surface friction forces (viscous) are important. Attempt to satisfy both the Froude and Reynolds laws simultaneously. Failing that, use the Froude law so as to properly model the wave forces and separately calculate the surface friction for both ships.

For surface drag, use the following equation from fluid mechanics:

$$F_f = C_f A \frac{\rho V^2}{2} \qquad\qquad\qquad\qquad \textbf{(9–13)}$$

In this equation, A is the surface area below the water line and C_f is a surface friction coefficient that is a function of the Reynolds number as given on page 348.

Laminar boundary layer (Re $\leqslant$ 500,000)

$$C_f = \frac{1.328}{Re^{1/2}} \tag{9-14}$$

Transition (500,000 < Re < 10^7)

$$C_f = \frac{0.455}{(\log Re)^{2.58}} - \frac{1700}{Re} \tag{9-15}$$

Fully turbulent boundary layer (Re $\geqslant$ 10^7)

$$C_f = \frac{0.455}{(\log Re)^{2.58}} \tag{9-16}$$

where Re = VL/ν (L = ship length).

When the model ship is towed according to the Froude law, the total force on the model is measured to be 6.7 lb. The prototype will operate in sea water (ρ = 1.99 slug/ft^3 and ν = 1.2 $\times$ 10^{-5} ft^2/s) while for convenience the model fluid is freshwater (ρ = 1.94 slug/ft^3 and ν = 1.05 $\times$ 10^{-5} ft^2/s).

Solution

We will first attempt to hold both Fr and Re constant. From Eq. 9–12b

$$\nu_m = \nu_p\left(\frac{L_m}{L_p}\right)^{3/2} = (1.2 \times 10^{-5})\left(\frac{16}{400}\right)^{3/2} = 9.60 \times 10^{-8}\ \text{ft}^2/\text{s}$$

There is no reasonable choice of liquid with this low a kinematic viscosity, and therefore both parameters cannot be held constant simultaneously. As an alternative, the Froude law will be used and only the wave (or gravity) forces will be scaled. On this basis, the model velocity is

$$V_m = V_p\left(\frac{L_m}{L_p}\right)^{1/2} = (30)(1/25)^{1/2} = 6\ \text{ft/s}$$

At that velocity

$$Re_m = \frac{(6)(16)}{1.05 \times 10^{-5}} = 9.14 \times 10^6$$

and the drag coefficient from Eq. 9–15 becomes

$$C_f = \frac{0.455}{(\log 9.14 \times 10^6)^{2.58}} - \frac{1700}{9.14 \times 10^6} = 0.00286$$

The below water line surface area is

$$A_m = (27,000)(\tfrac{1}{25})^2 = 43.2\ \text{ft}^2$$

and therefore the friction drag on the model can be calculated from Eq. 9–13,

$$F_{fm} = \frac{(0.00286)(43.2)(1.94)(6)^2}{2} = 4.31 \text{ lb}$$

Thus, the wave force on the model is

$$F_{wm} = F_m - F_{fm} = 6.7 - 4.31 = 2.39 \text{ lb}$$

The wave force on the prototype can now be found from the modeling equation,

$$F_{wp} = F_{wm}\left(\frac{L_p}{L_m}\right)^3 = (2.39)(25)^3 = 37,340 \text{ lb}$$

The friction force on the prototype can be found in the same way as it was for the model. At 30 ft/s, the Reynolds number is

$$\text{Re}_p = \frac{(30)(400)}{1.2 \times 10^{-5}} = 10^9$$

which leads to a drag coefficient of

$$C_f = \frac{0.455}{(\log 10^9)^{2.58}} = 0.00157$$

Whereupon the friction force on the prototype is

$$F_{fp} = \frac{(0.00157)(27,000)(1.99)(30)^2}{2} = 37,960 \text{ lb}$$

Finally, the total force on the prototype ship is the sum of the gravity and friction forces, or

$$F_p = F_{wp} + F_{fp} = 37,340 + 37,960 = 75,300 \text{ lb}$$

This is the required answer. Going a step further, the power requirement (in horsepower) for the prototype is

$$P_p = \frac{FV}{550} = \frac{(75,300)(30)}{550} = 4110 \text{ hp}$$

9–4 THE RIVER MODEL

The study of river behavior is an area in which hydraulic modeling has been used very successfully. Not only have hydraulic models been used to model the interaction of engineering works such as harbors, bridges, diversion structures, jetties, and dikes, but models of significant lengths of a river system have also permitted the study of flood waves, and the use of these models as flood predictors has been responsible for the saving of countless lives and untold millions of dollars in property.

In this section the modeling procedures will be considered for both fixed-bed and movable-bed river models.

Gravity and resistance forces are of consequence in either case. However, from our experience of the preceding section we can anticipate the impossibility of holding both Fr and Re constant simultaneously. As a consequence, we will disregard the Reynolds number for the present, considering it only to the extent that we ensure that the flow in both model and prototype are fully turbulent. It will usually be satisfactory to check that $\mathrm{Re}_m = VR/\nu$ is greater than, say, 1500 or 2000. Since the Reynolds number is ignored, another relationship must be found to handle the modeling of the resistance.

Fixed-Bed River Models

The fixed-bed model consists of a formed or molded channel, often constructed of plaster or concrete. The channel is shaped to resemble a portion of the river channel according to the scaling laws that follow. Although satisfactory modeling is far easier to achieve with the fixed-bed model than with the movable-bed model, nevertheless there are still a number of complicating considerations that set the model procedures apart from the modeling concepts treated previously.

As usual, similitude with respect to the effect of gravity requires that Fr be held constant. On this basis

$$V_r = L_r^{1/2} \tag{9–17}$$

Resistance effects can be introduced through the Manning equation. Since the Manning equation may be written for both the model and the prototype, it may also be expressed in terms of the various ratios. Thus, the velocity ratio may be written in either U.S. customary or SI units as

$$V_r = \frac{R_r^{2/3} S_r^{1/2}}{n_r} \tag{9–18}$$

In a geometrically scaled model, which will now be called an undistorted model, the hydraulic radius R can be treated the same as any length and R_r will equal L_r. Further, $S_r = 1$, and the above equations combine to yield

$$n_r = \frac{L_r^{2/3}}{L_r^{1/2}} = L_r^{1/6} \tag{9–19}$$

The use of two modeling relationships has again reduced the degrees of freedom so that the length ratio, rather than being arbitrary, is now constrained by n_r. Not only is this restrictive, but for relatively smooth prototype channels it may be impossible to satisfy Eq. 9–19 utilizing a reasonable length ratio. For example, if the river has an $n_p = 0.020$ and smooth concrete is used with $n_m = 0.012$, Eq. 9–19 requires that $L_r = 0.047$ or approximately 1:20. Normally, in a river model a much greater scale ratio is required. This leads to a second problem with an undistorted model. The width of a 400-ft-wide river could be modeled with a channel width of 2 ft if

$L_r = 1/200$, but a depth of 10 ft would reduce to only 0.05 ft in the model. Flow at this depth will be barely turbulent at best and shallower portions of the model channel will probably be laminar. In addition, the normally neglected surface tension effects may also influence the flow and satisfactory similitude becomes unlikely.

Both of the above problems can be circumvented through the use of a distorted model. Instead of using a single length ratio as required previously, different length ratios are chosen for the vertical and horizontal scales, thereby distorting the model. While this procedure may be expected to introduce some scale effects, it has been found that kinematic similitude is not seriously affected and the distorted model provides useful results.

The usual distortion is an exaggeration of the vertical scale relative to the horizontal. This causes a relative increase in depth, which eliminates the depth problem discussed above and in the process, perhaps surprisingly, solves the resistance problem as well. The vertical length ratio will be called Y_r. Once this scale ratio is determined all vertical dimensions, such as depths, will be scaled according to this ratio. Similarly, the horizontal length ratio will be X_r, which will provide the basis for the model channel widths as well as the longitudinal or downriver dimensions. The vertical exaggeration that results from the distorted model is shown in Fig. 9–1. The classic example is the Mississippi River model constructed by the U.S. Army Corps of Engineers Waterways Experiment Station that uses scale ratios of $Y_r = \frac{1}{100}$ and $X_r = \frac{1}{2000}$. Thus, a basin length of 2000 miles is duplicated by a model length of one mile at their outdoor facility. A portion of this model is shown in Fig. 9–2.

The various modeling ratios established in Section 9–2 can be readily modified for the distorted model. The velocity ratio based on Fr was $V_r = L_r^{1/2}$. Since the Froude number was originally defined using a depth dimension, the vertical scale is more appropriate for the distorted model, and we may write

$$V_r = Y_r^{1/2} \tag{9–20}$$

The acceleration ratio was given in Eq. 9–9. This is rarely important in the river model, but if necessary L_r would be defined in terms of X_r or Y_r, depending on whether the acceleration to be modeled was in the horizontal or vertical direction. The time ratio was first given in Eq. 9–10. In river modeling, time is generally considered with respect to distances traversed in the flow direction, and Eq. 9–10 usually becomes

$$t_r = \frac{X_r}{V_r}$$

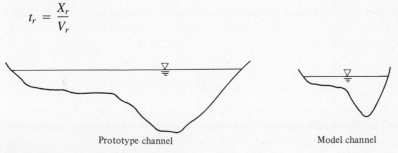

Prototype channel   Model channel

Figure 9–1 Distorted river model.

Figure 9–2 A portion of the Mississippi Basin Model located at the Jackson installation of the Waterways Experiment Station. This is the largest small-scale working model in the world. It covers 200 acres and reproduces 1,250,000 square miles of the Mississippi Basin. (Courtesy of the U.S. Army Engineer Waterways Experiment Station, Vicksburg, Mississippi).

When combined with V_r from the Froude number, we get

$$t_r = \frac{X_r}{Y_r^{1/2}} \tag{9-21}$$

The discharge is the product of the velocity and cross-sectional area. Consequently, in the discharge ratio given in Eq. 9–11, the cross-sectional area represented by L_r^2 must by replaced by the product of Y_r and X_r. When V_r is replaced by Eq. 9–20 as well, the discharge ratio becomes

$$Q_r = V_r X_r Y_r = X_r Y_r^{3/2} \tag{9-22}$$

The channel slope is based on the ratio of a vertical dimension to a horizontal dimension. Consequently, S_r will not be unity as it was for the undistorted model, but will be given by

$$S_r = \frac{Y_r}{X_r} \tag{9-23}$$

Because of the vertical exaggeration, the model slope will become greater than the prototype slope. This has the effect of tipping the model relative to the prototype and is largely responsible for the success of the distorted model. Since S_r directly measures the amount of distortion, it is also called the distortion ratio.

In addition, S_r will not drop out of the Manning relationship, Eq. 9–18, as it did previously, and the resistance equation now becomes

$$V_r = \frac{R_r^{2/3} Y_r^{1/2}}{n_r X_r^{1/2}}$$

Upon substituting Eq. 9–20 and solving for n_r, we get

$$n_r = \frac{R_r^{2/3}}{X_r^{1/2}} \tag{9–24}$$

In the event of a very wide channel in which the hydraulic radius equals the depth, the equation simplifies to

$$n_r = \frac{Y_r^{2/3}}{X_r^{1/2}} \tag{9–25}$$

Equation 9–25 can generally not be used in actual model studies, and Eq. 9–24 is unfortunately an inconvenient expression since it is not directly related to X_r and Y_r.[1] To obtain n_r, the actual model and prototype values of R must first be calculated from the respective channel geometries, as only then can R_r be calculated. Regardless, it is reasonable to assume that R_r is more closely related to Y_r than it is to X_r, since in many natural channels R is only a little less than the average depth. We can at least draw the qualitative conclusion that since Y_r (and therefore R_r) is greater than X_r, it follows that n_r is greater than one. This means that the model must be rougher than the prototype, which is just the opposite of the undistorted case. To illustrate this, assume that Eq. 9–25 is appropriate and substitute the scale ratios used for the Mississippi model. With $Y_r = \frac{1}{100}$ and $X_r = \frac{1}{2000}$, Eq. 9–25 becomes

$$n_r = \frac{(1/100)^{2/3}}{(1/2000)^{1/2}} = 2.08$$

and the model n value must be over twice that of the prototype Manning n.

Since the construction material for the model is likely to have a Manning n value in the range of 0.012 to 0.015, artificial roughness elements must be added. But this does not present any difficulty, as the model roughness can be increased almost indefinitely as needed. Although Eqs. 9–24 and 9–25 can be used as a guide, the usual procedure is to place roughness elements throughout the model on a primarily trial-and-error basis until past events such as major floods can be reproduced satisfactorily in the model. The roughness elements themselves can take many different forms, but small pieces of thin aluminum plate or wire mesh have often been used.

In summary, we see that the use of the distorted model overcomes the disadvantages of the undistorted model. The flexibility of independently choosing the vertical scale provides a means of achieving an adequate model depth so as to avoid severe scale problems. In addition, the exaggerated vertical scale results in a model slope

---

[1] In the problems at the end of the chapter, R_r will usually be assumed equal to Y_r to simplify the calculations.

in excess of the prototype slope, which in turn requires an increased roughness to retard the flow and maintain similitude. Instead of attempting to create an extremely smooth model as required in the undistorted case, the simpler task of adding roughness to the model is all that is necessary. It should be pointed out, however, that the entire process of calibration, verification, and application requires a considerable amount of patience, skill, and experience.

EXAMPLE 9–7

A river model is to be constructed using $Y_r = \frac{1}{100}$ and $X_r = \frac{1}{500}$, which provides a vertical exaggeration of 5 to 1. The prototype channel has an average depth of 5 m, an average width of 120 m, a discharge of 1500 m³/s, a slope of 0.00032, and a Manning $n = 0.021$.

Determine the corresponding geometry of the model channel and the required model discharge. Verify that turbulent flow will occur in the model, and calculate both model and prototype Froude numbers (based on the average depth).

Solution

Using the vertical scale, the model depth will be

$$y_m = y_p Y_r = (5)\left(\frac{1}{100}\right) = 0.05 \text{ m}$$

while from the horizontal scale

$$b_m = b_p X_r = (120)\left(\frac{1}{500}\right) = 0.24 \text{ m}$$

The slope ratio is determined from Eq. 9–23

$$S_r = \frac{Y_r}{X_r} = \frac{(1/100)}{(1/500)} = 5$$

and the model slope must be

$$S_m = (5)(0.00032) = 0.0016$$

Thus, the model channel has an average depth of 0.05 m, an average width of 0.24 m, and a slope of 0.0016.

Using Eq. 9–22, we find that the necessary model discharge is

$$Q_m = (1500)\left(\frac{1}{500}\right)\left(\frac{1}{100}\right)^{3/2} = 0.003 \text{ m}^3/\text{s}$$

Therefore, the average velocity in the model is

$$V_m = \frac{(0.003)}{(0.05)(0.24)} = 0.25 \text{ m/s}$$

and the corresponding Reynolds number becomes

$$\text{Re}_m = \frac{(0.25)(0.05)}{1.1 \times 10^{-6}} = 11{,}400$$

using 1.1×10^{-6} m^2/s as the kinematic viscosity. This ensures that turbulent flow will prevail in the model. Using the same velocity, the model Froude number is

$$\text{Fr}_m = \frac{0.25}{\sqrt{(9.81)(0.05)}} = 0.357$$

The average prototype velocity is

$$V_p = \frac{1500}{(5)(120)} = 2.5 \text{ m/s}$$

which gives a prototype Froude number of

$$\text{Fr}_p = \frac{2.5}{\sqrt{(9.81)(5)}} = 0.357$$

In conclusion, note that the two Froude numbers are identical as expected. This demonstrates that the similitude requirement was not lost in the channel distortion process.

Movable-Bed River Models

Just as the sediment transport equations have a much lower accuracy than the fixed-bed hydraulic equations, movable-bed similitude is much more difficult to establish than the rigid channel counterpart. It also follows naturally that the results of such a study are likely to be far less reliable. Nevertheless, many projects have been successfully designed on the basis of such a study, and it is appropriate for us to discuss at least some aspects of the problem. There are two major problems that complicate movable-bed modeling. The first is that the roughness is no longer independent as in the fixed-bed model, but rather depends on the flow itself. A change in discharge may result in a change in the bed form, say from dunes to a flat bed, which often results in a significant change in resistance and Manning n.

The second problem associated with movable-bed modeling is that the time scale of the sediment movement is different from that of the water itself. Whereas the suspended sediment movement has a time scale similar to the transporting water, other factors such as bed load, formation and movement of bed forms, and scour and deposition all occur at dissimilar rates. The result of these complicating factors is at least a lower degree of similitude between model and prototype. Often, the model results can only be used in a qualitative way to predict general trends in the prototype. To study movable-bed similitude, we will first examine one more-or-less theoretical approach. This will be followed by a look at the way one research facility has successfully operated a movable-bed model.

The sediment similitude will be based on the Shields diagram given in Fig. 8–10.[2] The Shields curve, which is labeled on the diagram, represents the limiting condition between transport and no transport. The two primary parameters τ_* and Re_* can be calculated from the hydraulic and sediment characteristics of a given flow. The farther above the Shields curve (as measured along the diagonal lines) the operating point lies, the greater will be the transport. It will be assumed that a point on the diagram fully describes the characteristics of a given flow with respect to the sediment transport, bed forms, and resistance. This provides a basis for sediment similitude that we assume will be satisfied if model and prototype share a common point on the diagram. That is, sediment similitude will be assumed satisfied it $\tau_{*m} = \tau_{*p}$, or

$$\left[\frac{\tau_0}{(\gamma_s - \gamma)d_s} \right]_m = \left[\frac{\tau_0}{(\gamma_s - \gamma)d_s} \right]_p \tag{9-26}$$

and in addition $Re_{*m} = Re_{*p}$, or

$$\left(\frac{u_* d_s}{\nu} \right)_m = \left(\frac{u_* d_s}{\nu} \right)_p \tag{9-27}$$

where $u_* = \sqrt{\tau_0/\rho} = \sqrt{gRS}$. To simplify the notation, Eq. 9–26 will be rewritten in terms of the sediment specific gravity and further simplified by introducing α for the submerged specific gravity, leading to

$$\left[\frac{\tau_0}{\gamma(\gamma_s/\gamma - 1)d_s} \right]_m = \left[\frac{\tau_0}{\gamma(\gamma_s/\gamma - 1)d_s} \right]_p$$

and finally

$$\left(\frac{\tau_0}{\gamma\alpha d_s} \right)_m = \left(\frac{\tau_0}{\gamma\alpha d_s} \right)_p$$

where $\alpha = (\gamma_s/\gamma) - 1$. Therefore, in terms of model-to-prototype ratios, Eq. 9–26 becomes

$$\frac{\tau_{0r}}{\gamma_r \alpha_r d_{sr}} = 1$$

On the basis of Eq. 7–2

$$\tau_{0r} = \gamma_r R_r S_r = \gamma_r R_r \frac{Y_r}{X_r}$$

and

---

[2]H. A. Einstein, and N. Chien, "Similarity of Distorted River Models with Movable Beds," *Trans. ASCE*, Vol. 121, 1956. Also see the reference to Henderson at the end of the chapter.

$$\frac{R_r Y_r}{\alpha_r d_{sr} X_r} = 1 \tag{9-28}$$

Squaring Eq. 9–27 and again introducing the appropriate ratios gives

$$\frac{R_r S_r d_{sr}^{\,2}}{v_r^{\,2}} = \frac{R_r Y_r d_{sr}^{\,2}}{X_r v_r^{\,2}} = 1$$

Assuming that $v_r = 1$, we get

$$\frac{R_r Y_r d_{sr}^{\,2}}{X_r} = 1 \tag{9-29}$$

Equations 9–28 and 9–29 are the two modeling equations that must be satisfied so that both model and prototype will plot at the same point on the Shields diagram. From the n versus d_s equations in Table 7–1, we get the additional equation,

$$n_r = d_{sr}^{\,1/6} \tag{9-30}$$

Whereas the original equations for n in Table 7–1 assumed a flat bed, the above assumption of similitude establishes equivalent bed form and resistance conditions. Thus, the ratios in Eq. 9–30 should be applicable under all conditions. If we now require the same hydraulic similitude conditions as with a fixed-bed model, namely similitude based on the Froude law and the Manning equation, then Eq. 9–24 must be satisfied. If Eq. 9–30 is now combined with Eq. 9–24 we get,

$$d_{sr}^{\,1/6} = \frac{R_r^{\,2/3}}{X_r^{\,1/2}} \tag{9-31}$$

The three equations (Eqs. 9–28, 9–29, and 9–31) contain four independent ratios, α_r, d_{sr}, X_r, and Y_r. As with the fixed-bed model, R_r is not independent, but depends on the X_r and Y_r. The difficulty, as discussed previously, is that R_m and R_p must be determined directly from each of the two channel cross sections. The first two ratios pertain to the sediment material and size, and the remaining two ratios provide the model size and distortion. The effort required to satisfy all three equations is somewhat laborious, but $R_r Y_r / X_r$ can be eliminated between Eq. 9–28 and Eq. 9–29 with the result that

$$\alpha_r d_{sr} = \frac{1}{d_{sr}^{\,2}}$$

or

$$d_{sr} = \alpha_r^{\,-1/3} \tag{9-32}$$

If the hydraulic engineer chooses a given model bed material (so that α_r is established), then Eq. 9–32 determines the required sediment size for the model. At this point, the three modeling equations can be used to determine the vertical and horizontal length scales. Note that neither of the length scales is arbitrary. In addition, according to the inverse relationship of Eq. 9–32, if the model sediment is lighter than that of the prototype, the model sediment size must be larger than the size of the prototype sediment.

If a wide channel is assumed so that $R_r = Y_r$, then the three modeling equations reduce to

$$\frac{Y_r{}^2}{\alpha_r d_{sr} X_r} = 1 \qquad (9\text{--}33)$$

$$\frac{Y_r{}^2 d_{sr}{}^2}{X_r} = 1 \qquad (9\text{--}34)$$

and

$$d_{sr}{}^{1/6} = \frac{Y_r{}^{2/3}}{X_r{}^{1/2}} \qquad (9\text{--}35)$$

Combining of the first two leads once again to Eq. 9–32. However, the equations can also be solved to give explicit equations for Y_r and X_r in terms of either the sediment size or specific gravity. The equations that result are

$$Y_r = d_{sr}{}^{-7/2} \qquad (9\text{--}36a)$$

or

$$Y_r = \alpha_r{}^{7/6} \qquad (9\text{--}36b)$$

and

$$X_r = d_{sr}{}^{-5} \qquad (9\text{--}37a)$$

or

$$X_r = \alpha_r{}^{5/3} \qquad (9\text{--}37b)$$

The application of these equations is essentially the same as with the more general equations discussed previously, except that the disappearance of the hydraulic radius makes them much easier to use. Their use will be illustrated in Example 9–8. Although both sets of modeling equations were developed in a logical process, the initial assumption that similitude can be based on the Shields diagram is not fully established. Consequently, the model results are a priori and require careful verification. As with the fixed-bed model, this is usually accomplished by reproducing one or more documented events.

Some of the constraints may need to be relaxed to verify the model. To get sediment movement in the model, for example, it may be required to tilt the model beyond that specified by the distortion ratio Y_r/X_r. This can be at least partially justified at relatively low Froude numbers (say Fr < 0.5) where gravitational effects are less. In the same way as a variation in Reynolds number is tolerated in a fully turbulent flow, a modest variation in Fr can be accepted here. A more extreme assumption proposed by Bogardi[3], but in the same vein, states that since the Shields

[3]J. Bogardi, "Hydraulic Similarity of River Models with Movable Bed," *Acta Tech. Acad. Sci. Hung.*, Vol. 24, 1959.

curve becomes horizontal at values of Re_* above 400, this parameter can be deleted. As a practical, but less conservative, limit, Bogardi suggests that Re_* be ignored when $Re_* > 100$. The effect of these various assumptions can be readily determined by modifying the development of the modeling laws accordingly.

EXAMPLE 9–8

The deposition in, and dredging requirements of, a navigational channel in a wide river are to be studied using a movable-bed model. The bed material in the river is sand (sp. gr. = 2.65) with a median size of 0.4 mm. A model bed with sp. gr. = 1.25 is proposed. Determine the necessary median diameter for the model bed and the required length scales. Assume that both prototype and model are sufficiently wide so that $R_r = Y_r$.

Assume also that the model is constructed and operated using the results obtained below. One sediment transport equation that is not included in Chapter 8 states that the dimensionless sediment transport q_s/u_*d_s is a function of the dimensionless τ_* from the Shields diagram. As given here, q_s is the volumetric sediment discharge per unit width. Use this equation to obtain the sediment transport ratio q_{sr}, and from this result determine the time ratio for sediment modeling. If the model study indicates that dredging of the model channel is required every 100 hours, calculate the frequency with which the prototype channel will require dredging.

Solution

The α_r is

$$\alpha_r = \frac{\alpha_m}{\alpha_p} = \frac{1.25 - 1}{2.65 - 1} = 0.152$$

Using this value, the sediment size ratio can be determined from Eq. 9–32, with the result that

$$d_{sm} = d_{sp}\alpha_r^{-1/3} = (0.4)(0.152)^{-1/3} = 0.75 \text{ mm}$$

The length ratios are

$$Y_r = \alpha_r^{7/6} = (0.152)^{7/6} = 0.111$$

and

$$X_r = \alpha_r^{5/3} = (0.152)^{5/3} = 0.0433$$

This means that the distortion ratio is

$$\frac{Y_r}{X_r} = \frac{0.111}{0.0433} = 2.56$$

This is liable to cause some trouble if the actual dredging is modeled, because side slopes that remain at the angle of repose in the prototype may be impossibly steep in the model.

Substituting $u_* = \sqrt{gRS}$ and

$$\tau_* = \frac{\tau_0}{(\gamma_s - \gamma)d_s} = \frac{\gamma RS}{\gamma\alpha d_s} = \frac{RS}{\alpha d_s}$$

and expressing the entire sediment transport equation in terms of ratios gives

$$\frac{q_{sr}}{R_n^{1/2}S_r^{1/2}d_{sr}} = \frac{R_r S_r}{\alpha_r d_{sr}}$$

or

$$q_{sr} = \frac{R_r^{3/2}S_r^{3/2}}{\alpha_r}$$

Equating $R_r = Y_r$ and $S_r = Y_r/X_r$ leads to

$$q_{sr} = \frac{Y_r^3}{\alpha_r X_r^{3/2}}$$

Eliminating Y_r and X_r using Eqs. 9–36b and 9–37b yields an equation for the sediment transport ratio,

$$q_{sr} = \frac{(\alpha_r^{7/6})^3}{\alpha_r(\alpha_r^{5/3})^{3/2}} = 1$$

Thus on a per-unit-width basis, the volumetric transport ratio $q_{sr} = 1$. In addition, this result can also be expressed in terms of g_s, the sediment discharge rate used in Chapter 8 based on weight. Then $g_{sr} = \alpha_{sr}q_{sr}$.

A sediment time ratio can be obtained by dividing a volume by the transport rate, an interpretation of which would be the time required for deposition to fill a specified volume. Using the transport per unit width, this becomes

$$t_r = \frac{X_r Y_r}{q_{sr}} = X_r Y_r$$

which for this example becomes

$$t_r = (0.111)(0.0433) = 0.00481$$

If dredging in the model is required every 100 hours, then the required prototype time would be given by

$$\frac{t_m}{t_p} = \frac{100}{t_p} = 0.00481$$

whereupon $t_p = 20{,}800$ hours $= 867$ days.

The previous procedures provide a basis for rational movable-bed model design, but they also provide a lot of constraint on the modeler. Most agencies that regularly conduct model studies develop their own procedures, which at least to a degree are based on trial and error. The Mead Hydraulics Laboratory of the Missouri River Division of the Corps of Engineers will serve as an example.[4] Their primary modeling interests are the improvement of hydraulic characteristics and navigation on the Missouri River. They have found that commercially available ground walnut shells with a specific gravity of 1.3 (or $\alpha = 0.3$) satisfactorily simulate the Missouri River sand, which has a specific gravity of 2.65 (and $\alpha = 1.65$). The walnut shells are similar in gradation and shape to the Missouri River sand, and they also demonstrate similar fall velocity characteristics.

From this point on, the methods employed by the Mead Hydraulics Laboratory to achieve similitude are altogether different from the previous, more theoretical analysis. The major criteria that they attempt to satisfy include, first, an adequate sediment movement with at least some suspended load to better simulate depositional patterns and, second, a reasonable sediment time scale. To attain these goals, they are willing to allow distortion not only of the vertical scale, but also of the sediment size, the transport ratio, the resistance, and the Froude number.

To illustrate the Mead Laboratory approach, their study of the Kansas City reach of the Missouri River will be examined. The purpose of this particular study was to determine the necessary control structures to ensure a navigation channel with minimum width and depth of 300 ft and 9 ft, respectively, at the design discharge of 40,000 cfs. In addition, it was also necessary to verify that the flood flows could be carried within the existing levees.

The lightweight walnut shells result in an α ratio of

$$\alpha_r = \frac{\alpha_m}{\alpha_p} = \frac{1.33 - 1}{2.65 - 1} = 0.20$$

The d_{50} sizes for the model and prototype material are 0.23 mm and 0.26 mm, demonstrating that Eq. 9–32 is completely ignored. With a distortion of the sediment size, it is apparent that the sediment time scale will be affected. However, they use this to their advantage since the size distortion will shorten the time required for channel adjustments to take place. The next divergence from the similitude requirements considered earlier in the section is in the choice of length scales. Scales based on Eqs. 9–36b and 9–37b for $\alpha = 0.20$ are $Y_r = 0.153$ and $X_r = 0.068$ for a vertical distortion of 2.25. The length of river reach to be modeled was 5.3 miles, and a scale of 1:150, or $X_r = 0.0067$, was chosen primarily on the basis of available space.

Rather than paying strict attention to the Froude law or Manning relationships, the vertical scale is determined as part of the model verification process. Once the

[4]Missouri River Division Report no. 1, "Operation and Function of the Mead Hydraulic Laboratory," U.S. Army Corps of Engineers, 1969. Also, D. J. Sveum, and W. J. Mellema, "Navigational Model Tests for Straight Reach of Missouri River," *Journal* Waterways, Harbors and Coastal Div. ASCE, May 1972.

width is established, a series of preliminary tests are run in which the depth and velocity are adjusted until a combination is obtained that best duplicates prototype conditions. In the Kansas City study, the result was $Y_r = 0.0185$. As a consequence of this, $Fr_m = 0.169$, while the prototype value was $Fr_p = 0.257$. These values are consistent with the allowable variation in the Froude number discussed previously. This choice of scales results in a vertical distortion of 2.76, which is not greatly out of line with the theoretical value based on α_r.

Instead of a slope matching the vertical distortion, the model was further tilted. The river slope is $S_p = 0.00019$ and the model slope $S_m = 0.00072$ resulted in $S_r = 3.79$. The effect of the various distortions is to distort the sediment transport ratio and increase the sediment time scale. Using the sediment transport and time scales from Example 9–8 leads to $t_r = X_r Y_r = (0.153)(0.068) = 0.010$, while the laboratory transport rate gave a scale $t_r = 0.033$. This means that the net result of the various distortions used in the model study accelerated the sediment behavior.

In summary, the lightweight-model bed material was chosen because of its favorable characteristics. The horizontal scale was determined by space constraints and the vertical scale was chosen more or less independently as part of the model verification rather than to meet prescribed similitude conditions. Consequently a number of distortions are present, but their main effect is to speed up the sediment processes, thereby reducing the time required to establish equilibrium and perform the desired studies.

 PROBLEMS

Section 9–2

9–1. Determine a relationship for the torque ratio of a turbine as a function of (1) the length and flow velocity ratios and (2) the length and discharge ratios. The model and prototype fluids are both water.

9–2. Determine a relationship for the torque ratio of a turbine as a function of the length ratio assuming (1) the Froude law and (2) the Reynolds law applies. The model and prototype fluids are both water.

9–3. Determine a relationship for the shear stress ratio as a function of (1) length, velocity, and fluid property ratios and (2) length, discharge, and fluid property ratios.

9–4. Express the volume ratio as a function of the area ratio.

9–5. Express the power ratio as a function of length, time, and fluid property ratios.

9–6. Express the discharge ratio as a function of length and time ratios.

9–7. Determine the power ratio if the velocity ratio is 2 in a full-scale model in which the prototype fluid is water and the model fluid is (1) water and (2) air.

9–8. Determine the discharge ratio if the velocity ratio is $\frac{1}{10}$ and the area ratio is $\frac{1}{25}$.

9–9. Determine the discharge ratio if the velocity ratio is $\frac{1}{5}$ and the volume ratio is $\frac{1}{10}$.

Section 9–3

9–10. The spillway for a dam is to be modeled with a $\frac{1}{100}$ model. If velocities at the crest and toe of the model spillway are determined to be 1 ft/s and 8 ft/s, respectively, what are the corresponding prototype velocities? Use the Froude law.

9–11. The maximum velocity on a 1-m-high model spillway is 2.2 m/s. If the prototype spillway has a height of 48 m, what is the corresponding prototype velocity?

9–12. The Froude law is to be used to study the operation of tide gates on a large river. If the closure time for the prototype gates is 20 min, determine the model closure time if the length ratio is (1) 1:50 and (2) 1:100.

9–13. Determine the horsepower required to close one of the tide gates in Prob. 9–12(1) if the model power requirement is 0.25 ft-lb/s.

9–14. Determine the wave forces on a 4-m-diameter circular tower if under conditions of Froude law similitude the wave force on a 0.2-m-diameter-model tower is 25 N.

9–15. A large dam must be designed to pass a design project flood of 200,000 cfs. What is the largest-scale model that can be used in a laboratory that has a maximum water supply of (1) 3 and (2) 10 cfs?

9–16. The flow over a two-dimensional ogee spillway with a crest length of 300 ft and design discharge of 107,000 cfs is to be modeled in a 2-ft-wide laboratory flume. The length ratio is $\frac{1}{100}$ (but note that the flow is two dimensional and the direction along the crest is not scaled). What is the model discharge required for dynamic similitude?

9–17. If the force on the 2-ft-wide model spillway of Prob. 9–16 is 65 lb, what is the force on the prototype spillway?

9–18. The flow over a two-dimensional ogee spillway with a crest length of 200 m and design discharge of 5000 m³/s is to be modeled in a 1-m-wide laboratory flume. The length ratio is $\frac{1}{100}$ (but note that the flow is two dimensional and the direction along the crest is not scaled). What is the model discharge required for similitude?

9–19. If the force on the 1-m-wide model spillway in Prob. 9–18 is 485 N, what is the force on the prototype spillway?

9–20. A hydraulic jump with upstream depth and velocity of 3 ft and 22 ft/s is to be studied using a $\frac{1}{10}$ model. Determine the upstream and downstream depths in the model. What are the upstream Froude numbers in the model and the prototype?

9–21. A hydraulic jump is to be studied using a $\frac{1}{25}$ model. If the discharge in the prototype is 160 m³/s, what is the required model flow rate? If the head loss in the model is 0.06 m, determine the head loss and energy dissipation in the prototype.

9–22. What velocity of water at 20°C in a 10-cm-diameter pipe would be dynamically similar to the flow of 30°C air (at atmospheric pressure) at 5 m/s in a 50-cm pipe? Use the Reynolds number for modeling.

9–23. Determine the pipe wall shear stress ratio in Prob. 9–22.

9–24. Determine the pressure gradient (dp/dx) ratio in Prob. 9–22.

9–25. Determine the air velocity in a 1-ft-diameter duct that is dynamically similar to the flow of water at 5 ft/s in a 1-in.-diameter pipe. Both fluids are at 70°F and the air is at atmospheric pressure. Use the Reynolds number as the modeling law.

9–26. Determine the discharge of water in a 6-in.-diameter pipe that is dynamically similar to a water discharge of 40 cfs in a 4-ft-diameter pipe. Both are at the same temperature. Use the Reynolds number as the modeling law.

9–27. Determine the ratio for the pressure gradient dp/dx in Prob. 9–26.

9–28. If the 6-in.-pipe in Prob. 9–26 has a pressure drop of 0.2 psi in a given length, what is the pressure drop in the 4-ft-diameter pipe in a length that is 1000 times that of the smaller pipe?

9–29. Determine the discharge of water at 15°C in a 1-m-diameter pipe that is dynamically similar to a water velocity of 8 m/s at 20°C in a 10-cm-diameter pipe.

9–30. If the pipes in Prob. 9–29 are of identical length, determine the ratio for the power requirement between the two pipes.

9–31. Assume that the kinematic viscosity ratio (v_m/v_p) is equal to 2 and the density ratio (ρ_m/ρ_p) is equal to 1 for a model study in which both the Froude and Reynolds modeling laws are required. Determine the resulting ratios for the length, velocity, force, and power.

9–32. Assume that the kinematic viscosity ratio (v_m/v_p) is equal to ½ and the density ratio (ρ_m/ρ_p) is equal to 1 for a model study in which both the Froude and Reynolds modeling laws are required. Determine the resulting ratios for the length, area, pressure difference, and torque.

9–33. Redo Example 9–6 if the total force on the model is 10 lb.

Section 9–4

For all problems in this section that involve a distorted model, assume that $R_r = Y_r$ except as otherwise indicated.

9–34. Assuming an undistorted river model in which $n_p = 0.022$ and $y_p = 5$ ft, determine the possible range of model depths that could be used if n_m can be varied over the range from 0.010 to 0.015.

9–35. Determine the discharge ratio Q_r in an undistorted model in which $n_r = 0.7$.

9–36. A river model is to be constructed using a vertical scale of ¹⁄₂₀ and a horizontal scale of ¹⁄₂₀₀. The prototype channel has an average depth of 10 ft, an average width of 400 ft, a Manning $n = 0.026$ and a slope of 0.0007. Assuming that the river is approximately rectangular, determine the model and prototype discharges. In addition, determine the distortion ratio, the model roughness, and both the model and prototype Reynolds and Froude numbers.

9–37. In Prob. 9–36, determine the velocity ratio.

9–38. If a flood wave passes through the model channel of Prob. 9–36 in 15 min., what time will be required for a similar flood to pass through the corresponding length of the river?

9–39. Determine n_m for the model channel of Example 9–7.

9–40. A rectangular channel with a depth of 2 m and a width of 30 m is to be modeled using a 0.1-m-deep-by-0.3-m-wide channel. Determine the length ratios and the actual ratio for the hydraulic radius. If $n_p = 0.018$, determine n_m based on the actual R_r and compare with a value calculated assuming that $R_r = Y_r$.

9–41. A trapezoidal channel has a depth of 10 ft and a bottom width of 100 ft. The side slopes are 2:1 ($z = 2$). Length ratios of $Y_r = \frac{1}{10}$ and $X_r = \frac{1}{25}$ are selected. Determine the geometry of the model channel and the actual R_r. Calculate n_r using the actual R_r and compare with the assumption that $R_r = Y_r$.

9–42. A river model is to be constructed using a vertical scale of $\frac{1}{50}$ and a horizontal scale of $\frac{1}{400}$. The prototype channel has an average depth of 2.5 m, an average width of 280 m, a Manning $n = 0.024$, and a slope of 0.0002. Assuming that the river is approximately rectangular, determine the model and prototype discharges. In addition, determine the distortion ratio, the model roughness, and both the model and prototype Reynolds and Froude numbers.

9–43. In Prob. 9–42, determine the velocity and time ratios.

9–44. Repeat Prob. 9–42 using the actual value of R_r whenever appropriate.

9–45. Determine the Reynolds number ratio as a function of the length ratios in a distorted model. Assume that fluid and temperature are identical for model and prototype in Prob. 9–42 and determine Re_r.

9–46. Sediment movement in a wide river is to be studied using a movable-bed model. The prototype bed is composed of equal portions of MS, CS, and VCS, and the river slope is 0.00027. The model bed material has a specific gravity of 1.35. Determine the required model sediment sizes, the length ratios, the water discharge per unit width ratio, and the shear stress ratio.

9–47. If the depth ranges from 1 ft to 6 ft in the protoype channel of Prob. 9–46, what is the corresponding depth range in the model?

9–48. Using the sediment transport expression introduced in Example 9–8, determine the sediment discharge ratio and the sedimentation time ratio in Prob. 9–46. What is the ratio of hydraulic time to sedimentation time?

9–49. Determine the ratio of hydraulic time to sedimentation time in Example 9–8.

9–50. If the prototype slope and depth in Prob. 9–46 are 0.00027 and 6 ft, respectively, determine from the Shields diagram whether or not sediment transport will occur in both the model and the protoype.

References

Allen, J., *Scale Models in Hydraulic Engineering,* London: Longmans Green, 1947.

ASCE Manual No. 25, *Hydraulic Models,* New York: ASCE, 1942.

Henderson, F. M., *Open Channel Flow,* New York: Macmillan, 1966.

Hickox, G. H., "Hydraulic Models," in *Handbook of Applied Hydraulics,* C. V. Davis (ed.), New York: McGraw-Hill, 1952.

Langhaar, H. L., *Dimensional Analysis and Theory of Models,* New York: Wiley, 1967.

Prasuhn, A. L., *Fundamentals of Fluid Mechanics,* Englewood Cliffs, NJ: Prentice-Hall, 1980.

Rouse, H. (ed.), *Engineering Hydraulics,* New York: Wiley, 1950.

Sharp, J. J., *Hydraulic Modelling,* London: Butterworths, 1981.

Yalin, M. S., *Theory of Hydraulic Models,* New York: Macmillan, 1971.

chapter
10
Hydraulic Machinery

Figure 10–0 Powerhouse at Oahe Dam on the Missouri River at Pierre, South Dakota. Note the tailrace in the foreground and three of the large-diameter surge tanks towering over the powerhouse (Prasuhn).

10

10–1 INTRODUCTION

Hydraulic machinery consists of those devices that transfer energy between a fluid and a mechanical system. The fluid of greatest interest to the hydraulic engineer is water, and the hydraulic machinery will be restricted to pumps and turbines. Specialized machines such as hydraulic couplers and torque converters that consist of a combined pump and turbine will not be considered. Neither will devices generally restricted to gas flows such as fans, blowers, and compressors. A *pump* converts mechanical energy, often from an electrical source, to fluid or flow energy. On the other hand, the *turbine* transforms the energy of the flowing water into mechanical energy. The turbine is frequently linked to a generator to produce electrical power.

Hydraulic machinery can be divided into two major classifications: positive displacement machinery and turbomachinery. The distinction is based on the mode of operation. The first category includes reciprocating or piston pumps and rotary or gear pumps. A typical one-stroke reciprocating pump is shown in Fig. 10–1. A hand or power-driven vertical motion is applied at the top, and the piston lifts the water on each upward stroke. The cylinder and valving can also be arranged so that water is lifted on the downward stroke as well, creating a two-stroke pump. This type of pump is ideally suited for deep-water wells, provided that the pulsating flow is acceptable. There were many types and uses for piston pumps; however, they have generally been replaced by other pump types.

A rotary pump detail is shown in Fig. 10–2. The two gears or cams are driven

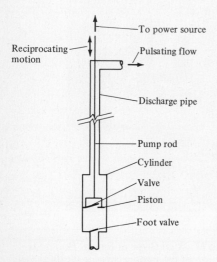

Figure 10–1 Reciprocating pump.

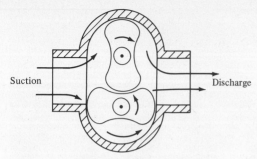

Figure 10–2 Two-lobe rotary pump.

in opposite directions as shown. A similar pump with two three-lobe rotors is also available. The pump is a positive-displacement pump, since the water is trapped between the lobes and subsequently forced out as they rotate. These are usually, but not always, restricted to low heads and discharges. Rotary pumps are self-priming, valveless, and have a flow rate that is directly proportional to the pump speed. Unlike the piston pump, the discharge is nearly steady. As an additional advantage, the rotary pump can usually be operated in either direction. The rotary pump is particularly well suited to the pumping of a prescribed quantity and may be used as a positive-displacement flow meter as well. Positive-displacement pumps will not receive further consideration.

The turbomachinery group is of primary interest to the hydraulic engineer. This classification contains nearly all turbines and most of the important pump types. There are three fundamental types of turbines. A *Pelton turbine*[1] or *Pelton wheel* is shown in Fig. 10–3. This turbine is an obvious outgrowth of the medieval waterwheel, particularly of the click-wheel type shown in Fig. 10–4. A jet of water is directed against the rotating ring of buckets on the periphery of the turbine. Pelton's contribution was primarily the introduction of a flow splitter in each bucket, but that simple expedient produced a highly efficient unit. The Pelton turbine is best suited for high heads and relatively low discharges; however, multiple jets are occasionally used to accommodate higher discharges. The use of additional jets is limited by the interference between the jets and the spent water.

The runner of a *Francis turbine*[2] is shown in Fig. 10–5. The Francis turbine generally operates under a moderate head and relatively high discharge. The runner was originally designed with a purely radial flow, but as the unit evolved into a more efficient machine, the blades were modified so that they now turn the flow from a radial to an axial direction. As the water passes through the blades, it loses its an-

---

[1]The Pelton turbine was developed by Lester A. Pelton (1829–1908) during the Gold Rush days in Camptonville, California. He allegedly received inspiration for the flow splitter (which increased the efficiency) from watching a jet of water squirt off the nose of a cow.
[2]Developed by James B. Francis (1815–1892), civil engineer in the industrial city of Lowell, Massachusetts.

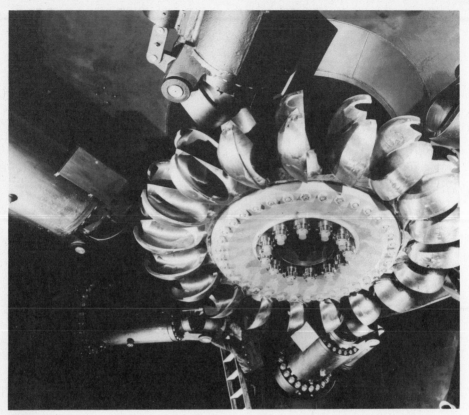

Figure 10–3 One of the six-nozzle Pelton turbines used in the Mont Cenis power house. Output 200 MW from a head of 869 m (courtesy of NEYRPIC, Grenoble, France).

gular momentum and in the process exerts a torque on the rotating runner. Because of this process the Francis turbine is also called a reaction turbine.

The *propeller turbine* shown in Fig. 10–6 is also a reaction turbine, since it likewise involves a change in the angular momentum. The propeller turbine is best suited for relatively low heads and high discharges. The pitch of the propeller blades is usually adjustable so that a high efficiency may be maintained over the range of operating discharges. In this case, the turbine is called a *Kaplan turbine*.[3]

Until the 1970s the trend was toward ever-larger turbines, with little emphasis on power extraction from limited discharges. The changing energy picture has forced a reassessment of the country's energy needs and previously unusable hydroelectric sites are now being considered and developed. This has created interest in the production of small turbines and considerable modification of the basic types is occurring. Often the turbine and generator are combined into a single unit called a *bulb*

[3]Named after the Austrian engineer Viktor Kaplan (1876–1934).

Figure 10–4 Early type of vertical axis turbine known as a click mill. This mill, preserved in the Orkney Islands, is typical of mills dating back to medieval times (Prasuhn).

turbine. Although the principles are the same, these units will not be considered further.

Unlike the early water wheels, turbines are rarely used for the direct production of mechanical power. In almost all cases the turbines are connected to an electrical generator. To produce the required constant-frequency alternating current, the generator (and turbine) must turn at a constant *synchronous speed*. This speed, based on the number of poles on the generator (an even number) and the frequency of the current, is given by

$$N = \frac{120f}{n_p} \qquad\qquad\qquad (10\text{--}1)$$

Here N is the synchronous speed in rpm, f is the frequency in hertz,[4] and n_p is the number of generator poles. In the U.S., the frequency is usually 60 Hz.

Typical layouts for the Pelton turbine and the two reaction turbines are shown in Figs. 10–7 and 10–8, respectively. In either case, water is usually drawn from an upper reservoir or *forebay*. Some care is required so that the flow out of the forebay does not interfere with navigational and recreational requirements. In addition, the

---

[4]The hertz, abbreviated Hz, equals one cycle per second.

Figure 10–5 Finishing of a large Francis turbine destined for the Agua Vermelha power house in Brazil. Output 250 MW under a head of 57 m (courtesy of NEYRPIC, Grenoble, France).

Figure 10–6 Kaplan turbine runner destined for the Oz-en-Oisans power house in France. Output 11.5 MW at a 40 m head (courtesy of NEYRPIC, Grenoble, France).

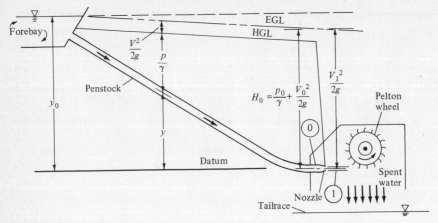

Figure 10–7 Installation layout for a Pelton turbine.

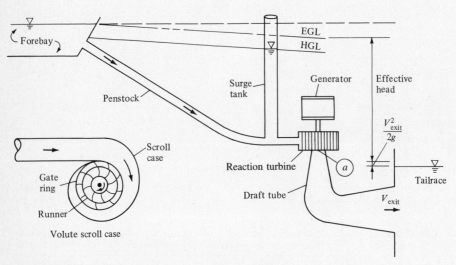

Figure 10–8 Installation layout for a reaction turbine.

forebay powerhouse arrangement may have to include consideration of the potential buildup of ice.

The water is carried from the forebay to the turbine through a large pipe or *penstock*. The entrance to the penstock is protected by a trash rack, which again may be subject to ice problems. Because of its size and the high velocity within, the penstock must also be protected from high water-hammer pressure. This can be accomplished by a *surge tank*, as shown in Fig. 10–8, or by an automatic bypass system that diverts the water past the turbine in the event of a sudden shutdown. The penstock is also susceptible to collapse under negative pressures due to a surge or rapid dewatering. If the penstock extends for a considerable length or contains changes in direction, massive thrust blocks may be required.

The turbine unit follows the penstock. From this point on, the installation differs significantly, as can be seen in Figs. 10–7 and 10–8. The Pelton or impulse turbine obtains all of its energy from the kinetic energy of the discharging jet of water. A nozzle is used to increase the velocity beyond that in the penstock, and a needle valve is often used to regulate the discharge. After striking the Pelton wheel, the spent water drops into the *tailrace* or outlet water below. Energy remaining in the water cannot be recovered and developed by the system. The wheel is usually mounted on a horizontal shaft and directly connected to the electrical generator. More than one wheel may be placed on a common shaft.

Referring now to Fig. 10–8 and the reaction turbine, the water from the penstock enters a spiral or volute scroll case (see the insert in Fig. 10–8). This chamber is designed to distribute the water uniformly around the circumference of the turbine while maintaining a constant velocity. The water then passes through a ring of hydraulically controlled wicket gates that simultaneously control the flow rate and direct the water in the proper rotational course so that it will smoothly and efficiently enter the turbine runner itself. Both Francis and propeller turbines are usually mounted on vertical axes with the generator positioned above, but on the same drive shaft. Horizontal runners are sometimes used. Also, other types of drives, such as belts, can be used, particularly if the vicinity of the turbine is subject to flooding and the generator must be raised to a higher elevation.

The runner of a reaction turbine is completely surrounded by water and the calculation of available energy involves changes in the pressure and potential energy as well as kinetic energy. The analysis is based on the energy equation, with a resulting effective head as shown. The effective head is increased by minimizing the exit velocity head, therefore the *draft tube* or passage from the runner to the tailrace becomes an important design consideration. An expanding section is usually included to reduce the exit velocity. This expansion is restricted, since an overall expansion angle in excess of 14 deg leads to flow separation and additional losses. (See the discussion of venturi meters in Section 6–4.) In addition, if the turbine is placed too far above the elevation of the tailrace, cavitation problems may result because of the low pressures within the turbine. This latter problem is somewhat offset by turning the draft tube in the horizontal direction as in Fig. 10–8. Various aspects of turbine design, including cavitation problems, will be considered in the following sections.

Although there are many kinds of pumps, the hydraulic engineer is usually most concerned with the turbomachine type. Turbopumps can be further divided into three groups, which are defined by the basic direction of flow through the unit: the centrifugal or radial-flow pump, the propeller or axial-flow pump, and the mixed-flow pump. The latter combines the two flow directions. There is no pump counterpart to the impulse turbine.

The operation of these three pump types is opposite to that of the reaction turbines. The motor or machine that drives the pump supplies mechanical energy that exerts a torque on the water and in the process increases its angular momentum. This results in a flow velocity, an increase in the kinetic energy, and an increased head.

The relative magnitude of the increases in the head and discharge depend on the individual pump characteristics and the system in which it operates.

A typical pump installation is shown in Fig. 10–9. The low-pressure inlet side is the *suction line* and the high-pressure outlet line is called the *discharge line,* as labeled. The pump chamber may require *priming* or prefilling, particularly if the pump is located some distance above the intake reservoir. However, a foot valve is sometimes located at the start of the suction line to keep the pump chamber full and primed. Regardless, the pump must not be located so far above the reservoir that the pressure near the pump drops to the water vapor pressure, making it impossible to lift the water. Further care is required to ensure that low pressures do not lead to cavitation problems within the pump. The head lines and the head added by the pump are shown in Fig. 10–9. Note that lowering the pump and positioning the suction line on the side of the reservoir would increase the pressure in the suction line.

The suction line must extend far enough below the reservoir surface to avoid air entrainment and the formation of intense vortices. This may require some care in design, and its importance cannot be overstated.

A centrifugal pump is shown in Fig. 10–10. The water usually enters at the axis and is thrown radially outward by the rotating impeller. The impeller vanes may have plates or shrouds on each side to improve the flow pattern and increase the efficiency, or the impeller may be left open to reduce clogging, particularly when solid material is transported with the water. A pump usually has a volute casing to improve the flow characteristics of the discharging water in the opposite manner to that used in the turbine scroll. This may or may not contain stationary guide vanes, depending on the size of the pump.

An axial-flow pump is shown in Fig. 10–11. This pump contains a propeller and works in an opposite manner to the similar propeller or Kaplan turbine. In between the two types are the mixed-flow pumps. The range of conditions encountered by pumps is far broader than that of turbines. Consequently, there is much more variation in pump types than is found with turbines. Only general design characteristics will be considered in the subsequent sections of this chapter.

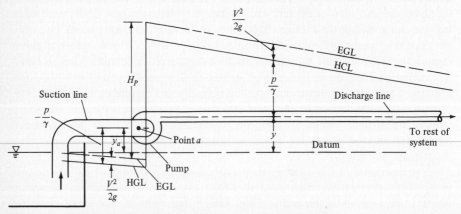

Figure 10–9 Typical installation layout for a centrifugal pump.

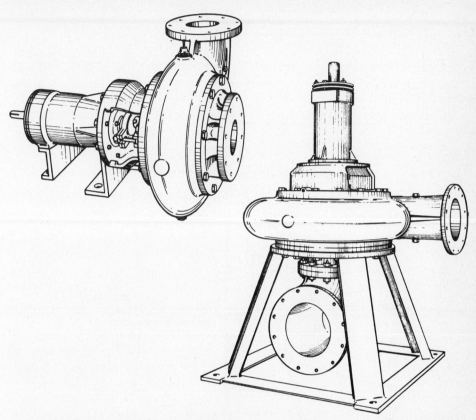

Figure 10–10 Horizontal and vertical centrifugal pumps (courtesy of Fairbanks Morse Pump Corporation, Kansas City, Kansas).

The similarity of reaction turbines to the corresponding pumps has led to reversible units and the concept of *pumped storage*. In this situation, the machine is used alternately as both a pump and a turbine. The gain through such an arrangement lies in the unique characteristic that hydroelectrical power generation, unlike steam-driven units, may be turned on and off with relative ease. The usual electrical power demand is one with periods of peak usage during the day followed by low demand during the night. Pumped storage works well in conjunction with steam generation because it permits the steam units to run continuously while the hydroelectric power plant provides the peaking power. During periods of peak demand, the water is directed down through the turbines to generate the additional power required. The water, or at least a portion of it, is subsequently pumped back up to an upper reservoir by the reversible units during the slack periods with the excess power from the steam plant. Although there is some energy loss through inefficiency, the net result is better utilization of generating equipment and lower overall power costs.

In addition to the two main pump classes of positive displacement and turbomachinery, there are a variety of other, usually special-purpose, pumps. They will

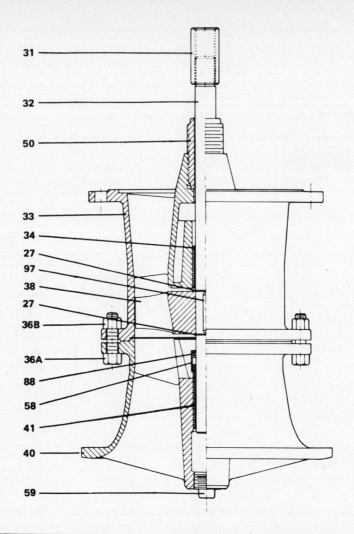

REF. NO.	DESCRIPTION	REF. NO.	DESCRIPTION
27	SNAP RING	41	SUCTION BELL BEARING
31	SHAFT COUPLING	50	CONNECTOR BEARING
32	PUMP SHAFT	58	SAND CAP
33	DISCHARGE BOWL	59	SUCTION BELL PIPE PLUG
34	DISCHARGE BOWL BEARING	88	SET SCREW
36A	CAP SCREW	97	PROPELLER KEY
36B	NUT		
38	PROPELLER		
40	SUCTION BELL		

Figure 10–11 Sectional drawing of an axial-flow or propeller pump (courtesy of Fair-banks Morse Pump Corporation, Kansas City, Kansas).

be briefly discussed at this point and will not be considered further. Included in the miscellaneous pump types are jet pumps and air-lift pumps. The details of a jet pump are shown schematically in Fig. 10–12. A relatively small quantity of water (or sometimes steam or air) is pumped at high velocity through the small pipe and nozzle into the throat of the main pipe. The suction pressure created by the jet induces the desired upward flow in the pipe. Applications include wells, dewatering of construction sites, and other low-capacity installations. Efficiencies on the order of 25 percent can be anticipated.

Air-lift pumps are frequently used for wells, particularly for domestic and farm uses. A typical arrangement is shown in Fig. 10–13. The air compressor pumps air

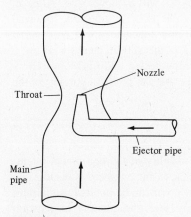

Figure 10–12 Jet pump.

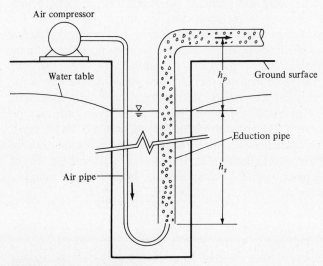

Figure 10–13 Air-lift pump.

through the small air pipe into the larger *eduction* or discharge pipe. The air-water mixture is lighter than the surrounding water and the imbalance in forces lifts the water upward in the eduction pipe. Efficiency is again low (25 to 50 percent), but relatively large amounts of water can be raised in this fashion. The lift is roughly limited to the vicinity of the ground level, and a second pump is usually required to distribute the water. Best operation occurs when the ratio h_p/h_s (see Fig. 10–13) is about 2 in deep wells where h_p approaches 500 ft, ranging to a ratio of 0.5 when h_p is only 50 ft.

10–2 ANALYSIS OF HYDRAULIC MACHINERY

In this section the theoretical analysis of turbomachines will be developed. After an introductory discussion of efficiency, the Pelton turbine will be analyzed, followed by reaction turbines and pumps. The simpler radial-flow machinery will be covered more thoroughly than the axial-flow machinery. The theoretical analyses allow a better understanding of the mechanics of turbomachine design and operation. However, the actual test results that are discussed later are far more important in the actual selection and use of the machinery.

Efficiency

The achieving of high efficiency is more the responsibility of the equipment manufacturer than that of the hydraulic engineer. Nevertheless, the engineer is responsible for selecting an efficient turbine or pump. The *efficiency* η can be defined as the ratio of an output quantity to the required input quantity. Here, the quantity in question will usually be the power, and the goal of high efficiency is simply that of obtaining the maximum output for a given input.

In a multiple-step process, there will generally be an efficiency associated with each step and the overall efficiency is the product of the several parts. Here, the turbine or pump efficiency may be expressed as the product of separate hydraulic, volumetric, and mechanical efficiencies. The hydraulic efficiency η_h will usually be restricted to the turbine or pump unit itself, that is, the turbine scroll case and draft tube or the pump casing. This term includes friction, eddy losses due to flow separation and secondary currents, and the unused kinetic energy leaving the draft tube. The other hydraulic losses associated with pipes, penstocks, and so forth may be included in an overall efficiency. The volumetric efficiency η_v accounts for the small amount of water that passes through clearances between the runner and the housing without doing any work. Mechanical efficiency η_m will always be present because of friction at the bearings and other moving mechanical parts. In general, the pump or turbine efficiency[5] is therefore

$$\eta = \eta_h\, \eta_v\, \eta_m$$

---

[5]The engineer is generally interested in an overall efficiency, but the above discussion indicates that even this term must be carefully defined to fit the specific circumstances.

There are also additional losses of energy associated with the generator or motor. These may or may not be included in a given analysis, depending on how the problem is formulated.

Without consideration of the component parts, the turbine or pump efficiency may be expressed as follows. The change in total head H through the turbine or pump results in a change in the power of magnitude $Q\gamma H$. The decrease in head through a turbine will be designated H_T, while the corresponding increase through a pump is H_P. When no confusion will result, the subscript will be deleted. Thus, the efficiency of a turbine may be written

$$\eta_T = \frac{P}{Q\gamma H_T} \tag{10–2a}$$

in consistent units or

$$\eta_T = \frac{P}{Q\gamma H_T/550} \tag{10–2b}$$

if the actual power output P is in horsepower. Likewise, the pump efficiency is

$$\eta_P = \frac{Q\gamma H_P}{P} \tag{10–3a}$$

or

$$\eta_P = \frac{Q\gamma H_P/550}{P} \tag{10–3b}$$

when horsepower units are specified. In the pump equations, the power P is the power required to drive the pump.

Pelton Turbine

A definitional sketch for the Pelton turbine is shown in Fig. 10–14. As the wheel in Part a turns, each bucket comes in line with the jet. The action against a particular bucket is shown in Part b. The analysis of the jet striking the bucket is based on the impulse-momentum equation. The jet velocity is indicated as V_1, and as the wheel rotates at a constant angular velocity ω, the peripheral velocity is u. Note that the wheel diameter D is defined relative to the centerline of the buckets as shown, and accordingly $u = D\omega/2$. The rotation (or actually translation as far as the derivation is concerned) is halted, for purposes of analysis, by superimposing a constant velocity $-u$ on both the bucket and jet.

The steady-flow control volume is shown in Fig. 10–14c. Writing the momentum equation in the flow direction,

$$-F = \rho Q(V_2' \cos\theta - V_1')$$

where the primes indicate steady-flow velocities. Since the rotation continuously brings the next bucket in line with the jet, all of the water is used, and Q is the

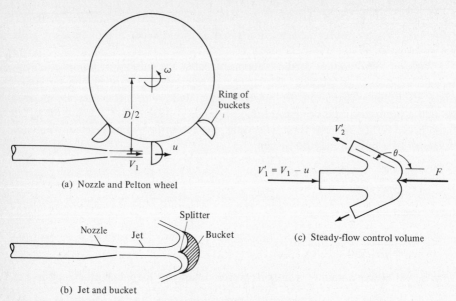

(a) Nozzle and Pelton wheel

(c) Steady-flow control volume

(b) Jet and bucket

Figure 10–14 Definitional sketches of the Pelton turbine.

actual discharge rather than the steady-flow discharge based on V'. The slight friction of the water as it flows along the surface of the bucket may be incorporated by

$$V_2' = kV_1'$$

where k is close to unity. Thus,

$$F = \rho Q V_1'(1 - k \cos \theta)$$

Rewriting in terms of the actual velocity

$$F = \rho Q (V_1 - u)(1 - k \cos \theta) \tag{10-4}$$

Since the power transferred from the jet to the wheel is $P = Fu$, the theoretical power that can be developed is

$$P = \rho Q (V_1 - u)(1 - k \cos \theta)u \tag{10-5}$$

Differentiating to determine the maximum power (assuming that k is not a function of u) and setting the result equal to zero yields

$$\frac{dP}{du} = \rho Q (1 - k \cos \theta)(V_1 - 2u) = 0$$

Thus, the theoretical maximum power corresponds to a wheel velocity $u = V_1/2$, whereupon

$$P_{\max} = \frac{\rho Q V_1^2 (1 - k \cos \theta)}{4} \tag{10-6}$$

A graph of Eq. 10–5 with $k = 1$ and $\theta = 180$ deg is given in Fig. 10–15. Under these conditions, the maximum power according to Eq. 10–6 is

$$P_{max} = \frac{\rho Q V_1^2}{2} \tag{10-7}$$

which corresponds to the available power in the jet. The friction coefficient for a polished bucket will be close to, but less than, unity. In addition, a deflection angle of 180 deg would leave the spent water in the way of the following bucket; an angle of 165 deg is closer to the standard. The power ratio of P in Eq. 10–5 to P_{max} in Eq. 10–7 provides a theoretical analysis of the hydraulic efficiency.

Although the bucket is made as small as possible to reduce air drag or windage, it generally has a width 3 to 4 times the jet diameter and the air drag along with mechanical friction further reduces the efficiency. Since the windage increases roughly with the square of the velocity, the optimum bucket speed is reduced to below 50 percent of the jet velocity and usually falls within the range of 43 to 48 percent. The ratio of bucket (or wheel) speed to jet velocity is designated as the relative speed.

The penstock and nozzle can be treated as in any other pipe system. The optimum penstock diameter is a trade-off between the increased cost of an excessively large diameter and the increased friction and accompanying loss of available energy if the diameter is too small. With reference to Fig. 10–7, the gross head at the forebay is y_0, while the head H_0 at the end of the penstock (section 0) is the sum of the pressure and velocity heads. The effective head at the jet itself is $V_1^2/2g$. Writing the energy equation between sections 0 and 1,

$$H_L = \frac{V_1^2}{2g} + C_L \frac{V_1^2 2g}{2g} \tag{10-8}$$

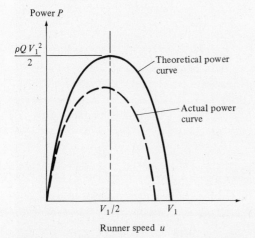

Figure 10–15 Power versus runner speed for Pelton turbine.

where C_L is a coefficient that may be as low as 0.04 to 0.05 at design conditions. The head loss in the nozzle is represented by the final term.

As a final consideration, the ratio of nozzle diameter to wheel diameter usually falls into the range of $1/14$ to $1/16$. In spite of all the contributing factors, the Pelton turbine attains an efficiency of 85 to 90 percent, with the higher values associated with the larger units.

EXAMPLE 10–1

A Pelton turbine is arranged as shown in Fig. 10–7. The water level in the forebay is 500 ft above the nozzle elevation and the available sustained discharge is 20 cfs. The steel penstock has a length of 4000 ft and a diameter of 1.5 ft. The loss coefficient at the entrance to the penstock is 0.5 and the loss coefficient of the nozzle is 0.04. Assume that the bucket friction coefficient is $k = 0.99$, the bucket angle $\theta = 165$ deg, and the combined shaft and generator efficiency is 92 percent. Finally, assume that the maximum turbine efficiency occurs when the wheel velocity is 46 percent of the water velocity. Ignore mechanical efficiency.

Determine the nozzle diameter and the resulting jet velocity. Then calculate the turbine diameter and speed if the unit is connected to a 60-Hz generator. What electrical power can be generated under the above conditions? What is the combined turbine and generator efficiency, and what is the overall efficiency of the entire system?

Solution

The energy equation between forebay and jet is

$$500 = \left(0.5 + \frac{fL}{d}\right)\frac{V^2}{2g} + (0.04)\frac{V_1^2}{2g} + \frac{V_1^2}{2g}$$

The entrance, friction, and nozzle losses have been included in addition to the energy terms. The velocity in the 1.5-ft-diameter penstock is 11.32 ft/s, leading to Re $= 1.4 \times 10^6$. Assuming $k = 0.00015$ ft for the absolute roughness of the steel penstock, $k/d = 0.0001$, whence from the Moody diagram, $f = 0.013$. Substituting into the above energy equation,

$$500 = \left[0.5 + \frac{(0.013)(4000)}{1.5}\right]\frac{(11.32)^2}{(2)(32.2)} + (1 + 0.04)\frac{V_1^2}{(2)(32.2)}$$

Solving, the jet velocity $V_1 = 163.2$ ft/s, which corresponds to a jet or nozzle diameter of 0.395 ft.

The corresponding runner velocity is

$$u = (0.46)(163.2) = 75.07 \text{ ft/s}$$

If we initially choose a wheel diameter 15 times that of the nozzle, then $D = 5.925$ ft. The angular velocity is obtained from

$$\omega = \frac{2u}{D} = \frac{(2)(75.07)}{5.925} = 25.34 \text{ rad/s}$$

or

$$N = \frac{60\omega}{2\pi} = \frac{(60)(25.34)}{2\pi} = 242.0 \text{ rpm}$$

The turbine must turn at a synchronous speed. Comparing with Eq. 10–1,

$$n_p = \frac{(120)(60)}{242.0} = 29.75 \text{ generator poles}$$

The number of generator poles is selected as the nearest even integer, or $n_p = 30$. Recalculating, the runner speed must be

$$N = \frac{(120)(60)}{30} = 240 \text{ rpm} = 25.13 \text{ rad/s}$$

To maintain the original peripheral velocity, the runner diameter must be adjusted slightly to

$$D = \frac{(2)(75.07)}{25.13} = 5.975 \text{ ft}$$

The power developed by the turbine is

$$P = (1.94)(20)(163.2 - 75.1)[1 - (0.99)(\cos 165°)](75.1)$$

$$= 502,000 \text{ ft-lb/s} = 913 \text{ hp}$$

and the generator output is $(913)(.92) = 840$ hp. The power available in the jet is

$$P = \frac{(1.94)(20)(163.2)^2}{2} = 517,000 \text{ ft-lb/s} = 939 \text{ hp}$$

Thus the turbine hydraulic efficiency is

$$\eta_T = \frac{913}{939} = 0.972 = 97.2 \text{ percent}$$

for a combined turbine and generator efficiency of $(.92)(.972)=0.894$ or 89.4 percent. Finally, the overall efficiency is

$$\eta = \frac{840}{(20)(62.4)(500)/550} = 0.740 \text{ or } 74 \text{ percent}$$

Keep in mind that the Pelton turbine analysis is essentially theoretical and, at the very least, does not include mechanical efficiency.

Reaction Machinery

Definitional sketches for radial flow turbines and pumps are shown in Figs. 10–16 and 10–17. The analysis in both cases is based on the torque/moment of momentum equation, which states that the external torque exerted on the water in the control volume equals the rate of flow of angular momentum from the control volume minus the rate of flow of angular momentum entering. The control volume consists of the region between two runner blades bounded by the inner and outer radii.

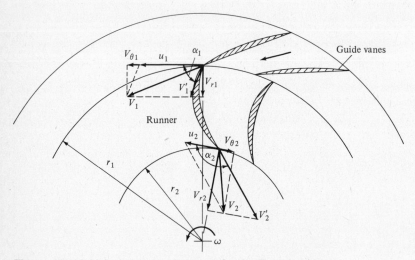

Figure 10–16 Velocity vector diagrams for the Francis turbine.

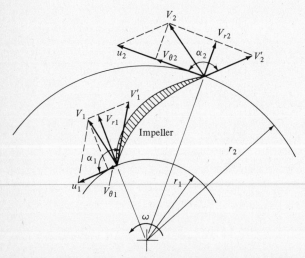

Figure 10–17 Velocity vector diagrams for the centrifugal pump.

With reference to the Francis turbine in Fig. 10–16, high efficiency requires that the water enters the runner smoothly. This is achieved by the stationary guide vanes, which must be positioned and matched to the turbine speed so that the water velocity is tangent to the rotating blade. The actual velocity entering the blade has a magnitude and direction V_1 with components $V_{\theta 1}$ and V_{r1} as shown. The velocity of the water relative to the blade V_1' must match the blade angle α_1, and the runner has a velocity $u_1 = r_1 \omega$. From relative motion the actual velocity, V_1 must equal the vector sum of the blade velocity u_1 and the velocity of the water relative to the blade V_1'. From the vector diagram

$$V_{\theta 1} = \omega r_1 + V_1' \cos \alpha_1$$

But the radial velocity $V_{r1} = V_1' \sin \alpha_1$. Thus,

$$V_{\theta 1} = \omega r_1 + V_{r1} \cot \alpha_1 \tag{10–9}$$

With exist velocities defined in a similar fashion, $u_2 = r_2 \omega$, and the corresponding vector diagram leads to

$$V_{\theta 2} = \omega r_2 + V_{r2} \cot \alpha_2 \tag{10–10}$$

It is assumed that these relationships hold for all of the water as it enters or exists the control volume. According to the previously stated torque/momentum equation, the torque T exerted by the blades on the water within the control volume equals the net rate of flow of angular momentum from the control volume, or

$$T = \rho r_2 V_{\theta 2} \, (V_{r2} A_2) - \rho r_1 V_{\theta 1} \, (V_{r1} A_1)$$

Here, $A = 2\pi r b$ is the area through which the water enters or exits.[6] The tangential momentum flux passing either section is the product of the mass discharge $\rho V_r A$ and the tangential velocity V_θ. Multiplying by the appropriate radius gives the moment of the momentum flux. From continuity,

$$Q = V_{r1} A_1 = V_{r2} A_2 \tag{10–11}$$

Substituting this equation into that for the torque and changing the sign on the torque to reflect the action of the water on the runner (rather than vice versa as formulated), we get

$$T = \rho Q (V_{\theta 1} r_1 - V_{\theta 2} r_2) \tag{10–12}$$

The foregoing equations provide a means of calculating the theoretical torque or power (since $P = T\omega$) developed by a Francis turbine. The radial flow or centrifugal pump follows an identical development. Examination of Fig. 10–17 reveals that the tangential velocities are again given by Eqs. 10–9 and 10–10. Repeating the analysis, the torque that the pump exerts on the water differs from that in Eq. 10–12 only by its sign, and we may write

$$T = \rho Q (V_{\theta 2} r_2 - V_{\theta 1} r_1) \tag{10–13}$$

---

[6]The runner width b is measured in the direction normal to the plane of Fig. 10–16.

Multiplying both equations by the angular velocity in radians per second gives parallel equations for the power. The theoretical power developed by a Francis turbine is

$$P = T\omega = \rho Q(V_{\theta 1} u_1 - V_{\theta 2} u_2) \tag{10-14}$$

while the power delivered to the centrifugal pump is

$$P = T\omega = \rho Q(V_{\theta 2} u_2 - V_{\theta 1} u_1) \tag{10-15}$$

In either case, the various efficiencies ignored in the derivation must also be considered. As a rule, turbines have significantly higher efficiencies than pumps. This is partly due to scale, since efficiency tends to increase with size. Clearances, for example, do not increase directly with size, thereby increasing volumetric efficiency as the size increases. Whereas pumps are manufactured in all sizes, the majority of turbines are large units. A second factor is that the turbine scroll case converges while the pump housing increase in cross-sectional area in the flow direction. Flow separation and an accompanying energy loss is more likely in the second instance.

As a further practical consideration, the usual turbine is designed so that the water exists without swirl and the accompanying loss of usable energy. Thus, $V_{r2} = 0$ in most cases. Similarly, $V_{r1} = 0$ in centrifugal pumps, and the equations for torque and power may be simplified accordingly. The power equations become

$$P = \rho Q V_{\theta 1} u_1 \tag{10-16}$$

for a turbine and

$$P = \rho Q V_{\theta 2} u_2 \tag{10-17}$$

for a pump.

The impulse-momentum analysis of axial-flow turbines and pumps and mixed flow pumps involves more complicated vector diagrams and generally less-accurate theoretical analyses. In fact, the better approach for axial-flow machines is the use of airfoil and propeller theory. In either case, the analysis is considerably more complex and will not be presented herein.

Equations 10–14 and 10–15 (or 10–16 and 10–17 as appropriate) can be used to determine the hydraulic efficiency of a turbine or pump, respectively. Indicating the head removed by the turbine as H_T and that added by a pump as H_P, as discussed at the beginning of this section, the corresponding power becomes $Q\gamma H_T$ or $Q\gamma H_P$.[7] Thus, the hydraulic efficiency of a reaction turbine is

$$\eta_h = \frac{V_{\theta 1} u_1 - V_{\theta 2} u_2}{g H_T} \tag{10-18}$$

while the corresponding pump efficiency is

$$\eta_h = \frac{g H_P}{V_{\theta 2} u_2 - V_{\theta 1} u_1} \tag{10-19}$$

---

[7]These terms refer to the power actually removed or added by the unit. Losses in head within the pipeline or penstock are not included.

Pump-characteristic curves will be examined in more detail later. However, the above analysis leads to theoretical-characteristic head-discharge curves that will assist in introducing the subject. From Eqs. 10–10 and 10–17, the theoretical head added by the pump must be

$$H_P = \frac{V_{\theta2}u_2}{g} = \frac{u_2^2}{g} + \frac{u_2 V_{r2} \cot \alpha_2}{g}$$

Substitution of Eq. 10–11 leads to

$$H_P = \frac{u_2^2}{g} + \frac{u_2 Q \cot \alpha_2}{A_2 g} \tag{10–20}$$

Here, u_2 is the outer runner velocity, A_2 the peripheral area through which the water leaves the runner, and α_2 is the exit angle relative to the runner, as shown in Fig. 10–17. A radial blade would correspond to $\alpha_2 = 90$ deg. Assuming that the pump impeller rotates at a constant speed, the theoretical head must vary linearly with the discharge at a rate dependent solely on the blade angle. This relationship, called the head-discharge characteristic curve for a pump, is shown in Fig. 10–18a. The most common design is the backward-facing blade ($\alpha_2 > 90$ deg), in which the head decreases with increasing discharge. The less common forward-facing blade ($\alpha_2 < 90$ deg) results in the opposite trend, whereas the radial blade represents the limiting case between the two.

The theoretical results are affected by a number of modifying factors. As the flow rate differs from the optimum or design discharge, the entrance angle no longer is optimum and some head loss due to turbulent eddies is likely to occur. The finite number of blades in the impeller means that the fluid between and away from the blades may not follow exactly the same flow direction. In addition, there is friction along the runner, heretofore ignored, that influences the relationship. Consequently, the actual characteristic curves differ markedly from the theoretical, as shown in Fig. 10–18b.

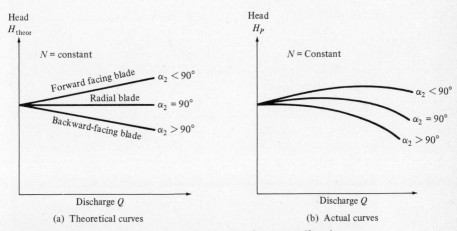

(a) Theoretical curves   (b) Actual curves

Figure 10–18 Head-discharge characteristics for a centrifugal pump.

EXAMPLE 10–2

A centrifugal pump has the following characteristics: $r_1 = 5$ cm, $r_2 = 20$ cm, the impeller width $b = 4$ cm, and the rotational speed is $N = 550$ rpm. The pump has a hydraulic efficiency of 85 percent and a mechanical efficiency of 80 percent. The pump must deliver a discharge of 0.03 m³/s against a head of 10.5 m. Determine the entrance and exit pump-blade angles and the shaft power required to drive the pump.

Solution

The angular velocity is

$$\omega = \frac{2\pi N}{60} = \frac{2\pi(550)}{60} = 57.60 \text{ rad/s}$$

and the runner velocities are

$$u_1 = \omega_1 r_1 = (57.60)(0.05) = 2.880 \text{ m/s}$$

and

$$u_2 = \omega_2 r_2 = (57.60)(0.20) = 11.520 \text{ m/s}$$

The radial water velocities at the entrance and exit to the impeller are

$$V_{r1} = \frac{0.03}{(2\pi)(0.05)(0.04)} = 2.387 \text{ m/s}$$

and

$$V_{r2} = \frac{0.03}{(2\pi)(0.20)(0.04)} = 0.597 \text{ m/s}$$

respectively.

The entrance blade angle will be based on the assumption that the actual water velocity is radial. For smooth entrance, Eq. 10–9 will be applied with the tangential velocity equal to zero.

$$0 = 2.880 + 2.387 \cot \alpha_1$$

Solving, $\alpha_1 = 140.3$ deg. This angle will provide a smooth entrance for the water at the design speed and discharge.

The exit angle must be based on the power transferred to the water. For the required head and the given hydraulic efficiency, Eq. 10–19 becomes

$$0.85 = \frac{(9.81)(10.5)}{V_{\theta 2}(11.520)}$$

Solving, $V_{\theta 2} = 10.52$ m/s. The exit blade angle required to achieve this tangential velocity is obtained from Eq. 10–10,

$$10.52 = 11.52 + 0.597 \cot \alpha_2$$

This yields the desired exit blade angle $\alpha_2 = 149.2$ deg.
The power transmitted to the water is

$$P = Q\gamma H_P = (0.03)(9810)(10.5) = 3090 \text{ W}$$

and the shaft power delivered to the pump must accordingly be

$$P_{\text{shaft}} = \frac{3090}{(0.85)(0.80)} = 4544 \text{ W}$$

or 4.544 kW.

10–3 TURBINE AND PUMP MODELING LAWS

Dimensional analysis can be used to determine a number of significant dimensionless pump and turbine variables. Rather than using this procedure, which can be found in many texts on fluid mechanics,[8] similitude considerations will be introduced directly. Regardless of approach, the results must be consistent with the procedures of Chapter 9 on general hydraulic modeling.

Pump and turbine manufacturers frequently produce model units for performance testing. The appropriate dimensionless terms can then be used as modeling or similitude laws to predict the performance characteristics of similar prototype units over a range of sizes. These results are usually presented to the user as a series of performance curves for each size unit.

Modeling Parameters

Kinematic similitude requires that all velocities be properly scaled. For example, the model velocity vectors at every point in Figs. 10–16 or 10–17, depending on whether the unit is a turbine or pump, must bear a constant relationship to the corresponding prototype velocities. That is, the direction of the velocity vector at any point in the model must be identical to that of the corresponding velocity vector in the prototype, and the velocity ratio between model and prototype must also be a constant at every point. This implies that

$$\frac{V_m}{V_p} = \frac{u_m}{u_p}$$

or

[8]The use of dimensional analysis is discussed in detail in A. L. Prasuhn, *Fundamentals of Fluid Mechanics,* Englewood Cliffs, NJ: Prentice-Hall, 1980, as well as other texts.

$$\frac{V_m}{u_m} = \frac{V_p}{u_p} = \text{constant}$$

where $\mathbf{V}$ is the velocity vector of the water and $\mathbf{u}$ is the velocity vector of the runner. The subscripts have the same meaning as in Chapter 9. The relationships that apply to the velocity vector must hold for the velocity components as well. Consequently, we can write

$$\left(\frac{V_r}{u}\right)_m = \left(\frac{V_r}{u}\right)_p$$

However,

$$V_r = \frac{Q}{A} \sim \frac{Q}{D^2}$$

In the radial flow machine, the area would actually be $\pi b D$, but the impeller width b may be replaced by the diameter D since all lengths must be proportional. The result is general for all turbomachinery. In a similar vein,

$$u = \frac{\pi D \omega}{2} \sim ND$$

When these substitutions are made, the velocity ratio becomes

$$\left(\frac{Q}{ND^3}\right)_m = \left(\frac{Q}{ND^3}\right)_p \qquad\qquad \text{(10–21a)}$$

or

$$\frac{Q}{ND^3} = \text{constant} \qquad\qquad \text{(10–21b)}$$

giving us a discharge equation based on the size and speed of the unit.

A second parameter can be generated by relating the average flow velocity V to the runner speed. The average velocity can be expressed by

$$V \sim g^{1/2}H^{1/2}$$

To demonstrate the basis for this relationship, note that the jet velocity in a Pelton turbine is essentially $V = \sqrt{2gH}$. Since the runner velocity is proportional to ND, the square of the velocity ratio becomes

$$\frac{gH}{N^2D^2} = \text{constant} \qquad\qquad \text{(10–22)}$$

Often, the gravitational acceleration g is dropped, but it will be retained here for dimensional consistency. This equilibrium relates the available head to the runner size and speed.

Assuming constant efficiency between model and prototype, the power can be expressed as proportional to $Q\gamma H$ or $Q\gamma V^2/2g$. In either case, when combined with the above we get

$$\frac{P}{\rho N^3 D^5} = \text{constant} \qquad (10\text{–}23)$$

as a third term involving power.

Equations 10–21 through 10–23 provide modeling parameters for discharge, head, and power, respectively. These equations are adequate for many modeling applications, but can be combined to create additional parameters as needed. However, these additional parameters will not be independent. Neither is the efficiency, which is itself a modeling parameter. Useful examples of additional terms include

$$\left(\frac{Q}{ND^3}\right)\left(\frac{N^2 D^2}{gH}\right)^{1/2} = \frac{Q}{D^2 g^{1/2} H^{1/2}} = \text{constant} \qquad (10\text{–}24)$$

relating discharge to head and size, and

$$\left(\frac{N^2 D^2}{gH}\right)^{3/2}\left(\frac{P}{\rho N^3 D^5}\right) = \frac{P}{\rho g^{3/2} H^{3/2} D^2} = \text{constant} \qquad (10\text{–}25)$$

which eliminates the speed while relating the power to head and diameter.

An alternative form of Eq. 10–22, called the *relative speed,* is sometimes used as a turbine parameter. It is defined as the ratio of runner velocity to water velocity:

$$\phi = \frac{u}{V} = \frac{(D/2)\omega}{\sqrt{2gH}} = \frac{(D/2)(2\pi N)/60}{\sqrt{2gH}}$$

or

$$\phi = \frac{1}{27.0}\frac{ND}{g^{1/2} H^{1/2}} \qquad (10\text{–}26)$$

Its relationship to Eq. 10–22 is obvious, but its importance is its ability to characterize turbine types. In any consistent set of units, relative speeds fall into the approximate ranges given in Table 10–1.

Note the consistency of the relative speed with the earlier theoretical analysis of the Pelton turbine. The use of similitude for turbines and pumps will be illustrated by the following examples.

The assumption of constant efficiency usually introduces a small error through scale effect. A number of factors are involved. For example, the tolerances and clearances will not be scaled between model and prototype. A small leakage past the

Table 10–1 APPROXIMATE RANGE OF RELATIVE SPEED

Turbine Type	Relative Speed Range
Pelton turbine	$0.43 < \phi < 0.48$
Francis turbine	$0.6 \ < \phi < 0.9$
Kaplan turbine	$1.4 \ < \phi < 2$

runner will be proportionally larger in the model. In addition, no attempt is made to hold the Reynolds number constant. As discussed previously in Chapter 9, the Reynolds number of the model must be at least sufficiently high to ensure turbulent flow.

Moody[9] proposed the following widely accepted equation to account for scale effect in both Francis and propellor turbines:

$$\frac{1 - \eta_p}{1 - \eta_m} = \left(\frac{D_m}{D_p}\right)^{1/5} \tag{10--27a}$$

A second turbine equation credited to Ackeret[10] that actually incorporates a Reynolds number is

$$\frac{(1 - \eta_h)_p}{(1 - \eta_h)_m} = \frac{1}{2}\left[1 + \left(\frac{Re_m}{Re_p}\right)^{1/5}\right] \tag{10--27b}$$

Here the Reynolds number Re is given by

$$Re = \frac{D\sqrt{2gH}}{\nu} \tag{10--27c}$$

An estimation of scale effect on pump efficiency is given by Wislicenus[11]

$$\frac{0.95 - \eta_p}{0.95 - \eta_m} = \left(\frac{\ln Q_m}{\ln Q_p}\right)^2 \tag{10--28}$$

All three equations are called step-up equations in that their primary function is to estimate the higher prototype efficiency based on the efficiency of a model.

EXAMPLE 10-3

A model of a particular style of propeller turbine is tested to determine optimum operating conditions. The model has a propeller diameter of 75 cm. At its maximum efficiency the model delivers 38,800 W at 650 rpm from a head of 15 m and a discharge of 0.30 m³/s.

If the design head and discharge are 50 m and 5 m³/s, respectively, determine the size, speed, and power capability of a prototype unit connected to a 60-Hz generator. What are the relative speeds of the model and prototype units?

[9]L. F. Moody and T. Zowski, Section 26, "Hydraulic Machinery," *Handbook of Applied Hydraulics*, third ed., C. V. Davis (ed.), New York: McGraw-Hill, 1984.
[10]Discussed in E. Muhlemann, "Zur Aufwertung des Wirkungsgrades von Ueberdruck-Wasserturbinen," *Schweizerische Bauzeitung*, vol. 66, June 12, 1948.
[11]G. F. Wislicenus, *Fluid Mechanics of Turbomachinery*, New York: McGraw-Hill, 1947.

Solution

A trial prototype diameter can be obtained directly from Eq. 10–24

$$D_p = D_m \left(\frac{Q_p}{Q_m} \right)^{1/2} \left(\frac{H_m}{H_p} \right)^{1/4}$$

$$= (0.75) \left(\frac{5}{0.3} \right)^{1/2} \left(\frac{15}{50} \right)^{1/4} = 2.266 \text{ m}$$

The optimum speed associated with this diameter according to Eq. 10–21 is

$$N_p = N_m \left(\frac{Q_p}{Q_m} \right) \left(\frac{D_m}{D_p} \right)^3$$

$$= (650) \left(\frac{5}{0.3} \right) \left(\frac{0.75}{2.266} \right)^3 = 392.8 \text{ rpm}$$

To match the prototype to a synchronous speed, we first check the approximate number of poles associated with this speed,

$$n_p = \frac{(120)(60)}{392.8} = 18.3$$

Choosing $n_p = 18$, the synchronous speed becomes

$$N_p = \frac{(120)(60)}{18} = 400 \text{ rpm}$$

Equation 10–23 cannot be used to calculate the power capability of the prototype because the slightly higher discharge required to drive the unit at maximum power and 400 rpm rather than the original 392.8 rpm is not available. Instead, and only as an approximation, the prototype power will be estimated on the basis of the model and prototype efficiencies. For the model (at maximum efficiency),

$$\eta = \frac{38,800}{Q\gamma H} = \frac{38,800}{(0.3)(9810)(15)} = 0.879 \text{ or } 87.9 \text{ percent}$$

The prototype efficiency may be estimated using Eq. 10–27a:

$$\frac{1 - \eta_p}{1 - 0.879} = \left(\frac{0.75}{2.266} \right)^{1/5}$$

Upon solving, $\eta = 0.903$ or 90.3 percent. With this efficiency, the prototype power may be evaluated from

$$0.903 = \frac{P}{(5)(9810)(50)}$$

Solving, $P = 2,210,000$ W or 2.21 MW. The relative speeds are

$$\phi_m = \frac{1}{27.0} \frac{(650)(0.75)}{(9.81)^{1/2}(15)^{1/2}} = 1.49$$

and

$$\phi_p = \frac{1}{27.0} \frac{(400)(2.266)}{(9.81)^{1/2}(50)^{1/2}} = 1.52$$

Although both are within the propeller turbine range, they differ slightly due to the required speed adjustment. This example illustrates the use of the modeling laws. It cannot completely evaluate the prototype behavior, because we don't know how the model itself performed at other than optimum conditions. This shortcoming will be addressed with the use of the efficiency hill later.

EXAMPLE 10–4

At optimum test conditions a model pump with a 6-in.-diameter impeller and speed of 1200 rpm requires 3.3 hp to discharge 500 gpm against a head of 20 ft. What are the diameter, speed, and power requirement of a homologous pump designed to discharge 6000 gpm against a head of 100 ft?

SOLUTION

From Eq. 10–24

$$D_p = (6)\left(\frac{6000}{500}\right)^{1/2}\left(\frac{20}{100}\right)^{1/4} = 13.90 \text{ in.}$$

For an impeller of this diameter, conditions of similitude require a speed given by

$$N_p = N_m\left(\frac{Q_p}{Q_m}\right)\left(\frac{D_m}{D_p}\right)^3$$

$$= (1200)\left(\frac{6000}{500}\right)\left(\frac{6}{13.90}\right)^3 = 1158 \text{ rpm}$$

Based on this speed, the power requirement is

$$P_p = P_m\left(\frac{N_p}{N_m}\right)^3\left(\frac{D_p}{D_m}\right)^5$$

$$= (3.3)\left(\frac{1158}{1200}\right)^3 \left(\frac{13.90}{6}\right)^5 = 198 \text{ hp}$$

This example illustrates the application of the modeling laws, but the selection procedure is incomplete. Except in very unusual circumstances where a large number of pumps are required or the pump is specially designed for a specific job, the pump will be selected from a catalog with only particular diameters available. This problem will be discussed later.

Specific Speed

The basic modeling parameters serve their purpose because they can be readily and easily used to determine the operating characteristics of all homologous units based on a single set of model tests. Further, as demonstrated in the previous examples, the size of a particular type of unit can be calculated on the basis of the same model tests to fit the given design conditions. One weakness in the foregoing procedures is that knowledge of the type of unit is required before the modeling laws can be applied. At this point an additional parameter is desirable, since a parameter that is independent of the runner size will better indicate the most suitable type of turbine or pump.

This parameter, known as the *specific speed* N_s, can be obtained by judiciously elminating the diameter between the various modeling equations. In turbine selection, the power output is of primary concern and the specific speed is developed by combining Eqs. 10–22 and 10–23.

$$\frac{\left(\dfrac{P}{\rho N^3 D^5}\right)^{1/2}}{\left(\dfrac{gH}{N^2 D^2}\right)^{5/4}} = \frac{N P^{1/2}}{\rho^{1/2}(gH)^{5/4}}$$

Since water is the sole fluid involved, it is customary to drop both ρ and g, leaving

$$\text{Turbine } N_s = \frac{N P^{1/2}}{H^{5/4}} \qquad\qquad (10\text{–}29)$$

The approximate range of satisfactory specific speeds for the major turbine types is given in Table 10–2 for both U.S. customary and SI units. Since the specific speed is dimensional, some care is required to ensure that the correct units are used.[12]

The efficiency of a turbine decreases when the limits of the specific speed range are exceeded. This is illustrated for U.S. customary units in Fig. 10–19.

---

[12]For the units specified in Table 10–2, the conversion equation is N_s (rpm, kW, m) = 3.813 × N_s (rpm, hp, ft).

Table 10-2 APPROXIMATE RANGE OF TURBINE SPECIFIC SPEEDS

Turbine Type	Specific Speed	
	U.S. Customary Units (rpm, hp, ft)	SI Units (rpm, kW, m)
Pelton turbine	3–7	11–27
Francis turbine	10–100	38–420
Propeller turbine	110–300	420–1100

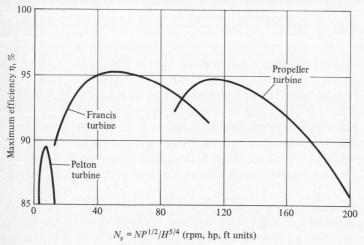

$$N_s = NP^{1/2}/H^{5/4} \text{ (rpm, hp, ft units)}$$

Figure 10–19 Approximate turbine efficiency based on specific speed.

Since the discharge is of greater interest in pump selection, the specific speed is based on Eqs. 10–21 and 10–22,

$$\frac{\left(\dfrac{Q}{ND^3}\right)^{1/2}}{\left(\dfrac{gH}{N^2D^2}\right)^{3/4}} = \frac{NQ^{1/2}}{(gH)^{3/4}}$$

Ignoring the acceleration due to gravity,

$$\text{Pump } N_s = \frac{NQ^{1/2}}{H^{3/4}} \tag{10–30}$$

Based on typical industry units,[13] the range of specific speeds for the various pump

---

[13]With units as given, the conversion equation from U.S. customary to SI units is N_s (rpm, l/s, m) = 0.6124 × N_s (rpm, gpm, ft). Discharge units of m³/s or cfs are also used, particularly for the larger pumps.

Table 10–3 APPROXIMATE RANGE OF PUMP SPECIFIC SPEEDS

Pump Type	Specific Speed	
	U.S. Customary Units (rpm, gpm, ft)	SI Units (rpm, l/s, m)
Radial flow	500–5000	300–3000
Mixed flow	3600–10,000	2200–6100
Axial flow	9000–15,000	5500–9200

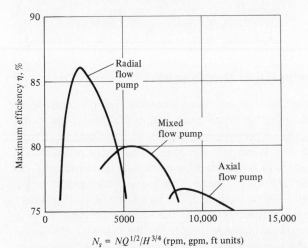

$N_s = NQ^{1/2}/H^{3/4}$ (rpm, gpm, ft units)

Figure 10–20 Approximate pump efficiency based on specific speed.

types is given in Table 10–3. The optimum efficiency range for each type of pump is also illustrated in Fig. 10–20 for U.S. customary units.

EXAMPLE 10–5

Determine the specific speeds for the model and prototype pumps in Example 10–4. What type of pump is indicated?

Solution

Based on rpm, gpm, and ft units, Eq. 10–30 yields the following results. For the model pump,

$$N_s = \frac{(1200)(500)^{1/2}}{20^{3/4}} = 2837$$

while for the prototype unit,

$$N_s = \frac{(1158)(6000)^{1/2}}{100^{3/4}} = 2837$$

As expected, the two calculations lead to identical results. The calculated value suggests that we are dealing with a radial flow pump. If the specific speed were calculated for the turbines of Example 10–3, slightly different values would result, since the speeds were not exactly matched.

10–4 CAVITATION IN HYDRAULIC MACHINERY

Both turbines and pumps can be subject to cavitation problems. The phenomenon of *cavitation* may occur when the pressure at any point within the flow drops to the vapor pressure of water. A significant pressure drop is most likely in regions of high velocity or at high points in a line. If the pressure drops to the vapor pressure in a siphon, a vapor pocket or cavity will form, breaking the siphon and disrupting the flow. In other cases, including turbines and pumps, the low pressure results in the formation of water vapor bubbles similar to those found in a kettle of boiling water. Their formation is followed by a sudden collapse as the bubbles are carried into a region of higher pressure. The sudden implosive collapse of the vapor bubbles in the vicinity of the runner may result in pitting and other structural damage. The manufacturers of hydraulic machinery have attempted to reduce potential cavitation and subsequent damage. However, additional care is required by the engineer in the selection of a particular unit and its placement in the system to ensure that cavitation does not occur.

With reference to Figs. 10–8 or 10–9, note that the critical point for low pressure in either pipe system is labeled point *a*. The pressure at point *a* must remain above the vapor pressure. However, the pressure will be even lower within the pump or turbine, since the rotating runner will cause locally high velocities and low pressures. The highest velocities and consequently the lowest pressures will tend to occur on the convex side of the runner blades of radial flow machines and toward the tip of the propeller of axial flow machines. Because of their high relative velocities,[14] propeller units are most susceptible to cavitation problems. Since the Pelton wheel does not operate in a completely confined flow field, it is far less susceptible to, but not completely free from, cavitation.

The difference in pressure between point *a* and the critical point within the unit cannot readily be determined without testing and must be supplied by the manufac-

[14]Consult Table 10–1.

turer. For the pump, which will be considered first, the difference in pressure head between point a and the critical point within the unit is called the *net positive suction head* and written NPSH. That is,

$$\text{NPSH} = \frac{p_a}{\gamma} - \frac{p_v}{\gamma} \tag{10–31}$$

Specifically, the NPSH is the piezometric head required to accelerate the water from the average velocity at point a in the suction line to its maximum velocity in the vicinity of the runner. In Eq. 10–31 and in the equations that follow, the pressures should be expressed in absolute units.

If we choose the elevation of the reservoir surface in Fig. 10–9 as the datum and work with absolute pressures, we may write the energy equation between the water surface and point a

$$\frac{p_{\text{atmos}}}{\gamma} = y_a + \frac{p_a}{\gamma} + \frac{V_a^2}{2g} + H_L \tag{10–32}$$

Here y_a, known as the *suction lift*, is the elevation of point a relative to the datum, and H_L includes all head losses between the reservoir and the pump inlet. Although y_a is shown positive, it may be negative as well. Substituting Eq. 10–31 into this equation and rearranging,

$$y_a = \frac{p_{\text{atmos}}}{\gamma} - \text{NPSH} - \left(\frac{p_a}{\gamma} + \frac{V_a^2}{2g} + H_L \right) \tag{10–33}$$

Equation 10–33 can be used by the pump manufacturer to determine experimentally the NPSH for a pump. With a pump in place, the suction lift can be continually increased until the onset of cavitation. This will be apparent because of the increased noise and vibration and reduced efficiency. Since all of the other terms in Eq. 10–33 are known, the pump manufacturer can calculate the NPSH.

Given the NPSH along with the other pump data, the engineer can now design a pump and piping system so as to avoid cavitation. Once a tentative layout is selected, y_a can be calculated. This will be the maximum elevation above the reservoir at which the pump can be located so that cavitation will not occur within the pump. Note that the greater the NPSH for a pump, the smaller the allowable y_a becomes.

Although best supplied by the manufacturer, the NPSH for a pump can be estimated using a cavitation parameter σ_c defined by

$$\sigma_c = \frac{\text{NPSH}}{H_p} \tag{10–34}$$

The value of σ_c at incipient cavitation is approximately related to the specific speed by the following relationships:

$$N_s \text{ in rpm, gpm, ft units} \qquad \sigma_c = \frac{N_s^{1.44}}{421,000} \tag{10–35a}$$

or

SANDBOX

$$N_s \text{ in rpm, l/s, m units} \qquad \sigma_c = \frac{N_s^{1.44}}{207{,}800} \qquad\qquad \textbf{(10–35b)}$$

Thus, Eqs. 10–34 and 10–35 together determine a value of NPSH that can be used in the calculation of y_a in Eq. 10–33. It should also be noted that susceptibility to cavitation increases with specific speed. Therefore, high pump heads can only be achieved using pumps with low specific speeds, that is, centrifugal pumps.

EXAMPLE 10–6

A pump has been selected to deliver water from a reservoir at the rate of 0.15 m³/s against a pump head of 10 m. The pump layout is similar to that of Fig. 10–9. The reservoir elevation is approximately at sea level and the water temperature is 20°C. The suction line has a diameter of 20 cm, a length of 12 m, and an absolute roughness height of 0.1 mm. It contains a check valve with a loss coefficient of 2.0, one 90 deg medium radius elbow with a loss coefficient of 0.8, and a rounded entrance with a loss coefficient of 0.08. Determine the maximum elevation above the reservoir at which the pump may be located and still avoid cavitation problems. The pump rotates at 1200 rpm.

Solution

The average velocity in the suction line is

$$V = \frac{0.15}{(\pi/4)(0.2)^2} = 4.775 \text{ m/s}$$

With the discharge in l/s the specific speed is

$$N_s = \frac{(1200)(150)^{1/2}}{(10)^{3/4}} = 2614$$

Using Eq. 10–35b, the critical value of the cavitation parameter is

$$\sigma_c = \frac{(2614)^{1.44}}{207{,}800} = 0.4011$$

Consequently

$$\text{NPSH} = (0.4011)(10) = 4.011 \text{ m}$$

The atmospheric pressure is 101.3 kN/m² (abs) and the vapor pressure at 20°C is 2.340 kN/m² (abs). The head loss in the suction line is based on a relative roughness $k/D = 0.0005$ and an $Re = 9.48 \times 10^5$, for which $f =$

0.017 from the Moody diagram. The maximum pump elevation is calculated by writing the energy equation as given in Eq. 10–33,

$$y_a = \frac{101{,}300}{9810} - 4.011 - \left\{ \frac{2340}{9810} + \frac{(4.775)^2}{(2)(9.81)} + \left[0.08 + 2.0 + 0.8 + \frac{(0.017)(12)}{0.2} \right] \frac{(4.775)^2}{(2)(9.81)} \right\}$$

This yields $y_a = 0.382$ m. To avoid cavitation, the pump must be located at an elevation less than 0.382 m above the reservoir surface. The low suction head in this example is at least partly due to the relatively high specific speed of the pump.

The analysis of cavitation in a reaction turbine is quite similar to the foregoing pump analysis. As stated previously, point a in Fig. 10–8 represents the critical location. The elevation y_a of this point, called the *draft head,* must be kept below some critical level. As with the pump, the pressure within the turbine will be less than that at point a. Although the term net positive suction head is not universally used in reference to turbines, it will be introduced to illustrate the similarity of analysis. Equation 10–31 defines the NPSH for a turbine. This quantity is related to the head removed by a turbine according to

$$\sigma_c = \frac{\text{NPSH}}{H_T} \tag{10–36}$$

The critical cavitation parameter may again be expected to depend on specific speed, and lacking better information, the following may be used as first approximations:

Francis turbine:

N_s in rpm, hp, ft units $\quad \sigma_c = \dfrac{N_s^2}{16{,}000}$ (10–37a)

N_s in rpm, kW, m units $\quad \sigma_c = \dfrac{N_s^2}{232{,}600}$ (10–37b)

Propeller turbine:

N_s in rpm, hp, ft units $\quad \sigma_c = \dfrac{N_s^{1.8}}{10{,}000}$ (10–38a)

N_s in rpm, kW, m units $\quad \sigma_c = \dfrac{N_s^{1.8}}{111{,}200}$ (10–38b)

Taking the datum at the elevation of the surface of the tailrace in Fig. 10–8 and writing the energy equation between point a and the tailrace gives

$$y_a + \frac{p_a}{\gamma} + \frac{V_a^2}{2g} = \frac{p_{atmos}}{\gamma} + H_L \tag{10-39}$$

The head loss term H_L includes all losses between the two points. Further, we continue to use absolute units to describe the pressures. Introducing Eq. 10-31 and solving for the draft head results in

$$y_a = \frac{p_{atmos}}{\gamma} - \text{NPSH} - \left(\frac{p_v}{\gamma} + \frac{V_a^2}{2g} - H_L\right) \tag{10-40}$$

The draft head can be determined using either a calculated or manufacturer-supplied value of NPSH. As with the pump, the turbine is positioned so as to not exceed the draft head. Note that if the head loss is ignored, Eqs. 10-33 and 10-40 become identical.

10-5 PERFORMANCE CURVES AND HYDRAULIC MACHINERY SELECTION

In the previous sections, the results of a model turbine or pump test were used with the similitude laws to determine specific prototype characteristics. The test results can also be used to develop general turbine or pump curves known as *performance curves* or *characteristic curves*. These curves were introduced theoretically in Fig. 10-18 at the end of Section 10-2.

Efficiency Hill and Performance Curves

The generation of performance curves is typically a two-step process. The first step is the plotting of a graph such as Fig. 10-21, creating an *efficiency hill*. For illustrative purposes, a turbine with a runner diameter of 30 in. has been chosen. The discharge of the model turbine, like that of the prototype, is controlled by a ring of wicket gates. At different gate openings, shown in Fig. 10-21 as full gate, 83 percent gate, and so forth, and usually under a nearly constant head, the discharge and power output are measured at different turbine speeds. The efficiency associated with each set of measurements is obtained from Eq. 10-2. Thus, a data set is created consisting of gate opening, head, runner speed, discharge, power, and efficiency.

U.S. customary units are used in Fig. 10-21, and the data has further been reduced to a unit head. To illustrate, consider one set of data that has the following values: 67 percent gate, a head of 17.5 ft, a speed of 193.3 rpm, a discharge of 84.9 cfs, and a power of 145.0 hp. To calculate the unit speed N_1, use Eq. 10-22 with the diameter constant,

$$\frac{1}{N_1^2} = \frac{17.5}{(193.3)^2}$$

giving $N_1 = 46.3$ rpm. Likewise, the unit power P_1 may be obtained from Eq. 10-25. For constant diameter,

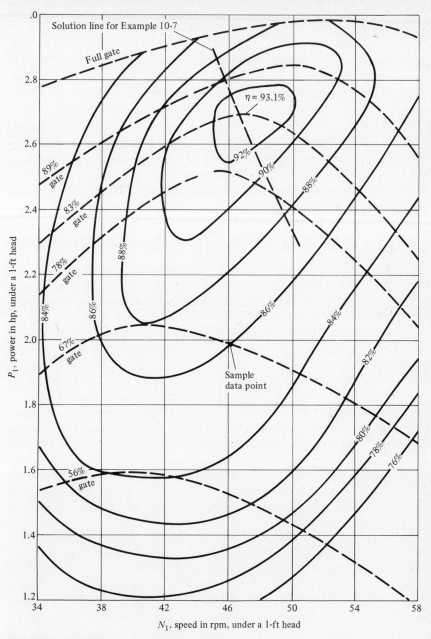

Figure 10–21 Efficiency hill for a turbine.

$$\frac{P_1}{(1)^{3/2}} = \frac{145.0}{(17.5)^{3/2}}$$

from which $P_1 = 1.98$ hp. Finally, the efficiency from Eq. 10–2b is

$$\eta = \frac{145.0}{(84.9)(62.4)(17.5)/550} = 0.860 \text{ or } 86.0 \text{ percent}$$

This data point is indicated on Fig. 10–21 as "sample data point." The remainder of the data points are not plotted, but they will of necessity all plot along the dashed lines of constant gate opening. Once all of the points are plotted and labeled with the specific values of efficiency, the lines of constant efficiency can be constructed. Their characteristic form gives the graph the name "efficiency hill." Note that the full gate position provides the maximum power output. However, this is generally not the design point of maximum efficiency. In the turbine test associated with Fig. 10–21, the maximum efficiency of 93.1 percent is associated with a gate opening of 83 percent.

The entire turbine test results are now available in usable form on the efficiency hill. This figure may be used in several ways. For example, it may be used to determine performance capability and characteristics for any homologous prototype unit given the operating conditions, or it may be used to determine the size of unit required to operate under given conditions. The efficiency hill may also be used to generate less comprehensive, but more convenient, performance curves for various size prototype units.

Both types of calculations depend on the previous model laws as adapted for the specific efficiency hill condition of unit head. The calculations must also be consistent with the synchronous speed given by Eq. 10–1. The direct use of the efficiency hill will be considered in the example that immediately follows. The second example will illustrate development of typical performance curves.

EXAMPLE 10–7

Use the efficiency hill in Fig. 10–21 for a 30-in. model turbine to determine the size, speed, and efficiency of a homologous turbine that will generate 40,000 hp from a head of 105 ft. The turbine will be connected to a generator that produces 60-hertz current. What discharge is required to obtain this output?

Solution

At maximum efficiency of 93.1 percent, $P_1 = 2.69$ hp and $N_1 = 46.8$ rpm. Through the use of Eq. 10–25,

$$40,000 = (2.69)\left(\frac{D_p}{30}\right)^2 \left(\frac{105}{1}\right)^{3/2}$$

Solving, $D_p = 111.53$ in. The corresponding turbine speed for maximum efficiency can be obtained through use of Eq. 10–22,

$$N_p = (46.8)\left(\frac{30}{111.53}\right)\left(\frac{105}{1}\right)^{1/2} = 128.99 \text{ rpm}$$

This is matched against a synchronous speed by first calculating the number of generator poles according to Eq. 10–1. For $f = 60$ Hz,

$$n_p = \frac{(120)(60)}{128.99} = 55.8 \text{ poles}$$

This is most readily rounded to 56 poles, whereupon

$$N_p = \frac{(120)(60)}{56} = 128.57 \text{ rpm}$$

This speed is so close to the optimum speed that the final turbine characteristics will be very close to optimum conditions, with an efficiency very close to 93.1 percent. This is a desired, but not always achieved, result.

The turbine speed is now fixed. To determine the actual runner diameter, we can combine Eqs. 10–22 and 10–25 as follows:

$$D_p = (30)\left(\frac{N_1}{128.57}\right)\left(\frac{105}{1}\right)^{1/2} = 2.391 \, N_1$$

and

$$D_p^2 = (30)^2 \left(\frac{1}{105}\right)^{3/2} \left(\frac{40,000}{P_1}\right) = \frac{33,460}{P_1}$$

When D_p is eliminated, we get

$$P_1 = \left(\frac{76.50}{N_1}\right)^2$$

This equation has been plotted on Fig. 10–21 and labeled as the "solution line for Example 10–7." By inspection, the best point can be selected as the point on the solution line that passes closest to the peak of the efficiency hill. This point has the coordinates $P_1 = 2.69$ hp and $N_1 = 46.6$ rpm. The efficiency, which will be the final efficiency of the prototype (ignoring scale effects), is approximately 93 percent.

Substituting into the appropriate equations above,

$$D_p = (2.391)(46.6) = 111.4 \text{ in.}$$

or

$$D_p = \left(\frac{33,460}{2.69}\right)^{1/2} = 111.5 \text{ in.}$$

verifying, within the accuracy of the graph, a diameter of 111.5 in. Because the speed adjustment was minor, there is essentially no change from the original runner diameter in this example.

The final calculation of discharge is obtained from Eq. 10–2.

$$0.93 = \frac{(550)(40,000)}{Q(62.4)(105)}$$

Solving, $Q = 3610$ cfs.

EXAMPLE 10–8

Develop performance curves of power output versus efficiency for heads of 90 ft, 105 ft, and 120 ft using the efficiency hill of Fig. 10–21 and the turbine of Example 10–7.

Solution

The 105-ft head was used to select the turbine in Example 10–7. It should be expected to yield the highest efficiency. The other two curves will show the results of operating the turbine at lower and higher heads. The higher head should result in a greater power output than the design head, but both will have lower efficiencies than the 105-ft head. As before, Eq. 10–25 can be used to calculate the prototype power, given P_1. For the three specified heads we may write:

$$H = 90 \text{ ft} \qquad P = P_1\left(\frac{111.5}{30}\right)^2\left(\frac{90}{1}\right)^{3/2} = 11,800\, P_1$$

$$H = 105 \text{ ft} \qquad P = P_1\left(\frac{111.5}{30}\right)^2\left(\frac{105}{1}\right)^{3/2} = 14,870\, P_1$$

and

$$H = 120 \text{ ft} \qquad P = P_1\left(\frac{111.5}{30}\right)^2\left(\frac{120}{1}\right)^{3/2} = 18,160\, P_1$$

As determined in Example 10–7, the unit speed for a 105-ft head is $N_1 = 46.6$ rpm. The performance curve for this head can be determined by taking selected values of efficiency and power along the vertical line $N_1 = 46.6$ rpm. The values of P_1 can be converted to values of prototype power using the equation for $H = 105$ ft. As examples, $P_1 = 2.003$ hp and $P = 29,800$ hp at $\eta = 86$ percent, and $P_1 = 1.662$ hp and $P = 24,700$ hp at $\eta = 84$ percent. The entire curve is plotted on Fig. 10–22 and labeled with $H = 105$ ft.

Values of N_1 are required for each of the other two heads. Using Eq. 10–22,

$$128.57 = N_1\left(\frac{30}{111.5}\right)\left(\frac{90}{1}\right)^{1/2}$$

yielding $N_1 = 50.4$ rpm for a head of 90 ft. In the same way, a head of 120 ft corresponds to $N_1 = 43.6$ rpm. As with the 105-ft head, data pairs of efficiency versus P_1 can be obtained from the efficiency hill for the particular value of N_1. After converting to prototype power, using the equations for $H = 90$ ft and $H = 120$ ft, respectively, these have also been plotted on Fig. 10–22.

The preceding examples illustrate a selection method based on the efficiency hill and the development of one type of turbine performance curve. Other types of performance curves such as discharge versus power can also be generated from the same turbine data. A model pump can be tested in the same way as the turbine, and a similar type of efficiency hill created. As was discussed when specific speed was

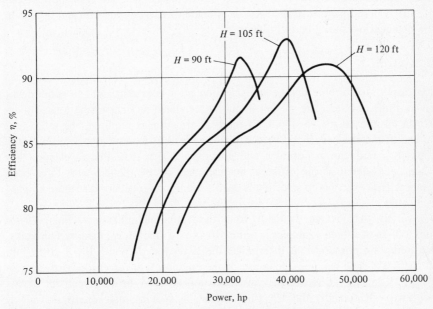

Figure 10–22 Turbine performance curves from Fig. 10–21.

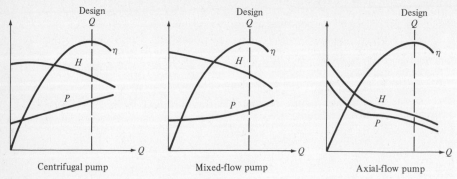

Figure 10–23 Pump performance curves.

considered, discharge becomes more important than power and the plotted parameters would generally be chosen to reflect this. A complete analysis will not be presented but typical performance curves will be introduced.

The performance curves are required because the turbine or pump cannot always be operated at optimum efficiency. These curves provide the additional information necessary to describe and analyze these other conditions. Although some pumps can be operated as variable speed pumps, we will consider the more usual case where the pump is connected to a constant-speed motor. Typical pump performance curves for the three common types of pumps are shown qualitatively in Fig. 10–23. Here, the various characteristics are given as functions of the discharge. The head at zero discharge is called the *shutoff head*. In each case, the design discharge is associated with the maximum efficiency. At higher or lower discharges, the efficiency is seen to fall off. A broad peak is desired on the efficiency curve so that efficient operation can be obtained over a range of discharges.

Pump Selection

Various aspects of turbine design and selection have been discussed in some detail throughout this chapter. The subject is quite specialized and cannot be treated thoroughly in an introductory textbook. If more information is required, the references at the end of the chapter should be consulted, as well as experts associated with the appropriate manufacturers, utilities and government agencies. Pump selection has also been considered as the different analysis and design requirements have been introduced. However, pump selection is such a common procedure that more emphasis is in order.

A variety of aids are available from the pump manufacturers. Figs. 10–24 through 10–26 are three examples. Figure 10–24 is a chart developed by Fairbanks, Morse Pump Corporation that determines the best type of pump for a given job. Once the pump type has been determined, the engineer can resort to charts such as Figs. 10–25 and 10–26 to complete the selection process. In the remainder of this section, we will consider the selection and analysis of a single pump as well as pumps that are arranged in series or parallel.

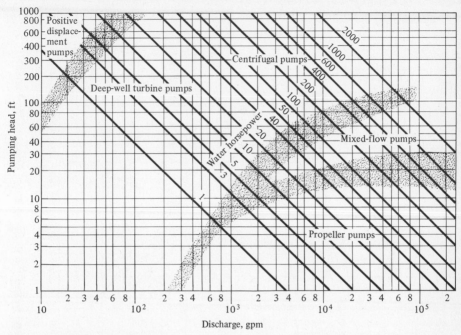

Figure 10–24 Pump selection chart (courtesy of Fairbanks Morse Pump Corporation).

A Single Pump in a System

Through an initial analysis of the pipe system using the energy equation, the engineer can calculate the pump head required to obtain the design discharge. Given this information, Fig. 10–24 immediately suggests the type of pump that should be used. For illustrative purposes, assume that a centrifugal pump is indicated.

The problem of pump selection still requires that consideration be given to a number of pump parameters. Although some pump/motor units are designed to operate at a select number or even an infinite number of speeds, most AC motors operate at a particular speed and the pump must be matched with the best motor (and speed) for given conditions. The next step in pump selection is best accomplished using a manufacturer-supplied coverage chart like that in Fig. 10–25 for a centrifugal pump. The calculated values of H_P and Q can be compared with the various regions or flags (so-called because of their characteristic shape) in Fig. 10–25. The possible choices may then be selected as illustrated in Example 10–9. The flag numbers refer to specific pumps, and the region covered by each flag is that of the actual head/discharge curves of that pump using different impeller diameters. The various regions designated by the same number refer to the same pump operated at different rotational speeds, with the upper region having the greatest speed.

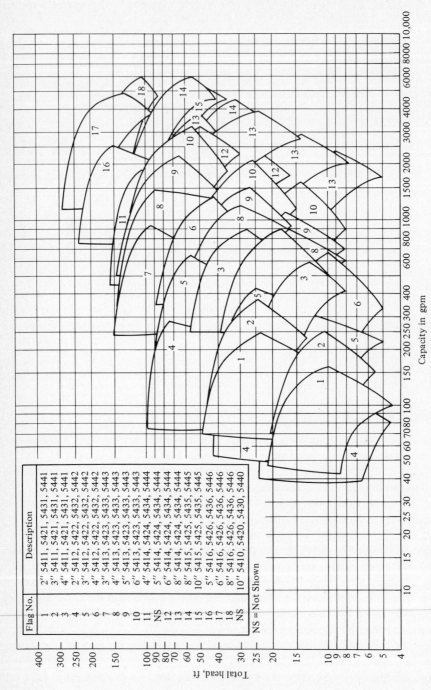

Flag No.	Description
1	2″ 5411, 5421, 5431, 5441
2	3″ 5411, 5421, 5431, 5441
3	4″ 5411, 5421, 5431, 5441
4	2″ 5412, 5422, 5432, 5442
5	3″ 5412, 5422, 5432, 5442
6	4″ 5412, 5422, 5432, 5442
7	3″ 5413, 5423, 5433, 5443
8	4″ 5413, 5423, 5433, 5443
9	5″ 5413, 5423, 5433, 5443
10	6″ 5413, 5423, 5433, 5443
11	4″ 5414, 5424, 5434, 5444
NS	5″ 5414, 5424, 5434, 5444
12	6″ 5414, 5424, 5434, 5444
13	8″ 5414, 5424, 5434, 5444
14	8″ 5415, 5425, 5435, 5445
15	10″ 5415, 5425, 5435, 5445
16	5″ 5416, 5426, 5436, 5446
17	6″ 5416, 5426, 5436, 5446
18	8″ 5416, 5426, 5436, 5446
NS	10″ 5410, 5420, 5430, 5440

NS = Not Shown

Figure 10–25 Centrifugal pump coverage chart (courtesy of Fairbanks Morse Pump Corporation, Kansas City, Kansas).

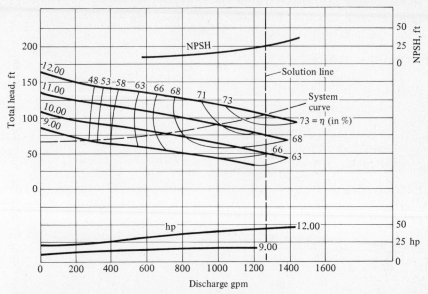

Figure 10–26a Pump flag number 8, 1750 rpm with 4-in. suction (courtesy of Fairbanks Morse Pump Corporation, Kansas City, Kansas).

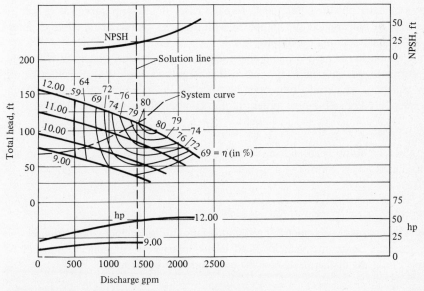

Figure 10–26b Pump flag number 9, 1750 rpm with 5-in. suction (courtesy of Fairbanks Morse Pump Corporation, Kansas City, Kansas).

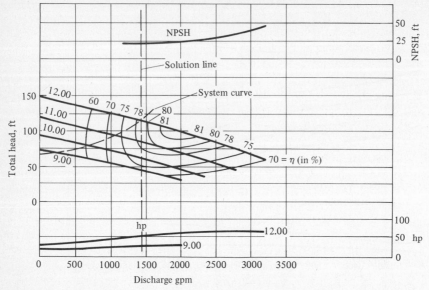

Figure 10–26c Pump flag number 10, 1750 rpm with 6-in. suction (courtesy of Fairbanks Morse Pump Corporation, Kansas City, Kansas).

We now refer to the characteristic curves for the selected pump and speed. Let us assume that regions 8, 9, and 10 are the best choices. These regions have been replotted in Fig. 10–26. Note that the various head/discharge curves are similar to the pump curves in Fig. 10–23. Additional curves include the NPSH for the largest diameter and the power requirement. In each case, the system curve, as described below, may be constructed and compared with the head characteristic curve to determine the actual operating conditions.

In addition to the plotted pump characteristics, the physical system of which the pump forms a part influences the actual discharge and head. Let us return to the energy equation, which may be written between the upstream reservoir (point 1) and a downstream point in the pipe system (point 2) in the following general form:

$$H_1 + H_P = H_2 + H_L$$

Solving for the pump head and separating the total head terms into piezometric and velocity heads,

$$H_p = h_2 - h_1 + \frac{V_2^2}{2g} - \frac{V_1^2}{2g} + C_L \frac{V^2}{2g}$$

where h represents a piezometric head and the final term includes both friction and minor losses.[15] Since the velocity head terms, regardless of number, may all be expressed in terms of the discharge, the equation may be written,

---

[15]With various size pipes within the system, several terms with different velocity heads may be required. Nevertheless, they may all be expressed as products of coefficients times velocity heads.

$$H_P = \Delta h + f(Q^2) \tag{10–41}$$

That is, the pump head is the sum of the increase in piezometric head plus a function of Q^2. If water were pumped from a lower reservoir to an upper, for example, then Δh would be the difference in water levels, or the static head. This equation, known as the *system curve*, may be superimposed on the head curves in Fig. 10–26. The actual pumping conditions will be those at the intersection of the characteristic head curve and the system curve. The development and use of the system curve will be further illustrated in Example 10–10.

The selection process must include consideration of the system curve, particularly if any variation in head is possible. Note that another important design consideration is that the pump should be selected so that the intersection corresponds as closely as possible to the point of maximum efficiency. (Contrast Fig. 10–26a with Figs. 10–26b and c.) As mentioned before, a pump with a flat efficiency peak is desirable so that small variations in operating conditions do not significantly reduce the efficiency.

In the event that the head developed by a single pump is inadequate, two or more pumps can be placed in series. If the pump choices do not provide the necessary discharge, two or more pumps may be placed in parallel. These two cases will be treated in the following subsections. The use of parallel pumps increases the flexibility of the system, because the pumps can also be operated in different combinations.

EXAMPLE 10–9

Select a pump to deliver a discharge of 1200 gpm from a lower to an upper reservoir. The elevation difference is 67 ft and the two reservoirs are connected by 200 ft of 6-in.-diameter pipe. Assume that the pump location avoids cavitation problems. Use $k = 0.00015$ ft for the absolute pipe roughness. Use Fig. 10–24 to verify that a centrifugal pump is the best choice. Then, from the pumps available in Fig. 10–25, determine possible pump choices. (Example 10–10, which involves the system curve, will illustrate the selection of the actual pump.) Assume that the minor losses between reservoirs comes to $5(V^2/2g)$.

Solution

The discharge in cfs is

$$Q = \frac{1200}{(7.48)(60)} = 2.674 \text{ cfs}$$

and the average velocity is

$$V = \frac{2.674}{(\pi/4)(0.5)^2} = 13.62 \text{ ft/s}$$

Assuming a temperature of 60°F

$$Re = \frac{(13.62)(0.5)}{1.217 \times 10^{-5}} = 560,000$$

and

$$\frac{k}{D} = \frac{0.00015}{0.5} = 0.0003$$

This leads to a resistance coefficient $f = 0.016$. This value will be assumed to remain constant. Writing the energy equation between the lower and upper reservoirs,

$$0 + H_P = 67 + \left[\frac{(0.016)(200)}{0.5} + 5\right]\frac{(13.62)^2}{(2)(32.2)}$$

or $H_P = 99.84$ ft. Consulting Fig. 10–24, the discharge of 1200 gpm and head of approximately 100 ft are best served by a centrifugal pump. Continuing to Fig. 10–25, the regions labeled 8, 9, and 10 appear to be good choices. The corresponding head/discharge curves along with efficiency, power, and NPSH are plotted in Fig. 10–26. These will be discussed further in the following example.

EXAMPLE 10–10

Refer to the sets of pump characteristics selected in Example 10–9. Determine and plot the system curve on each of the three charts. Select the best pump and determine the actual operating conditions for the pump.

Solution

The system was analyzed for a discharge of 1200 gpm in Example 10–9. If we introduce the discharge Q in lieu of the velocity we get

$$H_p = 67 + \frac{(11.4)\, Q^2}{(2)(32.2)(\pi/4)^2(0.5)^4}$$

which simplifies to

$$H_p = 67 + 4.592Q^2 \qquad (Q \text{ in cfs})$$

This must be converted into gpm units for comparison with Fig. 10–26, thus

$$H_p = 67 + \frac{4.592\,Q^2}{[(7.48)(60)]^2}$$

or

$$H_p = 67 + \frac{Q^2}{43{,}870} \qquad (Q \text{ in gpm})$$

This curve has been plotted on each of the three portions of Fig. 10–26 and labeled as the system curve. The operating conditions and pump characteristics that provide a minimum of 1200 gpm in each case are as follows:

Pump	Impeller diameter (in.)	Discharge (gpm)	Efficiency (%)	Power (hp)	Head (ft)
a	12	1275	73+	45	104
b	12	1400	79.7	47	109
c	12	1450	79	52	113

All three pumps operate at 1750 rpm. The solution for each pump provides a discharge somewhat in excess of the required 1200 gpm. Since pump *a* is closest to the desired discharge, it will require the least throttling and might be anticipated to provide the best solution. The other two pumps have higher efficiencies. However, these efficiencies are at excessive discharges and would not be achieved at the design discharge. Note also that pump *a* has the lowest power requirement. Of the three choices, pump *a* is the best. A control valve will have to be slightly closed to get the required discharge.

Arrangement of Pumps in Series

Two or more pumps may be connected in series as shown in Fig. 10–27. This becomes necessary when a high head makes a single impeller inefficient. As indicated in the diagram, the primary result is an increase in head. When the series pumps are located close together the increase in head becomes the sum of the individual head increases. This is the basis of the *multiple-stage pump,* in which several impellers are combined within a single housing, the water flowing through each in turn.

All pumps in series must be operated simultaneously. While the pumps do not have to be identical, it is best if they have similar characteristics. Individual head/discharge curves for two somewhat dissimilar pumps are given in Fig. 10–28. If the two pumps operate in series, the resulting head curve is the sum of the com-

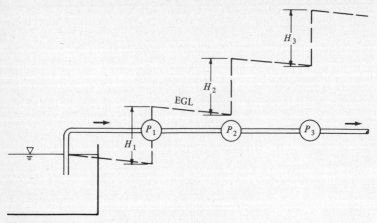

Figure 10–27 Pumps arranged in series.

ponent curves, as indicated. (The calculations will be demonstrated in the example that follows.) At this point, a system curve could be plotted as before to determine the actual operating conditions.

The efficiencies of the respective pumps are

$$\eta_1 = \frac{Q\gamma H_1}{P_1}$$

and

$$\eta_2 = \frac{Q\gamma H_2}{P_2}$$

while the overall efficiency of the two operating in series is

$$\eta = \frac{\gamma Q(H_1 + H_2)}{P}$$

Note that subscript P has been dropped from the head terms for convenience in this section.

Since the combined power requirement $P = P_1 + P_2$

$$\frac{\gamma Q(H_1 + H_2)}{\eta} = \frac{\gamma Q H_1}{\eta_1} + \frac{\gamma Q H_2}{\eta_2}$$

or

$$\eta = \frac{H_1 + H_2}{(H_1/\eta_1) + (H_2/\eta_2)}$$

In general, for any number of pumps in series,

$$\eta = \frac{\sum_i H_i}{\sum_i (H_i/\eta_i)} \tag{10–42}$$

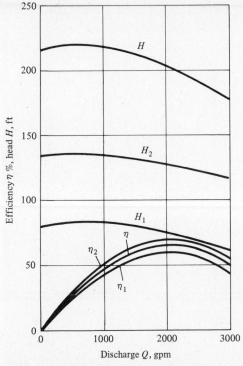

Figure 10–28 Performance curves for two pumps in series.

Figure 10–28 also includes the individual and combined efficiency curves. If two (or more) pumps are identical, the combined efficiency curve according Eq. 10–42 will be the same as the original.

EXAMPLE 10–11

The individual head/discharge and efficiency/discharge data for two pumps are tabulated below. Determine the head and efficiency curves when the two pumps are operated in series.

Solution

The data for selected points appear in the first five columns. This data are also plotted in Fig. 10–28 and identified by the appropriate subscripts. The values of combined head H in the sixth column are obtained by adding the pump heads at each discharge. The resulting efficiency η given in the final column is found from Eq. 10–42. A sample calculation follows the table.

Q (gpm) (1)	H_1 (ft) (2)	H_2 (ft) (3)	η_1 (%) (4)	η_2 (%) (5)	H (ft) (6)	η (%) (7)
0	80	134	0	0	214	0
300	82	136	15	18.5	218	17.0
700	83	136	32	39	219	36.0
1000	82.5	135	42	51	217.5	47.2
1500	79	132	54	65	211	60.4
1800	77	130	58	68	207	63.9
2000	75	128	60	70	203	65.9
2200	72.5	126	59.5	69	198.5	65.2
2500	68	122	57	67	190	63.0
3000	61	116	43	55	177	50.2

Sample calculations:

Discharge $Q = 2000$ cfs

$$H = 75 + 128 = 203 \text{ ft}$$

$$\eta = \frac{203}{\left(\dfrac{75}{60}\right) + \left(\dfrac{128}{70}\right)} = 65.9 \text{ percent}$$

The resulting head and efficiency curves have also been plotted in Fig. 10–28. The intersection of this combined head curve with a system curve would yield the actual operating conditions in a given situation.

Arrangement of Pumps in Parallel

Any number of pumps may be placed in parallel. The three pumps in Fig. 10–29 demonstrate one possible arrangement. The control valves are not shown, but following each pump there would normally be at least a check valve to prevent backflow. The parallel arrangement is necessary when a single pump will not efficiently provide the required discharge. When so arranged, the overall head is not materially affected, but the total discharge is the sum of the discharges of the individual pumps. As an additional advantage, at lower flow demands one or more of the pumps can be used independently. It is also possible to lay out the connective piping so that a number of pumps can be alternatively operated in either series or parallel, as desired.

Performance curves for two pumps are given in Fig. 10–30. When operated in parallel, the combined head/discharge curve is the sum of the respective discharges at each head. It is important that the individual pumps have reasonably similar characteristics. As Fig. 10–30 illustrates, the two pumps have significantly different head/discharge curves. When the head exceeds the shutoff head of the lower curve,

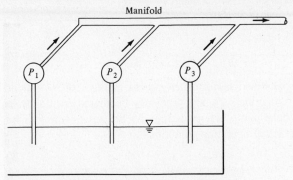

Figure 10–29 Pumps arranged in parallel.

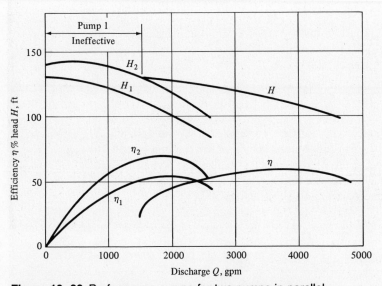

Figure 10–30 Performance curves for two pumps in parallel.

that pump becomes ineffective. Only when the operating head falls below the lower shutoff head does the second become effective and the head curve begin to reflect both pumps.

As with pumps in series, the combined efficiency is based on the power requirement, $P = P_1 + P_2$. In this case

$$\frac{\gamma(Q_1 + Q_2)H}{\eta} = \frac{\gamma Q_1 H}{\eta_1} + \frac{\gamma Q_2 H}{\eta_2}$$

or

$$\eta = \frac{Q_1 + Q_2}{(Q_1/\eta_1) + (Q_2/\eta_2)}$$

Generalizing to include any number of pumps,

$$\eta = \frac{\sum\limits_i Q_i}{\sum\limits_i (Q_i/\eta_i)} \tag{10–43}$$

The determination of the combined performance curves in Fig. 10–30 will be discussed in Example 10–12. As can be seen, the efficiency decreases drastically as the shutoff head for the lower head pump is approached. As before, a system curve can be plotted to determine the actual operating conditions.

EXAMPLE 10–12

The individual head/discharge and efficiency/discharge data for two pumps are tabulated below. Determine the head and efficiency curves when the two pumps are operated in parallel.

Solution

The data for selected points are given in the first five columns. These data are also plotted in Fig. 10–30 and identified by the appropriate subscripts. The values of combined discharge Q are obtained by adding the individual pump discharges at the selected heads. The resulting efficiency η is found from Eq. 10–43. A sample calculation follows the table.

H (ft) (1)	Q_1 (gpm) (2)	Q_2 (gpm) (3)	η_1 (%) (4)	η_2 (%) (5)	Q (gpm) (6)	η (%) (7)
140	—	0	—	0	0	—
143	—	400	—	27.5	400	—
130	0	1520	0	67.5	1520	—
129	350	1550	17	68	1900	43.8
125	770	1740	33	69.5	2510	51.9
120	1100	1930	43	69	3030	56.6
115	1380	2110	48	67.5	3490	58.2
110	1610	2280	53	64	3890	58.9
105	1830	2430	54	59.5	4260	57.0
100	2030	2570	54.5	53	4600	53.5

Sample calculations:

Head $H = 120$ ft

$$Q = 1100 + 1930 = 3030 \text{ gpm}$$

$$\eta = \frac{3030}{\left(\dfrac{1100}{43}\right) + \left(\dfrac{1930}{69}\right)} = 56.6 \text{ percent}$$

The resulting head and efficiency curves based on the final two columns have also been plotted in Fig. 10–30.

 PROBLEMS

Section 10–2

10–1. Determine the efficiency of a turbine that develops 6080 hp from an available head of 200 ft and a discharge of 300 cfs.

10–2. Determine the efficiency of a turbine that develops 21.5 MW from an available head of 50 m and a discharge of 50 m³/s.

10–3. If a turbine has an efficiency of 88 percent, determine the discharge required to develop 15,000 hp from a head of 300 ft.

10–4. If a turbine has an efficiency of 85 percent, determine the head necessary to produce 10 MW when the discharge is 20 m³/s.

10–5. Determine the efficiency of a 2-hp pump that delivers water at the rate of 6 ft/s through a 3-in. pipe against a head of 50 ft.

10–6. Determine the power required to pump 0.4 m³/s if the head is 500 m and the pump has an efficiency of 75 percent.

10–7. Determine the overall efficiency in Prob. 10–3 if the generator produces 9.84 MW.

10–8. If the pump in Prob. 10–6 has a volumetric efficiency of 98 percent and a mechanic efficiency of 90 percent, what is the hydraulic efficiency?

10–9. Redo Example 10–1 if the forebay level is 1000 ft above the nozzle elevation.

10–10. A Pelton turbine is arranged similar to Fig. 10–7. The available head at the nozzle is 550 ft and the discharge is 25 cfs. The nozzle has a loss coefficient of 0.055, the bucket angle is 162 deg, and the bucket friction coefficient is 0.98. The relative speed of the turbine will be 0.46. Determine the diameter, speed, power, and hydraulic efficiency if the turbine is connected to a 50-Hz generator.

10–11. A Pelton turbine is arranged similar to Fig. 10–7. The available head at the nozzle is 250 m and the discharge is 1 m³/s. The nozzle has a loss coefficient of 0.045, the bucket angle is 165 deg, and the bucket friction coefficient is 0.97. The relative speed of the turbine will be 0.455. Determine the diameter, speed, power, and hydraulic efficiency if the turbine is connected to a 50-Hz generator.

10–12. Write a computer program that will analyze the penstock and nozzle hydraulics for the arrangement shown in Fig. 10–7. Input the entrance loss coefficient, the penstock resistance coefficient f, the nozzle loss coefficient, and required discharge. Vary the penstock diameter and calculate and print out a table containing the nozzle diameter, the overall head loss, and the power in the jet as functions of the penstock diameter.

10–13. Append a program to that of Prob. 10–12 that will also calculate the turbine diameter, speed, power, and efficiency, given the relative speed, bucket friction coefficient, bucket angle, and generator frequency.

10–14. Repeat Prob. 10–12 changing the input to the absolute roughness and using Eq. 6–4 to calculate the resistance coefficient.

10–15. Repeat Prob. 10–13 substituting Prob. 10–14 for 10–12.

10–16. Plot a graph of hydraulic efficiency versus relative speed for the interval $0 < \phi < 1$, if a Pelton turbine has a wheel diameter of 2 m, a jet-to-wheel-diameter ratio of $\frac{1}{15}$, a bucket friction coefficient of 0.96, bucket angle of 165 deg, and a discharge of 1.05 m^3/s.

10–17. A Francis turbine has runner blade angles of $\alpha_1 = 135$ deg and $\alpha_2 = 150$ deg. The runner has radii of 0.65 m and 0.38 m and a constant width of 0.4 m. If the discharge is 10.5 m^3/s, determine the rotational speed necessary to avoid swirl as the water leaves the runner. If the available head at the turbine is 95 m, what power and efficiency are attained at this speed?

10–18. Under design conditions, a Francis turbine runs at 30 rpm when the discharge is 700 cfs. The inner and outer radii are 7.5 ft and 4 ft, respectively, and the runner has a constant width of 2.2 ft. If the guide vanes are placed at an angle of 30 deg relative to a tangent to the outer circumference, what blade angle is required for a smooth entrance? What angle α_2 is required to eliminate exit swirl? What power is developed?

10–19. Redo Example 10–2 if a discharge of 0.05 m^3/s is necessary.

10–20. A discharge of 5 cfs is required from a centrifugal pump with radii of 4 in. and 9 in. rotating at 700 rpm. The impeller width is 3 in. What entrance angle is required for smooth entry with a radial flow? What power is required if the exit angle is 165 deg?

Section 10–3

10–21. At maximum efficiency, a model turbine with a diameter of 90 cm develops 40.5 kW from a head of 12 m and a discharge of 0.4 m^3/s. Its speed is 234 rpm. Determine the power developed by a homologous unit 3.0 m in diameter if the head is 40 m. At what speed will the prototype operate with greatest efficiency? Use the relative speed to determine the type of turbine. Estimate the prototype efficiency.

10–22. Repeat Prob. 10–21 if the model speed was 460 rpm.

10–23. Repeat Example 10–3 if the turbine drives a 50-Hz generator.

10–24. Determine the necessary speed for the turbine in Prob. 10–21 if the generator produces a 60-Hz current.

10–25. Repeat Prob. 10–24 if the generator produces a 50-Hz current.

10–26. A model turbine with a 25-in.-diameter produces 61 hp at maximum efficiency from a head of 60 ft and a discharge of 10 cfs. The speed during the test was 1100 rpm. Determine the size, speed, and approximate power of a homologous unit if the available head is 200 ft and the discharge is 200 cfs. The turbine is connected to a 60-Hz generator. What is the best estimate of prototype efficiency?

10–27. Repeat Prob. 10–26 if the model speed is 550 rpm. Based on the relative speed, what type of turbine is indicated?

10–28. A 12-in.-diameter pump delivers 1200 gpm against a pump head of 70 ft when rotating at 1750 rpm. Determine the discharge and head that will result from a similar pump at that speed if the diameter is reduced to (1) 10 in. and (2) 8 in.

10–29. Determine the discharge and head that will result from the 12-in. pump in Prob. 10–28 if the pump speed is reduced to (1) 1150 rpm and (2) 550 rpm.

10–30. If a 10-in.-diameter pump delivers 800 gpm against a head of 50 ft with an efficiency of 65 percent, what is the power requirement? Assuming a constant efficiency, determine the discharge and head if the pump speed is increased from the original speed of 1100 rpm to (1) 1200 rpm and (2) 1800 rpm.

10–31. Determine the diameter required for a pump similar to that in Prob. 10–30 if 5 cfs are to be pumped against a head of 60 ft. At what speed must it rotate and what is the power requirement, assuming constant efficiency? What is the best estimate of efficiency for this pump?

10–32. If a 25-cm-diameter pump requires 200 kW to deliver 0.33 m^3/s of water against a head of 42 m when rotating at a speed of 1200 rpm, what is its efficiency? Assuming a constant efficiency, what would be the discharge of a similar 20-cm-diameter pump operated at 1800 rpm?

10–33. A model study is to be performed of a large 10-ft-diameter pump that is to deliver a discharge of 400 cfs against a head of 100 ft when rotated at 250 rpm. Determine the size and speed requirements for the model if the laboratory head and discharge must be 45 ft and 2.5 cfs, respectively.

10–34. Does the Pelton turbine in Prob. 10–10 have an acceptable specific speed?

10–35. Calculate the specific speeds for the model and prototype turbines in Prob. 10–21. What type of turbine is indicated?

10–36. Calculate the specific speeds for the model and prototype turbines in Prob. 10–26. What type of turbine is indicated?

10–37. Determine the specific speeds for the three pump diameters in Prob. 10–28. What type of pump is indicated?

10–38. What is the specific speed for the pump in Prob. 10–31? What type of pump is indicated?

10–39. Determine the specific speeds at both rotational speeds for the pumps in Prob. 10–32. What type of pump is indicated?

10–40. Determine the specific speeds for both the model and prototype pumps in Prob. 10–33. What type of pump is indicated?

10–41. Determine the conversion factor for pump specific speed between U.S. customary units in gpm and cfs.

10–42. Determine the conversion factor for pump specific speed between SI units in l/s and m³/s.

10–43. Verify the stated conversion factor for turbine specific speed between U.S. customary and SI units.

Section 10–4

10–44. Repeat Example 10–6 if the pump rotated at 600 rpm, all other conditions remaining unchanged.

10–45. Repeat Example 10–6 assuming a pump head of 25 m, all other conditions remaining unchanged.

10–46. A pump with a specific speed of 3000 (in gpm units) lifts water with a temperature of 70°F from a reservoir. Determine the maximum elevation at which the pump may be placed above the reservoir surface and still avoid cavitation. The pump head is 50 ft and the discharge is 3 cfs. The suction line has a diameter of 10 in. and total head loss of 3 ft.

10–47. Repeat Prob. 10–46 assuming the pump has a specific speed of 7000.

10–48. A pump with an NPSH = 20 ft is used to pump water from a reservoir at the rate of 4 cfs. The pump head is 40 ft and the water temperature is 60°F. The suction line has a rounded entrance with a loss coefficient of 0.06. It contains a check valve with a loss coefficient of 2.5 and a long-radius 90 deg elbow. Further, it has a diameter of 10 in. and a length of 30 ft. Assume that the pipe has an absolute roughness of 0.0001 ft. Determine the maximum elevation relative to the reservoir surface at which the pump can be located.

10–49. Repeat Prob. 10–48 assuming that the pump has an NPSH = 40 ft.

10–50. Repeat Prob. 10–48 assuming that the pump has an NPSH = 10 ft.

10–51. Determine the maximum level at which a Francis turbine may be located above the tailrace and still avoid cavitation problems. The water temperature is 40°F, the effective head is 100 ft and the flow rate is 300 cfs. In addition, the turbine has a specific speed of 45 and the velocity entering the draft tube is 20 ft/s. Assume that the total head loss through the draft tube is 3 ft.

10–52. Repeat Prob. 10–51 for a water temperature of 70°F.

10–53. Repeat Prob. 10–51 if the turbine has a specific speed of 70.

10–54. Repeat Prob. 10–51 if the effective head was only 50 ft and a propeller turbine with a specific speed of 160 was selected.

Section 10–5

10–55. Construct the efficiency hill based on a unit head for the model turbine data given on the facing page. The model has a diameter of 0.8 m.

10–56. Use the efficiency hill of Prob. 10–55 to select the diameter and speed of a homologous turbine that will produce 5000 kW from a head of 35 m. The generator will

Model Turbine Data

Run Number	Gate Position (%)	Head (m)	Speed (rpm)	Power (kW)	Efficiency (%)
1	100	5.80	127.6	154.9	79.5
2	100	5.80	154.1	159.2	82.0
3	100	5.80	175.8	162.0	84.0
4	100	5.75	204.1	161.5	89.0
5	100	5.75	225.4	160.1	86.0
6	100	5.70	243.5	152.3	80.0
7	89	5.85	130.6	144.5	81.0
8	89	5.85	149.7	149.0	83.0
9	89	5.80	180.6	151.8	89.3
10	89	5.80	195.3	152.3	92.4
11	89	5.75	216.1	147.0	87.3
12	89	5.75	244.6	134.0	80.2
13	84	5.85	132.8	128.8	82.1
14	84	5.85	159.1	134.6	86.4
15	84	5.85	181.2	137.2	90.2
16	84	5.85	189.1	137.4	89.8
17	84	5.80	216.7	132.0	85.1
18	84	5.80	247.1	119.7	79.0
19	79	5.85	128.2	110.6	81.0
20	79	5.85	157.2	116.6	85.7
21	79	5.85	186.5	119.3	86.2
22	79	5.80	204.9	115.9	83.8
23	79	5.80	224.2	111.2	80.9
24	79	5.80	245.4	100.0	77.3
25	69	5.85	133.0	91.7	79.5
26	69	5.85	145.1	94.9	81.8
27	69	5.80	171.0	97.2	82.8
28	69	5.80	192.9	94.8	81.6
29	69	5.80	226.9	85.1	78.0
30	69	5.80	244.7	78.1	76.0
31	58	5.90	131.2	76.1	78.0
32	58	5.90	143.3	78.8	79.4
33	58	5.85	157.5	78.8	80.5
34	58	5.85	184.5	76.4	80.0
35	58	5.85	208.2	72.0	78.1
36	58	5.85	244.5	62.8	74.8

produce a 60-Hz current. What is the best estimate of prototype efficiency? What discharge will be required?

10–57. Determine and plot performance curves (i.e., power versus efficiency) for the turbine in Prob. 10–56. Use heads of 30 m, 35 m, and 40 m.

10–58. Use the efficiency hill of Prob. 10–55 to select the diameter and speed of a homologous turbine that will produce 40,000 kW from a head of 50 m. The generator will produce a 60-Hz current. What is the best estimate of prototype efficiency? What discharge will be required? What type of turbine is indicated?

10–59. Determine and plot performance curves (i.e., power versus efficiency) for the turbine in Prob. 10–58. Use heads of 45 m, 50 m, and 55 m.

10–60. Determine the optimum size and speed for a turbine homologous to that in Prob. 10–55. The power requirement is 110 MW and the available head is 80 m. Ignore the synchronous speed, and also note that a solution line need not be plotted on the efficiency hill. What are the best estimates of prototype efficiency and discharge? What type of turbine is indicated?

10–61. Repeat Prob. 10–60 if the available head is 120 m.

10–62. Assume that the model turbine used to develop Fig. 10–21 has a diameter of 25 in., and use the efficiency hill to select the diameter and speed of a homologous turbine that will produce 5000 hp from a head of 70 ft. The generator will produce a 60-Hz current. What is the best estimate of prototype efficiency? What discharge will be required? What type of turbine is indicated?

10–63. Determine and plot performance curves (i.e., power versus efficiency) for the turbine of Prob. 10–62. Use heads of 60 ft, 70 ft, and 80 ft.

10–64. Select a pump to deliver 1700 gpm from a lower to an upper reservoir. The elevation difference is 50 ft, and the connecting pipe has a diameter of 10 in. and a length of 800 ft. Ignore possible cavitation problems. The pipe has an absolute roughness of 0.0001 ft and the minor losses total to $6(V^2/2g)$. Use Figs. 10–25 and 10–26 to select the best pump and determine the actual operating conditions.

10–65. The pump in Prob. 10–64 is located 20 ft from the inlet and the total head loss in the suction line equals two velocity heads. What is the maximum elevation of the pump relative to the lower reservoir surface?

10–66. Repeat Prob. 10–64 if a discharge of 1500 gpm is required.

10–67. Repeat Prob. 10–64 if a discharge of 900 gpm is required.

10–68. The head/discharge and efficiency/discharge curves for a centrifugal pump are tabulated below:

Discharge (m^3/s)	Head (m)	Power (kW)	Efficiency (%)
0	28.0	104.0	0
0.1	27.9	113.6	24.1
0.2	27.7	126.4	43.0
0.3	27.2	140.4	57.0
0.4	26.6	156.3	66.8
0.5	25.8	173.8	72.8
0.6	24.5	192.3	75.0
0.7	22.9	216.6	72.6
0.8	20.4	238.6	67.1
0.9	16.5	246.1	59.2

The pump is used to lift water up 20 m from a lower to an upper reservoir. The connecting pipe has a diameter of 0.5 m and a length of 100 m. Assume $f = 0.018$ and minor losses equal to 3 velocity heads. Determine the discharge, power requirement, and pump efficiency.

10–69. Redo Prob. 10–68 if the difference in reservoir levels is reduced to 10 m.

10–70. Determine the pipe diameter in Prob. 10–68 at which the pump will operate at maximum efficiency. What are the resulting discharge, head, power, and efficiency?

10–71. Repeat Prob. 10–68 if the connecting pipe has a length of 500 m.

10–72. To increase the head, two pumps from Fig. 10–26b using 10-in. impellers will be connected in series. The pumps will be used in a system where the overall lift is 105 ft. The pipe has a diameter of 6 in., a resistance coefficient $f = 0.018$, and a length of 400 ft. Assume that the minor losses total five velocity heads. Ignore the possibility of cavitation and determine the actual operating conditions of head, discharge, efficiency, and power required.

10–73. Redo Prob. 10–72 if three pumps are used to achieve an overall lift of 155 ft.

10–74. Plot the head, efficiency, and power curves if three identical pumps whose characteristics are given in Prob. 10–68 are used in series.

10–75. Plot the head, efficiency, and power curves if the 9-in. and 12-in. pumps in Fig. 10–26a are used in series.

10–76. Plot the head, efficiency, and power curves if the 9-in. and 12-in. pumps in Fig. 10–26c are used in series.

10–77. To increase the discharge, two pumps from Fig. 10–26b with 10-in. impellers will be connected in parallel. The overall lift of the system is 45 ft, the pipe diameter is 12 in., the pipe length is 400 ft, and the resistance coefficient is 0.018. The minor losses total five velocity heads. Ignore the possibility of cavitation and determine the expected head, discharge, efficiency, and power requirement.

10–78. Change the pipe diameter to 16 in. and redo Prob. 10–77 using three pumps.

10–79. Plot the head, efficiency, and power curves if three identical pumps whose characteristics are given in Prob. 10–68 are used in parallel.

10–80. Plot the head and efficiency curves if the 11-in. and 12-in. pumps in Fig. 10–26a are used in parallel.

10–81. Plot the head and efficiency curves if the 11-in. and 12-in. pumps in Fig. 10–26c are used in parallel.

References

Addison, H., *Centrifugal and Other Rotodynamic Pumps,* 3rd ed., London: Chapman and Hall, 1966.

Daily, J. W., "Hydraulic Machinery," Chapter 13 in *Engineering Hydraulics,* H. Rouse (ed.), New York: Wiley, 1950.

Dake, J. M. K., *Essentials of Engineering Hydraulics,* New York: Wiley Interscience, 1972.

Fair, G. M. and J. C. Geyer, *Water Supply and Waste-Water Disposal*, New York: Wiley, 1954.

Hicks, T. G. and T. W. Edwards, *Pump Application Engineering*, New York: McGraw-Hill, 1971.

Moody, L. F., and T. Zowski, "Hydraulic Machinery," Section 26 in *Handbook of Applied Hydraulics*, C. V. Davis and K. E. Sorensen (eds.), New York: McGraw-Hill, 1984 (reissue).

Norrie, D. H., *An Introduction to Incompressible Flow Machines*, New York: American Elsevier, 1963.

Parna, P. S., *Fluid Mechanics for Engineers*, London: Butterworth, 1957.

Prasuhn, A. L., *Fundamentals of Fluid Mechanics*, Englewood Cliffs, NJ: Prentice-Hall, 1980.

Shepherd, D. G., *Principles of Turbomachinery*, New York: Macmillan, 1956.

Wislicenus, G. F., *Fluid Mechanics of Turbomachinery*, 2 vols., New York: Dover, 1965.

chapter
11

Drainage Hydraulics

Figure 11–0 Debris accumulation against piers may lead to upstream flooding and/or increased scour problems (Prasuhn).

11

11-1 INTRODUCTION

This chapter will draw as much as possible from previous chapters and apply the hydraulic and hydrologic principles to the subject of drainage hydraulics. The material considered here will be limited to the estimation of surface drainage or runoff from urban or rural areas and the hydraulics of selected structures required to handle this runoff. However, neither the collection of urban runoff by gutter, storm sewer inlet, and storm sewer nor the subsurface collection of runoff by drains or tile fields will be covered. For reference to these subjects, the reader is directed to the literature in environmental and agricultural engineering, respectively.

What remains is the extremely important topic of drainage structures such as bridges and culverts. As they occur so frequently at the intersection of a water course with transportation systems, they are of interest not only to the hydraulic engineer but also to the bridge or transportation engineer. Although bridges and culverts are an important part of rail transportation systems, and to a lesser degree airport runways and taxiways, most applications are probably associated with highway crossings and the treatment will be slanted in that direction.

11-2 ESTIMATION OF DRAINAGE RUNOFF

For large projects such as river improvements for navigation, a dam, or major flood protection works, hydrologic data will frequently be available. In fact, for long-term projects a program of data collection will often be initiated while the project is in the early planning stages. In larger drainage basins or watersheds, there may be one or more precipitation recording stations that will provide a satisfactory estimation of rainfall. The runoff may then be calculated using the unit hydrograph method (discussed in the chapter on hydrology) or one of many computer-based mathematical modeling techniques. In a large basin with one or more stream gauging stations, the runoff record may actually be measured at or near the point—for example, a highway crossing—where the information is needed. If the gauging station is some distance away from the crossing, then stream flow routing may still provide the necessary information.

In many instances, a project site will be at a location where, at best, very limited data are available. For smaller projects, such as a culvert, small bridge, or bank protection for a small stream, this will usually be the case. The engineer may be faced with a complete absence of hydrologic data at even relatively large bridge crossings. The emphasis in this section will be placed on these ungauged basins and the estimation of discharge therefrom.

Estimating Precipitation in the Ungauged Basin

Actual hydrologic data should be used whenever possible. However, we will now look at alternatives to a satisfactory record of data at a project site. Our first concern will be with the estimation of precipitation. Various design aids are available to assist the engineer. One particularly important tool is the series of rainfall maps prepared by the U.S. National Weather Service.[1] Selected examples are included in Figs. 11–1 through 11–8. These maps give the expected rainfall throughout the continental United States during specified time intervals for a number of different recurrence intervals. They are based on a statistical analyses of extensive recorded rainfall data. The complete set of maps covers durations from 30 min to 24 hr and recurrence intervals from 1 yr to 100 yr. They can be used to estimate rainfall when actual data are unavailable. They have also been used by various agencies as the basis for the calculation of design discharges in ungauged basins.

Under the circumstances, it is logical to begin with an estimate of the precipitation, because the runoff depends on this as well as many additional parameters. Although a map of runoff per unit area could be drawn in much the same way as a precipitation map, its accuracy when applied to an ungauged basin would be far lower.

Estimating Runoff in the Ungauged Basin

The classic approach in this situation is to use the *rational equation* introduced previously by Eqs. 3–11a and 3–11b.

$$Q = ciA \qquad \text{(U.S. customary units)} \tag{11–1a}$$

and

$$Q = \frac{ciA}{360} \qquad \text{(SI units)} \tag{11–1b}$$

Only three parameters are needed to use this equation. The first is the area A of the drainage basin (in acres or hectares for U.S. customary or SI units, respectively). This area is readily obtained from the U.S.G.S. or other topographic maps. After identifying the watershed divide from the contour lines, the area can be measured using a planimeter.

The second variable is the rainfall intensity (in./hr or mm/hr). This may be estimated using a statistical analysis of measured precipitation data and a selected return period. Alternatively, the expected precipitation for that specific region of the country can be estimated from the U.S. National Weather Service rainfall maps. Depending on the significance of the structure to be designed and the associated risks of overtopping and upstream flooding, the rainfall return period may range from as little as 5 years to as much as 100 years or more.

---

[1] U.S. National Weather Service, *Rainfall Frequency Atlas for the United States,* Tech. Paper 40, 1961.

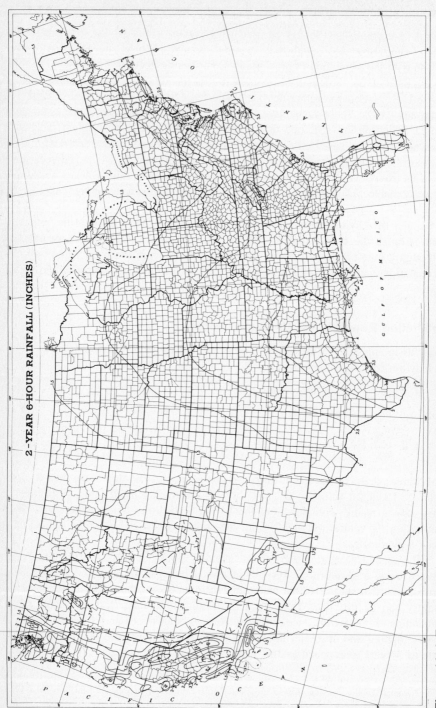

2-YEAR 6-HOUR RAINFALL (INCHES)

Figure 11–1 Two-year, 6-hr rainfall (in.) (U.S. National Weather Service NOAA).

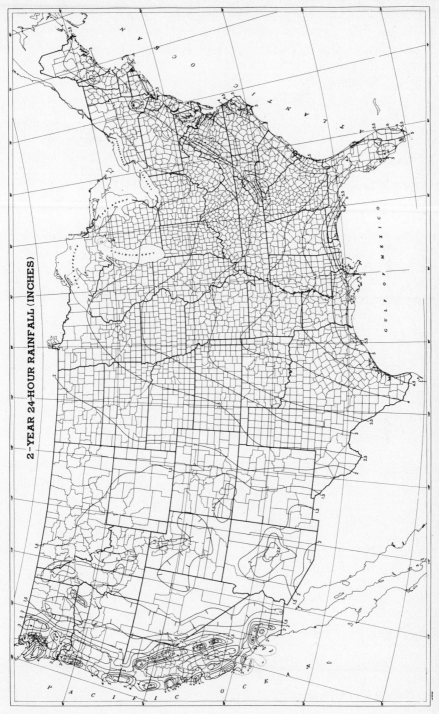

2-YEAR 24-HOUR RAINFALL (INCHES)

Figure 11–2 Two-year, 24-hr rainfall (in.) (U.S. National Weather Service NOAA).

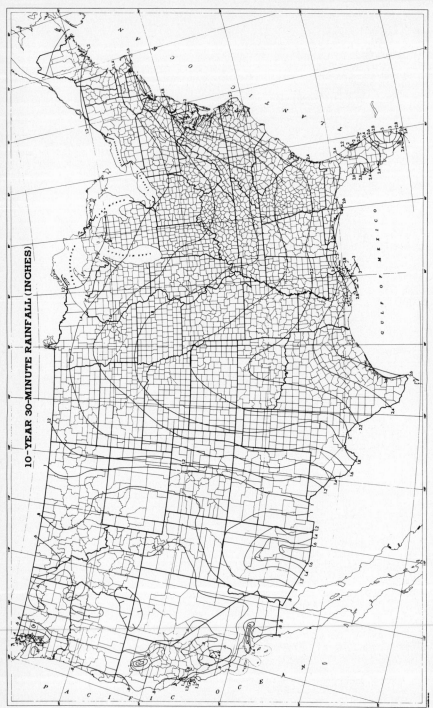

Figure 11–3 Ten-year, 30-min rainfall (in.) (U.S. National Weather Service NOAA).

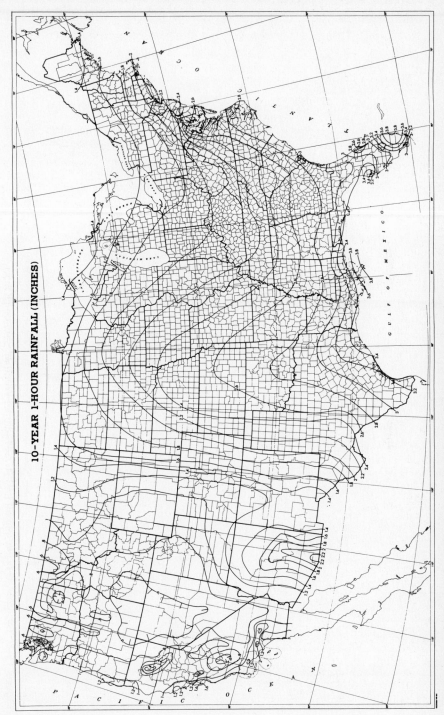

Figure 11-4 (Ten-year, 1-hr rainfall (in.) (U.S. National Weather Service NOAA).

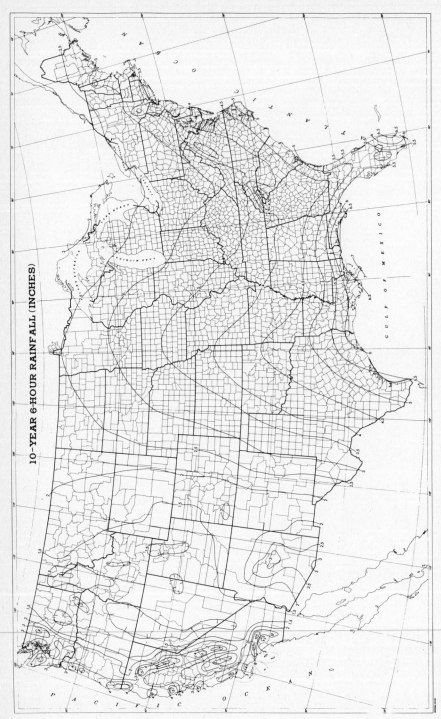

Figure 11–5 Ten-year, 6-hr rainfall (in.) (U.S. National Weather Service NOAA).

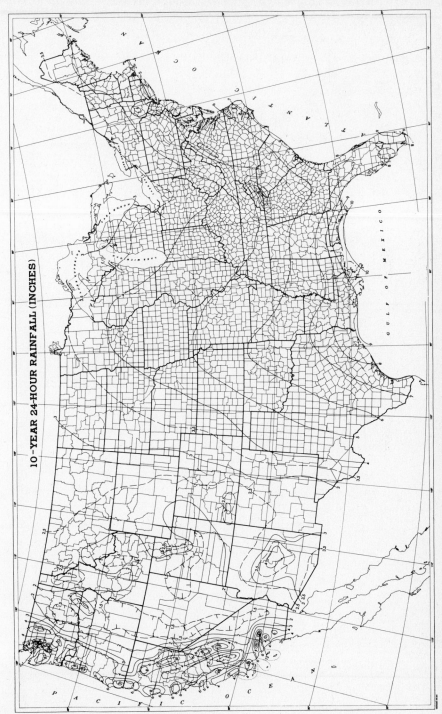

10-YEAR 24-HOUR RAINFALL (INCHES)

Figure 11–6 Ten-year, 24-hr rainfall (in.) (U.S. National Weather Service NOAA).

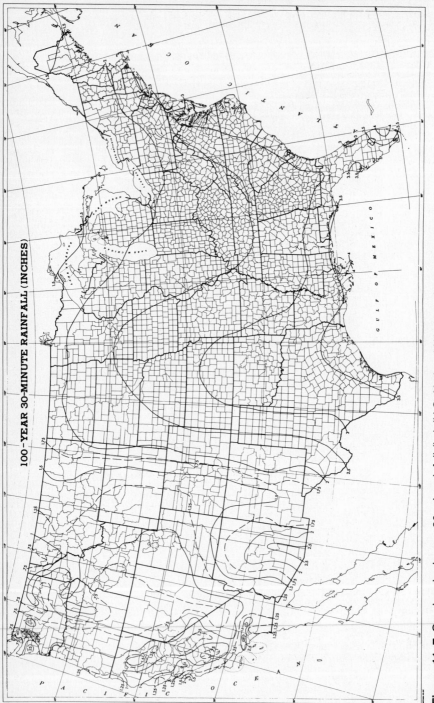

100-YEAR 30-MINUTE RAINFALL (INCHES)

Figure 11–7 One-hundred-year, 30-min rainfall (in.) (U.S. National Weather Service NOAA).

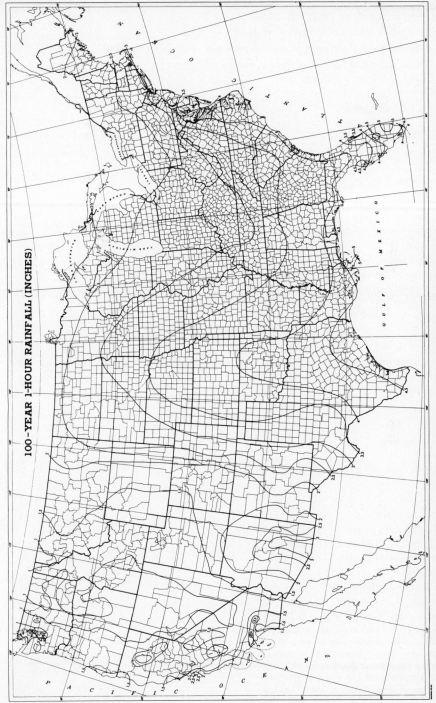

Figure 11-8 One-hundred-year, 1-hr rainfall (in.). (U.S. National Weather Service NOAA).

The final requirement is the value of the runoff coefficient c. A variety of typical values are included in Table 11–1.

The coefficients in the first portion of the table are selected on the basis of the land use that best describes the area in question. Judgment would have to be applied to estimate the most likely value of c within the specified range. The second portion

Table 11–1 VALUES OF C FOR THE RATIONAL EQUATION*

Description	c
Area	
Business	
Downtown	0.70–0.95
Neighborhood	0.50–0.70
Residential	
Single-family	0.30–0.50
Multiunits, detached	0.40–0.60
Multiunits, attached	0.60–0.75
Apartments	0.50–0.70
Industrial	
Light	0.50–0.80
Heavy	0.60–0.90
Parks, cemeteries	0.10–0.25
Playgrounds, railroad yards	0.20–0.35
Unimproved	0.10–0.30
Farmland, pasture	0.05–0.30
Forested area	0.05–0.20
Surface character	
Pavement	
Asphalt, concrete	0.70–0.95
Brick	0.70–0.85
Roofs	0.75–0.95
Lawns, sandy soil	
Flat, up to 2% slope	0.05–0.10
Average, 2–7% slope	0.10–0.15
Steep, over 7% slope	0.10–0.20
Lawns, heavy soil	
Flat, up to 2% slope	0.13–0.17
Average, 2–7% slope	0.18–0.22
Steep, over 7% slope	0.25–0.35

*Tabulated from *Design and Construction of Sanitary and Storm Sewers,* ASCE Manual and Report of Engineering Practice, No. 37, 1970. A. T. Hjelmfelt, and J. J. Cassidy, *Hydrology for Engineers and Planners,* Ames, IA: Iowa State University Press, 1975, and other sources.

of the table gives an estimated c value based on the description of the surface itself. In either case, an area that can be subdivided into identifiable types of subareas can be assigned an average runoff coefficient by weighing the various c values according to the relative magnitudes of the respective areas. The procedure is inherently inaccurate because of the large number of significant independent variables that are, at best, only implicitly included in the rational equation. However, ease of application, particularly when lacking an alternative, has led to its frequent use. Its accuracy is increased by selecting areas that are as small as practicable and limiting its use to storms with return periods of less than 10 years.

EXAMPLE 11–1

If the rainfall intensity is 2 in./hr, estimate the maximum rate of runoff from a 100-acre site consisting of lawns of which 10 percent of the area has a slope in excess of 7 percent, 35 percent of the area has a slope between 2 and 7 percent, and the remainder is flat. Assume that the soil is sandy.

Solution

A composite coefficient for the rational equation can be obtained

$(0.10)(0.18) = 0.018$

$(0.35)(0.12) = 0.042$

$(0.55)(0.08) = \underline{0.044}$

Sum $= 0.104$

Using an intermediate c value for each range of slope, a weighted value $c = 0.10$ results. Thus, the peak discharge to be expected on this basin from a rainfall intensity of 2 in./hr is

$Q = (0.10)(2.0)(100) = 20$ cfs

Various agencies have developed their own procedures for estimating local runoff. These are often based on the U.S. National Weather Service records as applied to a particular region, often a state, and depend on different parameters that are felt to be most reliable for the given region. The U.S. Department of Agriculture Soil Conservation Service has prepared an *Engineering Field Manual* on a state-by-state basis that includes the necessary charts and procedures to estimate volume, depth, and peak rate of runoff using parameters that include area, slope, soil type, land use, and the U.S. National Weather Service maps of 24-hr rainfall of different return periods. The list of available methods are far too numerous to cover exhaustively, but one example will be illustrated.

The U.S. Geological Survey conducted a study[2] to develop a procedure to estimate the peak discharge from an ungauged basin in South Dakota. By using data from 123 gauged sites, the peak discharge for return periods of 2, 5, 10, 50, and 100 years was found to be a function of the basin area A (sq. mi.), the slope S (ft/mi.), and the soil-infiltration index Si (in.). The resulting equations are

$$Q_2 = 21.2A^{0.48}S^{0.44}Si^{-1.16} \tag{11-2a}$$

$$Q_5 = 41.2A^{0.53}S^{0.42}Si^{-0.91} \tag{11-2b}$$

$$Q_{10} = 57.7A^{0.56}S^{0.43}Si^{-0.80} \tag{11-2c}$$

$$Q_{25} = 83.4A^{0.60}S^{0.44}Si^{-0.72} \tag{11-2d}$$

$$Q_{50} = 106A^{0.63}S^{0.45}Si^{-0.69} \tag{11-2e}$$

$$Q_{100} = 132A^{0.65}S^{0.46}Si^{-0.67} \tag{11-2f}$$

The area is readily determined from topographic maps. The slope is the average main channel slope between points 10 and 85 percent of the distance along the mainstream channel, as measured from the site (i.e., at the bridge or culvert) to the basin divide. The soil-infiltration index is a measure of the capacity of the soil to absorb moisture. Values are given in Fig. 11-9, a map based on data furnished by the U.S. Soil

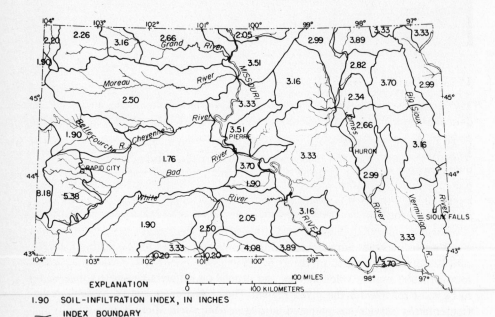

EXPLANATION

1.90 SOIL-INFILTRATION INDEX, IN INCHES

—— INDEX BOUNDARY

Figure 11–9 Map showing soil-infiltration indexes in South Dakota (U.S. Geological Survey; data provided by the U.S. Soil Conservation Service).

[2]L. D. Becker, ''Techniques for Estimating Flood Peaks, Volumes, and Hydrographs on Small Streams in South Dakota,'' U.S.G.S. Water Resources Investigations 80–80, Sept. 1980.

Conservation Service. The procedure is presumably valid for basins ranging from 0.05 to 100 square miles.

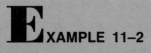XAMPLE 11–2

Estimate the 25-yr and 100-yr peak discharges to be expected on a small 5-square-mile basin tributary to the Moreau River in South Dakota (see Fig. 11–9). Assume the channel slope is 150 ft/mi.

Solution

With reference to Fig. 11–9, the value of $Si = 2.50$ in. This, along with $A = 5$ sq. mi. and $S = 150$ ft/mi. may be substituted into Eq. 11–3, parts d and f, to yield

$$Q_{25} = 83.4(5.00)^{0.60}(150)^{0.44}(2.50)^{-0.72} = 1030 \text{ cfs}$$

and

$$Q_{100} = 132(5.00)^{0.65}(150)^{0.46}(2.50)^{-0.67} = 2040 \text{ cfs}$$

for the 25-yr and 100-yr return periods, respectively.

11–3 CULVERT HYDRAULICS

The *culvert* is such a common structure that its analysis and design have become quite standardized. It may be manufactured from a range of materials, including precast concrete, reinforced concrete, cast iron, corrugated steel, and vitrified clay; each of the different pipe design manuals (the concrete pipe manual is included in the references at the end of the chapter) contains a section on the selection of the proper size and type of culvert. Careful choice of material and economical design are essential, because the drainage system and structures are a significant portion of the total cost of highways and transportation systems.

Most small- to moderate-size culverts are either circular in cross section or some other shape such as elliptical or arched, which the manufacturer can equate to a circular section. These latter shapes are similar in their analysis to the circular culvert, but recourse must be made to the specific design manual. The larger concrete culverts, called box culverts, are frequently rectangular in cross section. In this form, they differ little from the bridges considered in the next section. A hydraulic distinction might be made in that the box culvert, or any culvert, is designed for the possibility of flowing full. When water reaches a bridge deck, on the other hand, this is usually considered to be a flood condition.

The main component of the culvert is the pipe or *barrel,* which carries the drainage water under the roadway, usually through a fill embankment. The barrel may be constructed from any of the aforementioned materials. The other main structural features of the culvert include the *headwall* or *wingwall* at the entrance and the *endwall* at the exit. These are normally constructed of concrete. The headwall or wingwall is intended to protect the embankment from erosion and improve the hydraulic characteristics as the water enters the culvert. A flared wingwall is shown in Fig. 11–10a. For the smaller culvert, however, this would not be justified economically, and it would be simplified to a straight headwall perpendicular to the barrel or dispensed with altogether. The outlet structure serves primarily to prevent erosion and undercutting of the culvert. If used at all, this may also have a flared shape, or the less expensive U endwall of Fig. 11–10b may suffice.

Flooding upstream of a culvert is often the result of debris that has entered and clogged the culvert. To reduce this problem, a debris barrier is sometimes placed ahead of the entrance. It may take almost any form, although it is often constructed of metal bars or small-diameter pipes. Its design should allow for easy cleaning and should be so placed that debris accumulation on the barrier does not in itself result in either upstream flooding or plugging of the culvert.

Whenever possible, the axis of the barrel should follow the alignment of the natural stream. Not only is this more efficient hydraulically, but it is also less likely to lead to erosion problems. If the road alignment is skewed to the axis of the natural stream, then the alternative of a longer culvert that follows the natural channel must be weighed against the shorter culvert that can be used if the stream is realigned normal to the highway. If the option of stream realignment is chosen, care must be taken to protect both the embankment and the channel from the increased tendency for erosion.

The hydraulics of culvert design will be discussed with respect to Fig. 11–11. A number of interrelated factors are involved in the hydraulics of a culvert. These include the relative levels of the headwater *HW* and tailwater *TW,* the determination of whether the culvert flows full or not, and if it does not, it must be determined whether the slope of the culvert, which then becomes an open channel is mild or steep. To evaluate the above effects, the flow in a given culvert will be placed into

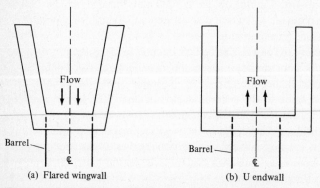

Figure 11–10 Typical culvert wingwall and endwall.

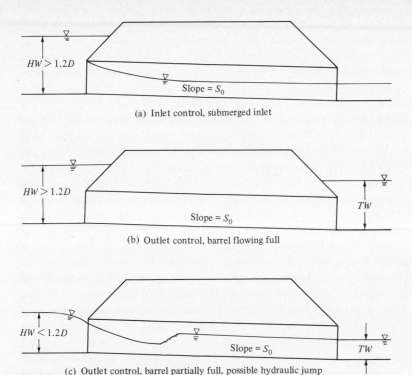

Figure 11–11 Culvert hydraulics.

one of two categories. In much the same way as open channel flow analysis depended on upstream versus downstream control, the flow/culvert system will be referred to as having inlet control or outlet control. In the first case, the discharge entering and passing through the culvert depends only on the entrance conditions, that is, the headwater elevation and the size, shape, and other details of the inlet geometry. The barrel can actually carry more water than is able to enter the culvert. Under these conditions, slope, friction, and barrel length have no effect on the discharge. In addition, the culvert will generally flow only partially full.

In the second case, where outlet control applies, the discharge through the culvert will be a function of all of the inlet factors listed above, as well as the tailwater elevation and the barrel factors such as length, diameter, roughness, and slope. This condition applies when water is able to enter the culvert more rapidly than it can flow through it. As the analysis will demonstrate, the culvert may flow full or only partially full.

In practice, the hydraulic analysis and subsequent selection of the proper culvert size is aided by charts and nomographs prepared for the specific shape and type of culvert. Such factors as the entrance loss coefficient for a particular pipe material may be incorporated directly into these procedures. The textbook presentation that follows is perforce more general and, therefore, more time consuming than direct recourse to design charts. However, comprehension of this material will result in a better understanding of the entire subject of culvert design and selection and will make the use of design manual procedures clearer and more reliable.

The analysis of culvert hydraulics requires the use of the energy equation along with a resistance equation, usually the Manning equation. To eliminate redundancy, the Manning equation will be specified in U.S. customary units. When applying SI units, the numerical value of 1.49 can always be replaced by unity.

If the headwater *HW* (see Fig. 11–11) is less than 1.2*D*, the entrance will not remain submerged and air will break into the culvert. Thus, free surface flow (with outlet control) will usually occur under this condition. For this to occur, the tailwater *TW* must also be low. The complete analysis would require determining whether the slope is mild or steep and evaluating the water surface profile by the open-channel methods of Chapter 7. If the slope is steep, a hydraulic jump will occur within or downstream of the culvert if the tailwater is subcritical. Economical culvert design normally requires the use of an allowable value of *HW* greater than 1.2*D*. Consequently, this case will not be pursued further.

The evaluation of whether the control is at the inlet or outlet is best made by assuming both cases and comparing the results. In the analysis of a given culvert, the resulting discharge for specified values of *HW* and *TW* or the value of *HW* (important because of upstream flooding) given *TW* and discharge would be the usual goals. If the discharge is determined by assuming both inlet and outlet control, the lower of the two values would occur. In the same way, if the value of *HW* is needed, then the higher value would be expected. On the other hand, the design problem of selecting a satisfactory culvert size, say a diameter, would be based on evaluating both types of control and choosing the larger diameter.

The analysis of inlet control shown in Fig. 11–11a is based on the similarity between the flow into a culvert and flow through an orifice. An orifice equation may be written

$$Q = C_d A \sqrt{2gh} \tag{11–3}$$

where *A* is the cross-sectional area of the culvert, *h* is the head relative to the center of the entrance (i.e., $h = HW - D/2$), and C_d is the discharge coefficient for the entrance. Obviously, much of the uncertainty is associated with this coefficient of discharge, which must reflect not only the characteristics of the end of the pipe or barrel and the type of wingwall or headwall, but also the direction and magnitude of the approaching flow. However, C_d generally ranges from a low of about 0.62 for a square-edged inlet up to approximate unity for a well-designed, well-rounded highly efficient entrance.

The analysis of the outlet control shown in Fig. 11–11b, requires application of the energy equation between points upstream and downstream of the culvert where the depths are *HW* and *TW*, respectively. In a reasonably general form, the energy equation may be stated as

$$HW + S_0 L + \frac{V_0^2}{2g} = TW + H_L \tag{11–4}$$

The total head upstream of the culvert is H_1:

$$H_1 = HW + S_0L + \frac{V_0^2}{2g}$$

The term S_0L is frequently small, particularly for short culverts, and V_0 is the approach velocity, which usually is negligible because of the inlet submergence at the design discharge. The downstream total head is

$$H_2 = TW$$

and the total head loss H_L is

$$H_L = C_L\frac{V^2}{2g} + \frac{V^2}{2g} + \frac{n^2V^2L}{2.22R^{4/3}}$$

where V is the average velocity in the culvert and the three terms on the right-hand side are respectively the entrance loss (see Fig. 6–3), the exit loss, and the friction loss using the Manning equation (Eq. 6–8). The coefficient of entrance loss, C_L is given approximately in Fig. 6–3, with values ranging from about 0.05 for a well-rounded entrance to 0.8 for a reentrant pipe. If there is a downstream submergence, the flow may continue downstream with more or less the same velocity, V. In this case $V^2/2g$ is an energy rather than loss term, but it doesn't affect the analysis. Manning n values for pipe materials are tabulated in Table 6–3. Ignoring the approach velocity, Eq. 11–4 becomes

$$H_L = HW + S_0L - TW \tag{11–5}$$

and the head loss from this equation may be related to the average velocity and discharge by the following:

$$H_L = \left(C_L + 1 + \frac{gn^2L}{1.11R^{4/3}}\right)\frac{V^2}{2g} \tag{11–6}$$

and

$$H_L = \left(C_L + 1 + \frac{gn^2L}{1.11R^{4/3}}\right)\frac{8Q^2}{\pi^2gD^4} \tag{11–7}$$

A circular barrel has been assumed, which further means that the hydraulic radius $R = D/4$.

In Fig. 11–11b, the barrel will continue to flow full even though the tailwater TW may drop below the soffit (or top) of the barrel, provided that the normal depth associated with the particular discharge is greater than the diameter D. Under these circumstances, the analysis remains unchanged. The actual value of TW depends upon downstream conditions and would have to be evaluated by extending a backwater analysis upstream to the culvert from some downstream control. Computer program 6 in Appendix E is based on the above procedures.

EXAMPLE 11–3

Determine the flow rate through a 50-ft-long, 18-in.-diameter steel culvert with a 1 percent slope if $HW = 5$ ft and $TW = 2.5$ ft. Assume $n = 0.015$, the entrance loss coefficient $C_L = 0.5$, and the orifice coefficient $C_d = 0.65$.

Solution

First, assuming inlet control with $h = 5 - (1.5/2) = 4.25$ ft,

$$Q = (0.65)\left(\frac{\pi}{4}\right)(1.5)^2\sqrt{(2)(32.2)(4.25)} = 19.0 \text{ cfs}$$

Next, on the basis of outlet control the head loss is

$$H_L = 5 + (0.01)(50) - 2.5 = 3.0 \text{ ft}$$

This may be substituted into Eq. 11–7 as follows:

$$3.0 = \left[0.5 + 1 + \frac{(32.2)(0.015)^2(50)}{(1.11)(1.5/4)^{4/3}}\right]\left[\frac{8Q^2}{\pi^2(32.2)(1.5)^4}\right]$$

Upon solving, the discharge is found to be $Q = 14.9$ cfs. Since this value is less than the previous value of 19.0 cfs, the outlet controls the flow through the culvert and the discharge is 14.9 cfs.

EXAMPLE 11–4

Determine the diameter of corrugated steel pipe required to convey a flow rate of 6.5 m³/s. The culvert will have a length of 20 m, a 0.5 percent slope, a maximum $HW = 3.8$ m, and the tailwater will not exceed the diameter of the pipe. Assume $n = 0.024$, $C_L = 0.5$, and $C_d = 0.62$.

Solution

We will start with the assumption of inlet control. However, with the yet un-known D we can only formulate the orifice head as a function of D, namely $h = 3.8 - D/2$. Thus Eq. 11–3 may be written as

$$6.5 = (0.62)\left(\frac{\pi}{4}\right)D^2 \sqrt{(2)(9.81)\left(3.8 - \frac{D}{2}\right)}$$

Upon solving this equation by trial and error we get $D = 1.30$ m. Continuing with the assumption of outlet control, we first write Eq. 11–5 in terms of the unknown D,

$$H_L = 3.8 + (0.005)(20) - D$$

$$= 3.9 - D$$

Then writing Eq. 11–7 in SI units

$$3.9 - D = \left[0.5 + 1 + \frac{(9.81)(0.024)^2(20)}{(0.5)(D/4)^{4/3}}\right]\left[\frac{(8)(6.5)^2}{\pi^2(9.81)D^4}\right]$$

Note that the numerical value of 1.49 in the Manning equation must be replaced by unity, which leads to a replacement of the value 1.11 by 0.5 in Eq. 11–7. This rearranges to

$$D + \left(1.5 + \frac{1.435}{D^{4/3}}\right)\left(\frac{3.49}{D^4}\right) = 3.9$$

Solving this equation by trial and error yields $D = 1.355$ m. As this is more than the previous value, outlet control must prevail and the minimum diameter is 1.355 m. The next larger standard culvert diameter would therefore be chosen.

In Example 11–4, the tailwater *TW* was assigned equal to the diameter. If *TW* drops below *D*, a more sophisticated procedure[3] is available. The procedure of Example 11–4, which errs slightly on the conservative side, is also used in Computer Program 6.

11–4 BRIDGE HYDRAULICS

Bridges are generally larger than culverts and, as such, are more expensive, must pass a greater discharge, and by and large require more care in optimizing their design. In addition to arriving at a safe, economical, and, one hopes, aesthetically pleasing design, there are a number of hydraulic considerations that are of the greatest interest here. These include the hydraulics of the flow at the bridge itself, the upstream effect of the bridge with regard to increased flooding potential, and the possibility of a failure due to scour at the bridge piers or abutments. This final concern was examined previously in the chapter on sediment mechanics and it remains now to discuss the other problems.

[3]See, for example, the Concrete Pipe Design Manual referenced at the end of the chapter.

Normally, the entire flow of the river passes through the bridge opening without impingement against any part of the bridge deck. Under this condition the bridge has some (usually small) effect on the water surface profile. The head loss or, more importantly, the surface elevation change created by the presence of the bridge translates into an increased stage upstream of the bridge. This increase in the water surface profile is potentially due to several factors, including (1) contraction and expansion of the flow as it first enters and then leaves the bridge section, (2) the impact and consequent change in momentum caused by the piers and abutments, and (3) conditions that make the bridge opening perform as a choke.

The first of these factors, the head loss due to contraction and expansion, occurs normally from river cross section to cross section. As the channel changes in area, the inability of the water to follow the boundary changes completely leads to flow separation and accompanying eddies. Between any two channel sections the head loss is expressed by

$$H_L = C_L \left| \frac{V_1^2}{2g} - \frac{V_2^2}{2g} \right| \tag{11-8}$$

where, for small changes in area, C_L is typically 0.1 and 0.3 for contractions and expansions, respectively (see Section 7–6). The absolute value is taken in Eq. 11–8, so that the equation can be applied directly to both contractions and expansions. If the bridge section is much smaller than the upstream and downstream sections, the coefficients can increase to as much as 0.6 and 1.0. The contraction coefficient is smaller than the expansion coefficient because the accelerating flow is less subject to separation than is the decelerating flow in the expansion.

Before analyzing the head loss specifically due to the bridge, the various types of flow conditions must be defined. Based on the chapter on open channel flow, the two distinct cases of subcritical and supercritical flow can be expected. The most common situation in natural rivers is for the flow to remain subcritical throughout. This condition is depicted in Fig. 11–12, and will be the only case subjected to analysis herein. The supercritical flow at a bridge crossing is relatively rare but can occur, particularly in mountain regions. Since a supercritical flow requires an upstream control, the bridge does not normally cause upstream backwater. The exception occurs when the bridge becomes a choke, backing up the stream and, in the process, forcing the upstream flow to become subcritical. An additional case (which we will not consider) can occur in which the otherwise subcritical flow is forced to pass through critical depth at the bridge. This condition will usually be followed by a hydraulic jump.

Return now to Fig. 11–12, where the water surface profile shown would be applicable between piers. If piers are not present, then the profile can be considered to apply between abutments. A momentum analysis[4] based on a control volume extending from Section 1 to Section 3 will be performed to obtain a theoretical estimate of the change in water surface elevation. The resulting Δy will not include head loss

[4]See reference to Henderson at the end of the chapter.

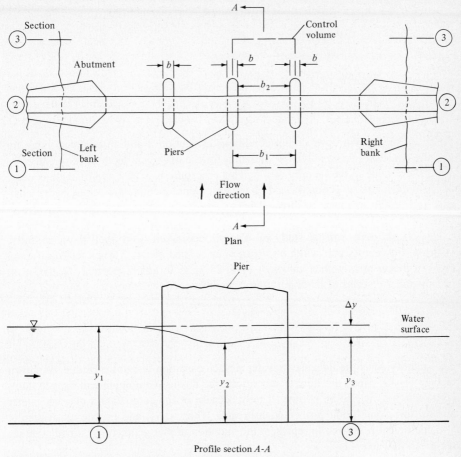

Figure 11–12 Bridge definitional sketch and water surface profiles.

due to contraction or expansion. However, the result of Eq. 11–8 can be added to reflect this head loss.

Referring to the control volume selected in Fig. 11–12, which is assumed representative of each region between piers, we may write

$$\frac{\gamma y_1^2 b_1}{2} - \frac{\gamma y_3^2 b_1}{2} - F_P = \rho q b_1 \left(\frac{q}{y_3} - \frac{q}{y_1} \right)$$

where the force terms consist of the usual hydrostatic forces and F_P is the force the pier exerts on the water. The discharge/unit width q is defined relative to the upstream and downstream sections, that is, $q = Q/b_1$. Introducing the drag force equation

$$F_P = C_D \, b y_1 \frac{\rho V_1^2}{2} \qquad\qquad\qquad\qquad \textbf{(11–9)}$$

where C_D is a drag coefficient leads after some rearrangement to

$$\frac{y_1^2}{2} + \frac{q^2}{gy_1} - \frac{C_D b y_1 V_1^2}{2gb_1} = \frac{y_3^2}{2} + \frac{q^2}{gy_3}$$

Letting $\alpha = b/b_1$ and $V_1 = q/y_1$ yields

$$\frac{y_1^2}{2} + \frac{q^2}{gy_1}\left(1 - \frac{C_D \alpha}{2}\right) = \frac{y_3^2}{2} + \frac{q^2}{gy_3} \tag{11-10}$$

Based on the known downstream conditions at Section 3 and an assumed value for C_D, the equation can be solved by iteration for the upstream depth y_1. Typically, C_D will have a value in the vicinity of 2.5. This procedure is at best approximate because of the uncertainty of geometry and alignment. Consequently, the actual bridge analysis and design are usually directed by experimental rather than theoretical procedures.

A classic experimental study by Yarnell[5] has often been applied successfully. Using α, the ratio of pier width b to span b_1 as defined above, Yarnell tested different types of model piers with α values of 11.7, 23.3, 35.0, and 50 percent. As the result of an extensive series of laboratory tests, he developed the following empirical equation:

$$\frac{\Delta y}{y_3} = K\mathrm{Fr}_3^2(K + 5\,\mathrm{Fr}_3^2 - 0.6)(\alpha + 15\,\alpha^4) \tag{11-11}$$

The quantity Δy is the increase in water surface elevation due to the bridge, as shown in Fig. 11–12. Since the flow is subcritical, the downstream depth and Froude number (Section 3) would be known (if only because of the computation of a backwater profile from some point further downstream). Thus α, y_3, and Fr_3 are all available. The coefficient K is used to represent the variation in pier geometry. Yarnell's values are given in Table 11–2.

Table 11–2 VALUES OF *K* FOR THE YARNELL EQUATION

Pier Shape	K
Semicircular nose and tail	0.90
Lens-shaped nose and tail	0.90
Twin-cylinder piers with connecting diaphragm	0.95
Twin-cylinder piers without diaphragm	1.05
90 deg triangular nose and tail	1.05
Square nose and tail	1.25

[5]D. L. Yarnell, "Pile Trestles as Channel Obstructions," U.S. Department of Agriculture, Tech. Bull. No. 429, July 1934, and "Bridge Piers as Channel Obstructions," U.S. Department of Agriculture, Tech. Bull. No. 442, November, 1934.

It is felt that the engineer can use this procedure with some confidence, providing a conservative estimate even in the range $\alpha < 0.117$. Other procedures are also available to estimate bridge backwater and more recent techniques are also recommended by the U.S. Department of Transportation. They were considered too lengthy for inclusion in an introductory textbook. For further information, the reader should refer to *Hydraulics of Bridge Waterways* as well as the other Department of Transportation references at the end of the chapter.

An additional problem can occur if the piers and abutments so constrict the channel (or the bridge approaches so constrict the floodplain) that the bridge section becomes a choke. This type of situation was discussed in some detail in the chapter on open channel hydraulics with respect to specific energy diagrams. At that point the emphasis was on the rectangular channel, but the same effect can occur in any channel. Assuming, as usual, a subcritical river flow, the contracted section causes a reduction in the water surface elevation. If the lower water surface just reaches critical depth, the section is a choke. A further contraction, in excess of that required to create a choke, would result in an increase in the upstream depth so as to provide the flow with the additional energy necessary to pass through the choke. The creation of a choke at a bridge section, which is illustrated in the following rather oversimplified example, should be avoided.

EXAMPLE 11-5

An approximately rectangular river channel has a width and depth of 40 m and 2.3 m, respectively. The bankfull discharge is 100 m³/s. A bridge crossing is proposed that will require abutments projecting into the channel, but no piers. Ignore head loss due to the bridge and determine the maximum encroachment into the channel that can be permitted without affecting the upstream depth.

Solution

The discharge/unit width is $q = 100/40 = 2.5$ m³/s/m. This leads to a critical depth of 0.861 m and indicates that the undisturbed flow is subcritical. The upstream specific head is

$$H_{01} = 2.3 + \frac{(2.5)^2}{(2)(9.81)(2.3)^2} = 2.360 \text{ m}$$

Since the contracted section is a choke, $y_{c2} = (\tfrac{2}{3})H_{02} = 1.573$ m. The corresponding unit discharge is obtained from

$$1.573 = \sqrt[3]{\frac{q_2^2}{9.81}}$$

Solving, $q_2 = 6.18$ m³/s/m and the contracted width is

$$b_2 = \frac{Q}{q_2} = \frac{100}{6.18} = 16.18 \text{ m}$$

Thus, the maximum distance which each abutment may extend into the channel is 11.91 m.

Under extreme conditions, the water surface may impact against the bridge deck or even flow over it. Although a rare flood discharge may strike or overtop the bridge, smaller discharges may unexpectedly lead to problems when a large amount of debris accumulates against the structure. In addition to the potential structural damage, particularly due to the debris, the increased stage increases the upstream flooding. This condition may or may not result in a pressure flow under the bridge. If it does not, but rather strikes the upstream face of the bridge and then flows under the bridge with a free surface, a sluice gate type of equation can be applied. If the flow remains in contact with the underside of the bridge deck, an orifice equation can be used. For details, reference should again be made to *Hydraulics of Bridge Waterways*.

If the flood discharge overtops the bridge or bridge approaches, a portion of the flow will pass under the bridge as described above. The remainder will pass over the bridge or approaches as a form of weir flow. Frequently, the bridge deck is placed at a higher elevation than the approaches so that if flooding occurs, the approach road will overtop first and thereby protect the more expensive bridge.

In conclusion, it should be noted that bridge scour, discussed in Chapter 8, increases the open area under the bridge and thereby tends to reduce the backwater upstream. This process is not recommended, however, because the resulting scour problems may be more serious than the additional flooding.

PROBLEMS

Section 11–2

11–1. Refer to the rainfall maps and by plotting the available data, estimate the 10-yr, 2-hr rainfall and the 10-yr, 12-hr rainfall for Columbus, Ohio.

11–2. Repeat Prob. 11–1 for Atlanta, Georgia.

11–3. If the rainfall intensity is 2.7 in./hr, estimate the rate of runoff in cfs from an area of 3 mi², of which 10 percent is forest and the remainder is farmland.

11–4. Assuming that Table 11–1 is accurate, determine the range of runoff rates that might occur in Prob. 11–3.

11-5. If the rainfall intensity is 10 cm/hr, estimate the rate of runoff in m^3/s from an area of 12 km^2. The area is composed of the following: 10 percent concrete pavement, 90 percent lawns with a heavy soil, of which 25 percent is flat, 60 percent has an average slope of 5 percent, and 15 percent has an average slope of 10 percent.

11-6. Repeat Prob. 11-5 if the lawns are composed of sandy soil.

11-7. Determine the maximum runoff in Prob. 11-3 due to a 10-yr 1-hr rainfall in central Illinois.

11-8. Determine the maximum runoff in Prob. 11-3 due to a 100-yr, 30-min. rainfall in southern Mississippi.

11-9. Determine the maximum runoff in Prob. 11-5 due to a 10-yr, 6-hr rainfall in central Maine.

11-10. Determine the maximum runoff in Prob. 11-5 due to a 2-yr, 6-hr rainfall in central Florida.

11-11. Use the U.S.G.S. method to estimate the 10-yr and 50-yr peak discharges from a 25 mi^2 basin just west of Huron, South Dakota. The channel slope is 10 ft/mi.

11-12. Determine the 2-yr and 5-yr peak discharges from the basin in Prob. 11-11.

11-13. Repeat Prob. 11-11 if the channel slope is (1) 5 ft/mi and (2) 50 ft/mi.

11-14. Write a computer program to estimate the peak discharge for any return period between 2 and 100 years based on the U.S.G.S. procedure for South Dakota. Input the basin area, the slope, and the soil-infiltration index. Base the computations on Eqs. 11-2 and linearly interpolate between return periods.

11-15. Repeat Prob. 11-14, but use logarithmic interpolation for the specific return period.

Section 11-3

In Probs. 11-16 through 11-34, assume $C_d = 0.62$, $C_L = 0.5$, and $n = 0.013$ unless otherwise instructed.

11-16. Determine the capacity of a 24-in. concrete culvert if *HW* and *TW* are 10 ft and 3 ft, respectively. The culvert has a slope of 1 percent and a length of 50 ft.

11-17. Repeat Prob. 11-16 if the culvert has a length of 500 ft.

11-18. Repeat Prob. 11-16 if the tailwater is not backed up.

11-19. Determine the capacity of a 50-cm-diameter, 12-m-long concrete culvert if the maximum head water is 1.5 m, the outlet discharges freely into the atmosphere, and the slope is 3 percent.

11-20. Repeat Prob. 11-19 if the culvert is 100 m long.

11-21. If the culvert in Prob. 11-16 must pass a discharge of 9 cfs, how much headwater is required?

11-22. If the culvert in Prob. 11-16 must pass a discharge of 10 cfs, how much headwater is required?

11-23. If the culvert in Prob. 11-19 must pass a discharge of 0.8 m^3/s, how much headwater is required?

11–24. If the culvert in Prob. 11–20 must pass a discharge of 0.75 m³/s, how much headwater is required?

11–25. A corrugated steel culvert ($n = 0.023$) must carry 50 cfs through a 600-ft-long embankment. The tailwater will not exceed the barrel diameter and the road surface is 50 ft above the outlet invert. Allow 3 ft of freeboard and use a culvert slope of 2 percent. Determine the required culvert diameter.

11–26. Repeat Prob. 11–25 if the culvert length is 120 ft.

11–27. Repeat Prob. 11–25 if the tailwater may extend to 5 ft above the outlet invert.

11–28. A corrugated steel culvert ($n = 0.022$) must carry 200 cfs through a 600-ft-long embankment. The tailwater may be as much as 12 ft and the road surface is 60 ft above the outlet invert. Allow 4 ft of freeboard and use a culvert slope of 3 percent. Determine the required culvert diameter. What discharge could the selected culvert carry before the road is overtopped?

11–29. Repeat Prob. 11–28 if the tailwater will never exceed the diameter of the barrel.

11–30. A 50-m-long culvert with $n = 0.018$ must transport 3 m³/s. The headwater cannot exceed 1 m above the inlet crown. Determine the diameter if the slope is 5 percent and the culvert discharges freely into the atmosphere.

11–31. Repeat Prob. 11–30 if the culvert is 150 m long.

11–32. Repeat Prob. 11–30 if the headwater cannot exceed 0.4 m above the inlet crown.

11–33. Determine the necessary culvert diameter if the embankment of Prob. 11–28 crosses the outlet of the drainage basin in Prob. 11–3.

11–34. Determine the necessary culvert diameter if the embankment of Prob. 11–29 crosses the outlet of the drainage basin in Prob. 11–3.

Section 11–4

In Prob. 11–35 through 11–43, assume that the channel is approximately rectangular, $k = 0.9$ and the pier drag coefficient $C_D = 2.5$, unless otherwise instructed.

11–35. Plot graphs of water level increase $\Delta y/y_3$ versus $\alpha = b/b_1$ over the range $0 < \alpha < 0.5$ for $Fr_3 = 0.1, 0.3$, and 0.5. Compare with selected calculations using Eq. 11–10.

11–36. A bridge crosses a 500-ft-wide channel on four piers. Each pier consists of two 10-ft-diameter cylindrical piers without connecting diaphragm. The undisturbed depth and discharge are 8 ft and 20,000 cfs, respectively. Estimate the increase in upstream depth using Yarnell's procedure.

11–37. Repeat Prob. 11–36 if the discharge is 30,000 cfs (and the same depth).

11–38. Repeat Prob. 11–36 using Eq. 11–10.

11–39. Repeat Prob. 11–37 using Eq. 11–10.

11–40. A bridge crosses a 300-m-wide river on nine 3-m-wide piers. The piers have a lens-shaped nose and tail. The undisturbed depth and velocity are 5 m and 3.5 m/s, respectively. Estimate the increase in upstrem depth using Yarnell's procedure.

11–41. Repeat Prob. 11–40 if the piers have a square nose and tail.

11–42. Repeat Prob. 11–41 if the velocity is only 2.5 m/s.

11–43. Repeat Prob. 11–40 using Eq. 11–10.

References

Bradley, J. N., *Hydraulics of Bridge Waterways,* Hydraulic Design Series no. 1, U.S. Department of Transportation, Washington, D.C.: Federal Highway Administration, 1970.

Chang, F. F. M. and H. W. Shen, *Debris Problems in the River Environment,* Report no. FHWA-RD-79-82, Washington, D.C.: Federal Highway Administration, 1979.

Chow, V. T., *Open Channel Flow,* New York: McGraw-Hill, 1959.

Concrete Pipe Design Manual, American Concrete Pipe Association, Arlington, VA, 1974.

Farraday, R. V., and F. G. Charlton, *Hydraulic Factors in Bridge Design,* Wallingford, England: Hydraulics Research Station, 1983.

Fletcher, J. E., A. L. Huber, F. W. Hawes, and C. G. Clyde, *Runoff Estimates for Small Rural Watersheds and Development of a Sound Design Method,* Report no. FHWA-RD-77-159, Washington, D.C.: Federal Highway Administration, 1977.

Henderson, F. M., *Open Channel Flow,* New York: Macmillan, 1966.

Hjelmfelt, A. T. and J. J. Cassidy, *Hydrology for Engineers and Planners,* Ames, IA: Iowa State University Press, 1975.

Linsley, R. K., M. A. Kohler, and J. L. H. Paulhus, *Hydrology for Engineers,* 3rd ed., New York: McGraw-Hill, 1982.

appendix
A

Notation, Units, and Fundamental Dimensions

Table A–1 PARTIAL LIST OF SYMBOLS

Symbol	Quantity	Usual Units		Dimensions	
		U.S. Customary	SI	F-L-T	M-L-T
a	Acceleration	ft/s^2	m/s^2	LT^{-2}	LT^{-2}
a	Skew coefficient				
A	Area	ft^2	m^2	L^2	L^2
b	Width, bottom width	ft	m	L	L
B	Top width	ft	m	L	L
c	Coefficient				
c	Sonic velocity	ft/s	m/s	LT^{-1}	LT^{-1}
C	Coefficient				
C	Concentration	lb/ft^3	N/m^3	FL^{-3}	$ML^{-2}T^{-2}$
C_c	Contraction coefficient				
C_d	Discharge coefficient				
C_D	Drag coefficient				
C_f	Surface friction coefficient				
C_H	Hazen-Williams coefficient				
C_L	Loss coefficient				
C_w	Weir coefficient				
d, d_s	Sediment diameter	ft	mm	L	L
D	Diameter	ft or in.	m or cm	L	L
e	Vapor pressure	in. Hg	cm Hg	L	L
E	Evaporation rate	in./day	cm/day	L	L
E	Modulus of elasticity	psi	N/m^2	FL^{-2}	$ML^{-1}T^{-2}$
f	Rate of infiltration	in./hr	cm/hr	LT^{-1}	LT^{-1}
f	Resistance coefficient				
F	Force	lb	N	F	MLT^{-2}
F_{gr}	Sediment mobility factor				
Fr	Froude number				
g	Gravitational acceleration	ft/s^2	m/s^2	LT^{-2}	LT^{-2}
g_s	Sediment transport per unit width	ton/day/ft	N/s/m	$FT^{-1}L^{-1}$	MT^{-3}
G_s	Sediment transport	tons/day	N/s	FT^{-1}	MLT^{-3}
G_{gr}	Dimensionless transport				
h	Vertical distance, piezometric head	ft	m	L	L

Table A–1 PARTIAL LIST OF SYMBOLS *(continued)*

Symbol	Quantity	Usual Units U.S. Customary	SI	Dimensions F-L-T	M-L-T
h_l	Head loss due to friction	ft	m	L	L
H	Head, total head	ft	m	L	L
H_L	Head loss	ft	m	L	L
H_0	Specific energy	ft	m	L	L
HW	Headwater	ft	m	L	L
i	Rainfall rate	in./hr	mm/hr	LT^{-1}	LT^{-1}
I	Moment of inertia	ft^4	m^4	L^4	L^4
I	Inflow	cfs	m^3/s	L^3T^{-1}	L^3T^{-1}
k	Absolute pipe roughness	ft	mm	L	L
K	Coefficient, constant				
K	Modulus of compressibility	psi	N/m^2	FL^{-2}	$ML^{-1}T^{-2}$
K	Muskingum coefficient	days	days	T	T
K	Permeability coefficient	ft/day	m/day	LT^{-1}	LT^{-1}
l	Length	ft	m	L	L
L	Length	ft	m	L	L
m	Rank of item				
M	Mass	slugs	kg	FT^2L^{-1}	M
n	Manning coefficient				
N	Normal force	lb	N	F	MLT^{-2}
N	Number of items				
N	Rotational speed	rpm	rpm	T^{-1}	T^{-1}
N	Time period	days or years	days or years	T	T
N_s	Specific speed	(see Section 10–3)			
NPSH	Net positive suction head	ft	m	L	L
O	Outflow	cfs	m^3/s	L^3T^{-1}	L^3T^{-1}
p	Pressure	psi or psf	N/m^2	FL^{-2}	$ML^{-1}T^{-2}$
p	Probability				
p_i	Fraction of ith size				
p_v	Vapor pressure	psi	N/m^2	FL^{-2}	$ML^{-1}T^{-2}$
P	Power	ft-lb/s	m-N/s	FLT^{-1}	ML^2T^{-3}
P	Precipitation	in.	mm	L	L
P	Wetted perimeter	ft	m	L	L
q	Discharge per unit width	cfs/ft	m^3/s/m	L^2T^{-1}	L^2T^{-1}
Q	Discharge	cfs	m^3/s	L^3T^{-1}	L^3T^{-1}
r	Radial distance	ft	m	L	L
R	Hydraulic radius	ft	m	L	L

Table A–1 PARTIAL LIST OF SYMBOLS *(continued)*

Symbol	Quantity	Usual Units U.S. Customary	Usual Units SI	Dimensions F-L-T	Dimensions M-L-T
R	Radius	in. or ft	cm or m	L	L
Re	Reynolds number				
s	Storage	sfd	m^3	L^3	L^3
S	Slope				
S_c	Storage coefficient				
S_f	Friction slope				
S_0	Channel slope				
Si	Soil-infiltration index	in.		L	L
t	Time	s, hr or days	s, hr or days	T	T
t	Thickness	in.	mm	L	L
TW	Tailwater	ft	m	L	L
T	Transmissibility	gpd/ft	m^3/day/m	L^2T^{-1}	L^2T^{-1}
T	Temperature	°F or °R	°C or °K		
T	Tangential force	lb	N	F	MLT^{-2}
T	Torque	lb-ft	N-m	FL	FL
t_p	Return period	years	years	T	T
u	Runner velocity	ft/s	m/s	LT^{-1}	LT^{-1}
u_*	Shear velocity	ft/s	m/s	LT^{-1}	LT^{-1}
V	Average velocity	ft/s	m/s	LT^{-1}	LT^{-1}
V_s	Surge velocity	ft/s	m/s	LT^{-1}	LT^{-1}
$\forall$	Volume	ft^3	m^3	L^3	L^3
w	Fall velocity	ft/s	cm/s	LT^{-1}	LT^{-1}
WS	Water surface elevation	ft	m	L	L
x	Muskingum weighting factor				
x	Horizontal distance	ft	m	L	L
X	Sediment flux				
y	Depth, elevation, vertical distance	ft	m	L	L
y_c	Critical depth	ft	m	L	L
y_n	Normal depth	ft	m	L	L
y_p	Distance to center of pressure	ft	m	L	L
Y	Aquifer thickness	ft	m	L	L
z	Concentration exponent				
z	Drawdown, head, elevation	ft	m	L	L
α	Angle				
α	Coefficient of thermal expansion	ft/ft°F	m/m°C		
γ	Specific weight	lb/ft^3	N/m^3	FL^{-3}	$ML^{-2}T^{-2}$
δ	Elongation	in.	mm	L	L

Table A–1 PARTIAL LIST OF SYMBOLS *(continued)*

Symbol	Quantity	Usual Units U.S. Customary	Usual Units SI	Dimensions F-L-T	Dimensions M-L-T
ϵ	Diffusion coefficient	ft^2/s	m^2/s	L^2T^{-1}	L^2T^{-1}
ϵ	Strain	in./in.	mm/mm		
η	Efficiency	%	%		
θ	Angle				
μ	Absolute viscosity	$lb\text{-}s/ft^2$	$N\text{-}s/m^2$	FTL^{-2}	$ML^{-1}T^{-1}$
μ	Poisson ratio				
ν	Kinematic viscosity	ft^2/s	m^2/s	L^2T^{-1}	L^2T^{1-1}
π	Dimensionless variable, constant				
ρ	Density	$slug/ft^3$	kg/m^3	FT^2L^{-4}	ML^{-3}
σ	Standard deviation				
σ	Tangential stress	psi	N/m^2	FL^{-2}	$ML^{-1}T^{-2}$
σ_c	Cavitation parameter				
τ	Shear stress	lb/ft^2	N/m^2	FL^{-2}	$ML^{-1}T^{-2}$
ϕ	Angle of repose				
ϕ	Relative speed				
Φ	Phi index	in./hr	cm/hr	LT^{-1}	LT^{-1}
ψ	Duboys transport function	ft/s	m/s	LT^{-1}	LT^{-1}
ω	Angular velocity	rad/s	rad/s	T^{-1}	T^{-1}

appendix
B

Fluid Properties

Table B–1a PROPERTIES OF WATER (U.S. CUSTOMARY UNITS)

Temperature °F	Density ρ slugs/ft³	Specific weight γ lb/ft³	Dynamic viscosity μ × 10⁵ lb-s/ft²	Kinematic viscosity ν × 10⁵ ft²/s	Surface tension σ × 10² lb/ft	Vapor pressure p_v psia	Modulus of compressibility E × 10⁻⁵ psi
32	1.940	62.42	3.746	1.931	0.518	0.087	2.93
40	1.940	62.43	3.229	1.664	0.514	0.12	2.94
50	1.940	62.41	2.735	1.410	0.509	0.18	3.05
60	1.938	62.37	2.359	1.217	0.504	0.26	3.11
70	1.936	62.30	2.050	1.059	0.500	0.36	3.20
80	1.934	62.22	1.799	0.930	0.492	0.51	3.22
90	1.931	62.11	1.595	0.826	0.486	0.70	3.23
100	1.927	62.00	1.424	0.739	0.480	0.96	3.27
110	1.923	61.86	1.284	0.667	0.473	1.28	3.31
120	1.918	61.71	1.168	0.609	0.465	1.69	3.33
130	1.913	61.55	1.069	0.558	0.460	2.22	3.34
140	1.908	61.38	0.981	0.514	0.454	2.89	3.30
150	1.902	61.20	0.905	0.476	0.447	3.72	3.28
160	1.896	61.00	0.838	0.442	0.441	4.75	3.26
170	1.890	60.80	0.780	0.413	0.433	5.99	3.22
180	1.883	60.58	0.726	0.385	0.426	7.51	3.18
190	1.876	60.36	0.678	0.362	0.419	9.34	3.13
200	1.868	60.12	0.637	0.341	0.412	11.52	3.08
212	1.860	59.83	0.593	0.319	0.404	14.69	3.00

Table B-1b PROPERTIES OF WATER (SI UNITS)

Temperature °C	Density ρ kg/m³	Specific weight γ N/m³	Dynamic viscosity $\mu \times 10^3$ N − s/m²	Kinematic viscosity $\nu \times 10^6$ m²/s	Surface tension $\sigma \times 10^2$ N/m	Vapor pressure p_v kN/m²	Modulus of compressibility $E \times 10^{-9}$ N/m²
0	999.8	9805	1.794	1.794	7.62	0.61	2.02
5	1000.0	9806	1.519	1.519	7.54	0.87	2.06
10	999.7	9802	1.308	1.308	7.48	1.23	2.11
15	999.1	9797	1.140	1.141	7.41	1.70	2.14
20	998.2	9786	1.005	1.007	7.36	2.34	2.20
25	997.1	9777	0.894	0.897	7.26	3.17	2.22
30	995.7	9762	0.801	0.804	7.18	4.24	2.23
35	994.1	9747	0.723	0.727	7.10	5.61	2.24
40	992.2	9730	0.656	0.661	7.01	7.38	2.27
45	990.2	9711	0.599	0.605	6.92	9.55	2.29
50	988.1	9689	0.549	0.556	6.82	12.33	2.30
55	985.7	9665	0.506	0.513	6.74	15.78	2.31
60	983.2	9642	0.469	0.477	6.68	19.92	2.28
65	980.6	9616	0.436	0.444	6.58	25.02	2.26
70	977.8	9588	0.406	0.415	6.50	31.16	2.25
75	974.9	9560	0.380	0.390	6.40	38.57	2.23
780	971.8	9528	0.357	0.367	6.30	47.34	2.21
85	968.6	9497	0.336	0.347	6.20	57.83	2.17
90	965.3	9473	0.317	0.328	6.12	70.10	2.16
95	961.9	9431	0.299	0.311	6.02	84.36	2.11
100	958.4	9398	0.284	0.296	5.94	101.33	2.07

Table B–2a PROPERTIES OF AIR (U.S. CUSTOMARY UNITS)

Temperature °F	Density $\rho \times 10^3$ slugs/ft^3	Specific weight $\gamma \times 10^2$ lb/ft^3	Dynamic viscosity $\mu \times 10^7$ lb-s/ft^2	Kinematic viscosity $\nu \times 10^4$ ft^2/s
0	2.68	8.62	3.38	1.26
10	2.63	8.46	3.45	1.31
20	2.57	8.27	3.50	1.36
30	2.52	8.11	3.58	1.42
40	2.47	7.94	3.62	1.46
50	2.42	7.79	3.68	1.52
60	2.37	7.63	3.74	1.58
70	2.33	7.50	3.82	1.64
80	2.28	7.35	3.85	1.69
90	2.24	7.23	3.90	1.74
100	2.20	7.09	3.96	1.80
120	2.15	6.84	4.07	1.89
140	2.06	6.63	4.14	2.01
160	1.99	6.41	4.22	2.12
180	1.93	6.21	4.34	2.25
200	1.87	6.02	4.49	2.40

Table B–2b PROPERTIES OF AIR (SI UNITS)

Temperature °C	Density ρ kg/m^3	Specific weight γ N/m^3	Dynamic viscosity $\mu \times 10^5$ N-s/m^2	Kinematic viscosity $\nu \times 10^5$ m^2/s
0	1.293	12.68	1.71	1.32
10	1.248	12.24	1.76	1.41
20	1.205	11.82	1.81	1.50
30	1.165	11.43	1.86	1.60
40	1.128	11.06	1.90	1.68
60	1.060	10.40	2.00	1.87
80	1.000	9.81	2.09	2.09
100	0.946	9.28	2.18	2.31

appendix
C
Conversion Factors

C

Table C–1a U.S. CUSTOMARY UNITS/SI UNITS[a]

Quantity	U.S. Customary Units	Conversion factor	SI Units
Acceleration	ft/s^2	0.3048	m/s^2
Area	ft^2	0.0929	m^2
Area	in.^{2}	645.2	mm^2
Area	mile^2	2.5898	km^2
Area	acre	0.4047	hectare
Area	acre	4047	m^2
Density	slugs/ft^3	515.38	kg/m^3
Discharge	cfs	0.02832	m^3
Discharge	gpm	0.06309	l/s
Dynamic viscosity	lb − s/ft^2	47.88	N-s/m^2
Energy, work	ft-lb	1.3558	N-m (joule)
Energy, work	Btu	1055.1	N-m (joule)
Force	lb	4.4482	N
Force	ton	8.8964	kN
Kinematic viscosity	ft^2/s	0.0929	m^2/s
Length	ft	0.3048	m
Length	in.	25.4	mm
Length	mile	1.6093	km
Length	mile	1609.3	m
Mass	slug	14.594	kg
Specific weight	lb/ft^3	157.09	N/m^3
Pressure	lb/ft^2	47.88	N/m^2 (Pa)
Pressure	psi	6894.8	N/m^2 (Pa)
Power	ft-lb/s	1.3558	N-m/s (watt)
Power	horsepower	745.69	N-m/s (watt)
Velocity	ft/s	0.3048	m/s
Velocity	miles/hr	0.44704	m/s
Volume	ft^3	0.02832	m^3
Volume	in.^{3}	16,387	mm^3
Volume	yard^3	0.7646	m^3
Volume	gallon	3.784	1 (liter)

[a]To convert a quantity having U.S. Customary units to SI units, multiply by the conversion factor. To convert from SI units to U.S. Customary units, divide by the conversion factor.

Table C–1b MISCELLANEOUS CONVERSION FACTORS[a]

Quantity (1)	(2)	Conversion Factor (3)	(4)
Area	acre	43,560	ft^2
Area	mi^2	640	acre
Discharge	cfs	448.8	gpm (U.S.)
Energy, work	Btu	778	ft-lb
Length	mile	5280	ft
Power	horsepower	550	ft-lb/s
Power	horsepower	0.708	Btu/s
Velocity	mph	1.467	ft/s
Velocity	knot	1.689	ft/s
Velocity	knot	1.152	mph
Volume	ft^3	7.48	U.S. gal
Volume	British gal	1.2	U.S. gal

[a]To convert a quantity having the units of column (2) to those of column (4), multiply by the conversion factor. To reverse the process, divide by the conversion factor.

appendix
D

Geometric Properties of Plain Surfaces and Volumes

Table D–1 PROPERTIES OF COMMON PLANE SURFACES

Shape	Sketch	Area	Centroid	Moment of inertia
Rectangle		$A = ab$	$\bar{y} = \dfrac{a}{2}$	$\bar{I} = \dfrac{ba^3}{12}$ $I_x = \dfrac{ba^3}{3}$
Triangle		$A = \dfrac{ab}{2}$	$\bar{y} = \dfrac{a}{3}$	$\bar{I} = \dfrac{ba^3}{36}$ $I_x = \dfrac{ba^3}{12}$
Circle		$A = \dfrac{\pi D^2}{4}$	$\bar{y} = \dfrac{D}{2}$	$\bar{I} = \dfrac{\pi D^4}{64}$
Semicircle		$A = \dfrac{\pi D^2}{8}$	$\bar{y} = \dfrac{4r}{3\pi}$	$\bar{I} = \left(\dfrac{1}{4} - \dfrac{16}{9\pi^2}\right)\dfrac{\pi r^4}{2}$ $I_x = \dfrac{\pi r^4}{8}$
Quarter circle		$A = \dfrac{\pi D^2}{16}$	$\bar{x} = \bar{y} = \dfrac{4r}{3\pi}$	$\bar{I} = \left(\dfrac{1}{4} - \dfrac{16}{9\pi^2}\right)\dfrac{\pi r^4}{4}$ $I_x = \dfrac{\pi r^4}{16}$
Ellipse		$A = \pi ab$	$\bar{y} = a$	$\bar{I} = \dfrac{\pi ba^3}{4}$

Table D–2 PROPERTIES OF COMMON VOLUMES

Shape	Sketch	Volume	Centroid
Cylinder		$V = \dfrac{\pi D^2 a}{4}$	$\bar{y} = \dfrac{a}{2}$
Cone		$V = \dfrac{\pi D^2 a}{12}$	$\bar{y} = \dfrac{a}{4}$
Sphere		$V = \dfrac{\pi D^3}{6}$	$\bar{y} = \dfrac{D}{2}$
Hemisphere		$V = \dfrac{\pi D^3}{12}$	$\bar{y} = \dfrac{3r}{8}$

appendix
E
Computer Program

The following typical hydraulic engineering computer programs are referenced within the text. Some have been selected because they provide useful subroutines for other programs. All of the programs may be run with little or no modification depending on the computer available. More importantly, they may also serve the student as a guide to similar programming. The programs are written in Basic for the Apple II +. However, they may be easily changed to run on any machine. Some documentation is provided in each case as well.

PROGRAM 1: KINEMATIC VISCOSITY OF WATER[1]

This program evaluates the approximate kinematic viscosity of water as a function of temperature specified in either U.S. or SI units. The program is written in Basic for the Apple II +, but may be readily modified for other micros or translated into Fortran. By deleting the appropriate lines of code, the program can be easily inserted as a subroutine into other programs requiring the kinematic viscosity. Since the program is based on prompts, no additional documentation should be required.

Program 1

```
10   REM  PROGRAM TO CALCULATE KINEMATIC VISCOSITY OF WATER
20   REM  CAN BE USED AS A SUBROUTINE IN OTHER PROGRAMS
30   REM  TEMPERATURE IS T
40   REM
50   HOME
60   PRINT "PROGRAM TO CALCULATE KINEMATIC VISCOSITY"
70   PRINT
80   PRINT "ARE YOU USING (1) U.S. CUSTOMARY OR (2) SI UNITS?"
90   INPUT SI
100  IF SI <  > 1 AND SI <  > 2 THEN  PRINT : PRINT "ENTER 1 OR 2 ONL
     Y": PRINT : GOTO 70
110  HOME
120  PRINT "ENTER THE TEMPERATURE IN DEGREES ";
130  IF SI = 1 THEN  PRINT "FAHRENHEIT": GOTO 150
140  PRINT "CELSIUS"
150  INPUT T
160  REM CALCULATE KINEMATIC VISCOSITY
170 TC = T
180  IF SI = 2 THEN 230
190  REM
200  REM  CALCULATIONS ARE IN SI UNITS
210  REM
220 TC = (T - 32) * (5 / 9)
230 XT = TC - 20
240  IF TC > 20 THEN 270
250 NU = 10 ^ (1301 / (998.333 + 8.1855 * XT + .00585 * XT * XT) - 3.
     30233)
```

[1]Referenced in Chapter 2.

Program 1 cont.

```
260   GOTO 290
270  TT = 20 - TC
280  NU = .01002 * (10 ^ (((1.3272 * TT - .001053 * XT * XT) / (TC + 10
     5)))
290  NU = NU / 10000: REM IN SQ M/S
300  IF SI = 1 THEN NU = NU * 10.7639: REM IN U.S. CUSTOMARY UNITS
310  PRINT "TEMPERATURE","VISCOSITY"
320  PRINT T,NU
330  REM  ADD RETURN IF IN A SUBROUTINE
340  END
```

PROGRAM 2: THE MUSKINGUM METHOD[2]

Inflow and outflow hydrographs are used to determine the parameters x and K in the Muskingum river routing procedure. Although presented in Basic, it was originally written in Fortran by Todd Connelly. Data requirements include the number of hydrograph points (both inflow and outflow hydrographs must have the same number of points), the time interval between subsequent hydrograph values (in hours), and the inflow and outflow hydrographs for the reach. The data is read, rather than inputed, and the included data (located at the end of the program) is that for Prob. 3–70. The program uses only U.S. Customary units. A companion flood routing program utilizing the results of this program is not included, but can be easily written.

The program follows the procedure in Section 3–8 using Eq. 3–19 to determine the incremental storage which is then summed and compared with the weighted discharge $[xI + (1 - x)O]$ for different values of x. The weighting factor x is varied from 0 to 0.50 in steps of 0.01. The storage versus the weighted discharge is analyzed for each x by using least-squares to determine the line of best fit (similar to Fig. 3–19). The x-value which gives the best line (based on the squares of the deviations) is selected and the K value calculated from the slope of the line.

Program 2

```
10   REM PROGRAM TO CALCULATE X AND K FOR THE MUSKINGUM METHOD
20   REM ORIGINALLY WRITEN IN FORTRAN
30   REM BY TODD CONNELLY WHILE A GRADUATE STUDENT
40   REM AT SOUTH DAKOTA STATE UNIVERSITY
50   REM
60   REM DATA INCLUDED AT END OF PROGRAM IS FOR PROB. 3-70
70   REM SEE REMARKS WITH DATA
80   REM A PROGRAM CAN BE EASILY WRITTEN TO USE X AND K IN THE MUSKING
     UM METHOD
90   REM
100   GOTO 180
110   REM
120   REM TURN ON AND FORMAT PRINTER
130   REM
140   PR# 1
150   PRINT  CHR$ (27);"L,528,432,$"
160   PRINT  CHR$ (27);"J,120,900,$"
170   RETURN
```

[2]Referenced in Chapter 3.

Program 2 cont.

```
180   HOME : VTAB (5)
190   PRINT "PROGRAM TO CALCULATE X AND K FOR THE MUSKINGUM METHOD"
200   PRINT : PRINT "THE PROGRAM USES U.S. CUSTOMARY UNITS ONLY"
210   PRINT : PRINT "DO YOU WANT TO USE THE PRINTER (Y/N)";
220   INPUT PT$
230   IF PT$ = "Y" THEN PT = 1
240   HOME : VTAB (5)
250   PRINT "DO YOU WANT (1) FINAL RESULTS OR (2) A DETAILED TRACE";
260   INPUT PL
270   IF PL < > 1 AND PL < > 2 THEN  PRINT : PRINT "ENTER 1 OR 2 ONL
      Y": PRINT : GOTO 250
280   IF PT = 1 THEN  GOSUB 140
290   REM
300   REM THE INFLOW AND OUTFLOW HYDROGRAPHS
310   REM MUST HAVE THE SAME NUMBER OF VALUES
320   REM INPUT THE NUMBER OF DATA POINTS
330   REM INPUT THE TIME INTERVAL BETWEEN POINTS IN HOURS
340   READ N,IT
350   DIM QI(N),QO(N),AI(N),AO(N),S(N),XX(52),X(N),QQ(52,N),XM(52),QM(
      52),XK(52),D2(52)
360   FOR I = 1 TO N: READ QI(I): NEXT I
370   FOR I = 1 TO N: READ QO(I): NEXT I
380 DT = IT / 24: REM CONVERT INTERVAL TO DAYS
390   IF PL = 1 THEN 460
400   PRINT : PRINT "TIME","INFLOW","OUTFLOW"
410   PRINT "IN DAYS","CFS","CFS"
420   FOR I = 1 TO N
430   PRINT (DT * (I - 1)),QI(I),QO(I)
440   NEXT I
450   PRINT : PRINT
460 SS = 0:S(1) = 0
470   REM
480   REM DETERMINE THE AVERAGE INFLOW AND OUTFLOW FOR EACH INTERVAL
490   REM CONVERT TO STORAGE
500   REM
510   FOR I = 2 TO N
520 AI(I) = (QI(I) + QI(I - 1)) / 2
530 AO(I) = (QO(I) + QO(I - 1)) / 2
540 DS = (AI(I) - AO(I)) * DT
550 SS = SS + DS
560 S(I) = SS
570   NEXT I
580   IF PL = 1 THEN 740
590   PRINT "TIME","STORAGE"
600   PRINT "DAYS","SFD"
610   FOR I = 2 TO N
620   PRINT (I - 1) * DT,S(I)
630   NEXT I
640   PRINT : PRINT
650   REM
660   REM THE X-VALUE IS VARIED FROM 0 TO 0.5
670   REM TO DETERMINE THE LINE OF BEST FIT
680   REM XX IS WEIGHTING FACTOR X FOR ROUTING
690   REM
700   PRINT "TREND OF VALUES FOR X AND K"
710   PRINT
720   PRINT "X-VALUES" TAB( 20)"K-VALUES" TAB( 40)"DEVIATIONS SQUARED"
730   PRINT
740   FOR J = 1 TO 51
750 XX(J) = (J / 100) - .01
760 SX = 0:SY = 0
770   REM
```

Program 2 cont.

```
780   REM A STATISTICAL ANALYSIS IS RUN ON THE FLOW CHARACTERISTICS
790   REM  VERSUS THE STORAGE FOR EACH X
800   REM
810   FOR K = 1 TO N
820  X(K) = S(K)
830  QQ(J,K) = XX(J) * QI(K) + (1 - XX(J)) * QO(K)
840  SX = SX + X(K)
850  SQ = SQ + QQ(J,K)
860   NEXT K
870   REM
880   REM DETERMINE THE MEANS FOR X AND QQ
890   REM
900  XM(J) = SX / N
910  QM(J) = SQ / N
920  SX = 0:SQ = 0
930  SZ = 0
940  Q2 = 0:X2 = 0
950   FOR IB = 1 TO N
960  X3 = X(IB) - XM(J)
970  Q3 = QQ(J,IB) - QM(J)
980  X2 = X2 + X3 * X3
990  Q2 = Q2 + Q3 * Q3
1000 SZ = SZ + X3 * Q3
1010 SX = SX + X3
1020 SQ = SQ + Q3
1030  NEXT IB
1040 XK(J) = SZ / X2
1050 SD = 0
1060  REM
1070  REM THE ACTUAL VALUE OF Q IS COMPARED WITH THE VALUE
1080  REM OF Q ON THE REGRESSION LINE
1090  REM THE DEVIATION IS SQUARED AND SUMMED FOR EACH TRIAL
1100  REM
1110  FOR LA = 1 TO N
1120 QH = QM(J) + XK(J) * (X(LA) - XM(J))
1130 DQ = (QQ(J,LA) - QH) ` 2
1140 SD = SD + DQ
1150  NEXT LA
1160  REM
1170  REM THE K-VALUE IS CALCULATED BY TAKING
1180  REM THE RECIPROCAL OF THE SLOPE
1190  REM
1200 XK(J) = 1 / XK(J)
1210 D2(J) = SD
1220  IF PL = 1 THEN 1240
1230  PRINT XX(J) TAB( 20)XK(J) TAB( 40)D2(J)
1240  NEXT J
1250 CC = D2(1)
1260  REM
1270  REM THE LOWEST SUM OF SQUARES IS TESTED FOR
1280  REM TO REPRESENT THE LINE OF BEST FIT
1290  REM THE X-VALUE AND K-VALUE ARE RECORDED
1300  REM
1310  FOR L = 2 TO 51
1320  IF D2(L) ) = CC THEN 1350
1330 CC = D2(L)
1340 LL = L
1350  NEXT L
1360  REM
1370  REM FINAL RESULTS
1380  REM
1390  PRINT
```

Program 2 cont.

```
1400  PRINT "THE SELECTED VALUES OF X AND K ARE ";XX(LL);" AND ";XK(L
      L)
1410  END
1420  REM
1430  REM DATA IS FOR PROB 3-70
1440  REM DATA ORDER IS AS FOLLOWS:
1450  REM FIRST LINE--NUMBER OF HYDROGRAPH VALUES
1460  REM AND TIME INTERVAL IN HOURS BETWEEN DATA POINTS
1470  REM SECOND LINE--INFLOW HYDROGRAPH
1480  REM THIRD LINE--OUTFLOW HYDROGRAPH
1490  REM
1500  DATA    15,12
1510  DATA 790,2620,6970,6370,5250,4100,3270,2500,1960,1510,1240,1020
      ,850,680,580
1520  DATA 790,800,1580,3830,6000,5560,4710,3780,3220,2650,2280,2060,
      1930,1680,1510
```

PROGRAM 3: MULTIPLE RESERVOIR PROBLEM[3]

This program will calculate the discharge in either U.S. Customary or SI units for any reasonable number of reservoirs and junctions arranged as in Fig. 6–8. The printout is based on a machine language program developed by Bongers[4] for the Apple II + . Line 2000 may be deleted and lines 2070 and 2180 replaced by conventional print statements.

The input is by prompts and should be self-explanatory. The user can determine whether to print to the monitor or printer. The results of each iteration can be printed or the user may choose just the final results. The geometric input requirements include the number of junctions and the number of reservoirs connected to each junction, the elevation of each reservoir, and the diameter, length and absolute roughness of each pipe. All length dimensions including the roughness must be specified in either feet or meters as appropriate. The starting piezometric head at each junction is based on the average water surface elevation of the reservoirs connected to that junction. The discharge calculations are based on Eq. 6–6.

The kinematic viscosity is required if known (a default value of 60° F is included). However, Program 1 could be easily substituted and the input requirement changed to the temperature. Lines 80 and 90 program the printer to set suitable margins, and may require modification depending on the available printer.

Program 3

```
10   GOTO 110
20   REM  TURN ON PRINTER
30   PR# 1
40   REM
50   REM  FORMAT INSTRUCTIONS FOR PRINTER
60   REM  TO SET PAGE LENGTH AND MARGINS
70   REM
```

[3]Referenced in Chapter 6.
[4]C. Bongers, "Amper Print-use Program," *Nibble Express,* Vol. 2, 1982.

Program 3 cont.

```
80    PRINT  CHR$ (27);"L,528,432,$"
90    PRINT  CHR$ (27);"J,120,$"
100   RETURN
110   PRINT "MULTIPLE RESERVOIR PROBLEM"
120 N = 0
130 PI = 3.14159265
140   PRINT
150   PRINT "PROGRAM TO CALCULATE THE DISCHARGE IN EACH PIPE"
160   PRINT : PRINT "DO YOU WANT RESULTS PRINTED (Y/N)?"
170   INPUT B$
180   IF B$ = "Y" THEN PT = 1
190   PRINT
200   REM
210   REM   INPUT THE PIPE DATA
220   REM
230   INPUT "ENTER THE NUMBER OF PIPE JUNCTIONS    ";M
240   DIM N(M): DIM SU(M)
250   FOR I = 1 TO M
260   PRINT
270   PRINT "ENTER THE NUMBER OF RESERVOIRS CONNECTED TO JUNCTION ";I
280   INPUT N(I)
290 N = N(I) + N: REM COUNT TOTAL NUMBER OF RESERVOIRS
300   NEXT I
310 Q2 = 1 + N: REM COUNTER TO INCLUDE PIPES BETWEEN JUNCTIONS
320   PRINT
330   PRINT "SELECT PRINTOUT LEVEL: (1) RESULTS ONLY, (2) ADDITIONAL D
      ETAILS ";
340   INPUT PL
350   PRINT
360   IF PL < 1 OR PL > 2 THEN 330
370   PRINT : PRINT "ENTER THE MIN. CORRECTION FOR HJ"
380   PRINT ".001 IS THE RECOMMENDED VALUE"
390   INPUT AL
400   PRINT "IS THE PROBLEM SPECIFIED IN (1) U.S. CUSTOMARY OR (2) SI
      UNITS? ";
410   INPUT SI
420   IF SI < 1 OR SI > 2 THEN 400
430   PRINT
440 G = 32.174
450   IF SI = 2 THEN G = 9.80665
460   PRINT "ENTER THE KINEMATIC VISCOSITY, OR IF NOT KNOWN ENTER 0 ";
470   INPUT NU
480   IF NU > 0 THEN 510
490 NU = .00001217
500   IF SI = 2 THEN NU = .000001141
510   PRINT
520 NP = M + N: REM  NUMBER OF PIPES
530   DIM D(NP): DIM A(NP): DIM K(NP): DIM HL(NP): DIM L(NP): DIM Q(NP
      )
540   DIM EL(NP): DIM HJ(M)
550 K = 1
560   FOR J = 1 TO M
570   PRINT "FOR THE RESERVOIRS CONNECTED TO JUNCTION NO. ";J
580   FOR I = 1 TO N(J)
590   PRINT
600   PRINT "ENTER THE WATER SURFACE ELEVATION OF RESERVOIR NO. ";K
610   INPUT EL(K)
620 SU(J) = SU(J) + EL(K)
630   REM
640   REM   SUM RES ELEV AT EACH JUNCTION TO GET STARTING HJ'S
650   REM   HJ IS PIEZOMETRIC HEAD AT JUNCTION
660   REM
```

Program 3 cont.

```
670 K = K + 1
680  NEXT I
690  PRINT
700  NEXT J
710 K = 1
720  FOR J = 1 TO M
730  FOR I = 1 TO N(J)
740  PRINT : PRINT "FOR PIPE NO. ";K
750  PRINT "RUNNING FROM RESERVOIR NO. ";K
760  PRINT "TO JUNCTION NO. ";J
770  PRINT "ENTER DIAMETER, LENGTH, AND ROUGHNESS K"
780  INPUT D(K),L(K),K(K)
790 K = K + 1
800  NEXT I
810  NEXT J
820  PRINT
830  IF M = 1 THEN 930
840  REM   ADD PIPES CONNECTING JUNCTIONS
850  FOR II = 1 TO (M - 1)
860  PRINT : PRINT "FOR THE PIPE FROM JUNCTION NO. ";II
870  PRINT "TO JUNCTION NO. ";(II + 1)
880  PRINT "ENTER DIAMETER, LENGTH, AND ROUGHNESS K"
890  INPUT D(K),L(K),K(K)
900 K = K + 1
910  NEXT II
920  REM
930  REM   GET STARTING HJ'S
940  REM   BASED ON AVE WS ELEV
950  REM   AT EACH RESERVOIR
960  REM
970  FOR I = 1 TO M
980 HJ(I) = SU(I) / N(I)
990  NEXT I
1000  IF PT = 1 THEN  GOSUB 30
1010  IF PL = 1 THEN 1070
1020  FOR I = 1 TO M
1030  PRINT "HJ(";I;") = ";HJ(I)
1040  NEXT I
1050 NI = 0: REM  NO. OF ITERATIONS FOR TRACE ONLY
1060  REM
1070  REM   FLOW DIRECTION DETERMINED BY TESTING WS EL AGAINST HJ
1080  REM   IF WS EL GREATER THAN HJ THEN FLOW TOWARD JUNCTION
1090  REM   THEN Q WILL BE NEGATIVE
1100  REM   SUM OF ALL Q'S AT JUNCTION MUST EQUAL ZERO
1110  REM   IF SUM IS NEGATIVE THEN RAISE HJ
1120  REM
1130 K = 1: REM  COUNTER FOR RESERVOIRS AND PIPES
1140  IF PL = 1 THEN 1160
1150 CI = CI + 1: PRINT : PRINT "ITERATION NO. ";CI: PRINT
1160  FOR J = 1 TO M
1170  FOR I = 1 TO N(J)
1180 MI = 1: REM   TO SET SIGN ON DISCHARGE
1190 HL(K) = EL(K) - HJ(J)
1200  IF HL(K) > 0 THEN MI =  - 1
1210 HL(K) =  ABS (HL(K))
1220  IF HL(K) > 0 THEN 1300
1230 Q(K) = 0: GOTO 1480
1240  REM
1250  REM   CALCULATE Q
1260  REM   USING COLEBROOK-WHITE EQUATION
1270  REM   COMBINED WITH THE DARCY WEISBACH EQUATION
1280  REM   SEE EQ. 6-6
```

Program 3 cont.

```
1290   REM
1300   T1 =   SQR (2 * G * D(K) * HL(K) / L(K))
1310   T2 = K(K) / (3.7 * D(K))
1320   T3 = 2.51 * NU / (D(K) * T1)
1330   T4 =   LOG (T2 + T3)
1340   T4 = T4 / 2.302585
1350   A(K) = PI * D(K) * D(K) / 4
1360   Q(K) =   - 2 * A(K) * T1 * T4 * MI
1370   IF PL = 1 THEN 1430
1380   PRINT "FLOW IN PIPE NO. ";K;" = ";Q(K)
1390   REM
1400   REM  CQ NEEDED FOR HJ CORRECTION
1410   REM CT SUM OF CQ
1420   REM
1430   CQ =   ABS (Q(K) / HL(K))
1440   CT = CT + CQ: REM   SUM OF Q/HL
1450   REM
1460   REM   SUM Q'S ALGEBRAICALLY
1470   REM
1480   Q = Q + Q(K)
1490   IF FG = 1 THEN   RETURN
1500   K = K + 1
1510   NEXT I
1520   IF M = 1 THEN 1670
1530   FG = 1: REM  SET FLAG FOR PIPES BETWEEN JUNCTIONS
1540   Q1 = K: REM   SAVE ACCUMULATED NO OF PIPES
1550   IF J = 1 THEN 1610
1560   K = Q2
1570   EL(K) = HJ(J - 1)
1580   GOSUB 1180
1590   IF J = M THEN 1640
1600   Q2 = Q2 + 1
1610   K = Q2
1620   EL(K) = HJ(J + 1)
1630   GOSUB 1180
1640   FG = 0
1650   K = Q1
1660   REM
1670   REM  MAKE CORRECTION TO HJ AT JUNCTION
1680   REM  CR IS CORRECTION TO HJ
1690   REM
1700   CR = 2 * Q / CT
1710   IF   ABS (CR) <  = AL THEN CC = 1
1720   SC = SC + CC
1730   CC = 0
1740   Q = 0:CT = 0
1750   HJ(J) = HJ(J) - CR
1760   IF PL = 1 THEN 1800
1770   PRINT : PRINT "THE CORRECTION TO HJ(";J;") = ";CR
1780   PRINT "THE NEW VALUE FOR HJ(";J;") = ";HJ(J)
1790   PRINT
1800   NEXT J
1810   IF PL = 1 THEN 1850
1820   PRINT "TEST FOR CONVERGENCE"
1830   PRINT "JUNCTIONS PASSING = ";SC;" OF ";M
1840   PRINT
1850   IF SC = M THEN 1890
1860   SC = 0
1870   Q2 = N + 1
1880   GOTO 1130
1890   PRINT : PRINT : PRINT : PRINT
1900   PRINT "RESULTS FOR MULTIPLE RESERVOIR PROBLEM"
```

Program 3 cont.

```
1910   PRINT : PRINT "THERE ARE ";N;" RESERVOIRS"
1920   PRINT "WITH ";M;" JUNCTIONS"
1930   PRINT
1940 K = 1
1950   REM
1960   REM   FORMAT PRINTOUT FOR APPLE II+ ONLY
1970   REM   REQUIRES SPECIAL UTILITY PROGRAM
1980   REM   DELETE OR REPLACE WITH FORMAT FOR PRINT USING
1990   REM
2000 A$ = "#####.###"
2010   FOR J = 1 TO M
2020   PRINT "FOR THE RESERVOIRS CONNECTED TO JUNCTION NO. ";J;":"
2030   FOR I = 1 TO N(J)
2040   REM
2050   REM   PRINT STATEMENT FOR APPLE II+ ONLY
2060   REM
2070   &  PRINT USEA$; ABS (Q(K));
2080   IF SI = 2 THEN  PRINT " CMS";: GOTO 2100
2090   PRINT " CFS";
2100   IF Q(K) < 0 THEN  PRINT " OUT OF RESERVOIR NO. ";K: GOTO 2120
2110   PRINT " INTO RESERVOIR NO. ";K
2120 K = K + 1
2130   NEXT I
2140   PRINT : NEXT J
2150   IF M = 1 THEN  END
2160   PRINT : PRINT "FLOW BETWEEN JUNCTIONS IS AS FOLLOWS:"
2170   FOR J = 1 TO (M - 1)
2180   &  PRINT USEA$; ABS (Q(K));
2190   IF SI = 2 THEN  PRINT " CMS";: GOTO 2210
2200   PRINT " CFS";
2210   IF Q(K) < 0 THEN  PRINT " FROM JUNCT. ";J;" TO JUNCT. ";(J + 1)
       : GOTO 2230
2220   PRINT " FROM JUNCT. ";(J + 1);" TO JUNCT. ";J
2230 K = K + 1
2240   NEXT J
2250   END
```

PROGRAM 4: WATER SURFACE PROFILES IN NATURAL CHANNELS[5]

This program uses Eq. 7–48 with the water surface elevation correction given by Eq. 7–49 to determine water surface profiles in natural channels. The program is written for the Apple II+ and the general comments are essentially the same as those for Program 3. The use of a printer is optional as is the level of printout and the use of U.S. Customary or SI units.

As written, the program assumes a constant Manning n from section to section and no provision is made for overbank flow. Any number of discharges can be run, but the downstream water surface elevation must be known or estimated for each. Thus, subcritical flow is assumed. All lengths must be entered in feet or meters as appropriate. Additional data requirements include the number of cross sections and the distance between them. The cross-sectional geometry is specified by *(x, y)* coor-

[5]Referenced in Chapter 7.

dinate points commencing with the left bank. The calculation of area and wetted perimeter are based on Eqs. 7–45 and 7–46.

A relatively large number of coordinate points may be required to define the channel geometry. This is particularly true if there are a large number of cross sections. The program was written with prompts for the data entry because it was felt that it would be easier to follow. The user will probably find it more convenient to replace the input statements with read statements and data lines. As a final note, the program will work for prismoidal channels as well as natural channels.

Program 4

```
10   REM WATER SURFACE PROFILES
20   REM NATURAL CHANNEL
30   REM ALAN L. PRASUHN
40   REM PROMPTS MAY BE EASILY REPLACED BY DATA AND READ STATEMENTS
50   REM RUNS ON APPLE II+
60   REM PRINT USE MAY BE REPLACED BY PRINT USING OR DELETED
70   REM
80   GOTO 780
90   REM
100  REM SUBROUTINE
110  REM DETERMINE INTERSECTION OF BOUNDARY AND WATER SURFACE
120  REM
130  I = 1
140  IF WS(K) >  = Y(K,I) THEN 170
150  I = I + 1
160  GOTO 140
170  IF I = 1 AND WS(K) > Y(K,I) THEN 380
180  LY = WS(K)
190  REM
200  REM STRAIGHT LINE INTERPOLATION
210  REM FOR LEFT BANK
220  REM
230  LX = X(K,I) - (X(K,I) - X(K,I - 1)) * (LY - Y(K,I)) / (Y(K,I - 1)
     - Y(K,I))
240  J = N(K)
250  IF WS(K) >  = Y(K,J) THEN 280
260  J = J - 1
270  GOTO 250
280  IF J = N(K) AND WS(K) > Y(K,J) THEN 380
290  RY = WS(K)
300  IF J < N(K) THEN 360
310  RX = X(K,J)
320  GOTO 460
330  REM
340  REM INTERPOLATION FOR RIGHT BANK
350  REM
360  RX = X(K,J) + (X(K,J + 1) - X(K,J)) * (RY - Y(K,J)) / (Y(K,J + 1)
     - Y(K,J))
370  GOTO 460
380  PRINT "WATER SURFACE ABOVE TOP ELEVATION AT SECTION (";K;")"
390  PRINT "REENTER GEOMETRIC DATA AFTER HITTING RUN": PRINT : END
400  REM
410  REM SUBROUTINE
420  REM CALCULATE A,P,R,B
430  REM I=FIRST COORDINATE BELOW WS
440  REM J=LAST COORDINATE BELOW WS
450  REM
```

Program 4 cont.

```
460  A(K) = LX * Y(K,I) - LY * X(K,I)
470  P(K) = SQR (((X(K,I) - LX) ^ 2) + (LY - Y(K,I)) ^ 2)
480  IF I > = J THEN 550
490  FOR L = I TO (J - 1)
500  T = X(K,L) * Y(K,L + 1) - Y(K,L) * X(K,L + 1)
510  A(K) = A(K) + T
520  PP = SQR (((X(K,L + 1) - X(K,L)) ^ 2) + (Y(K,L + 1) - Y(K,L)) ^
     2)
530  P(K) = P(K) + PP
540  NEXT L
550  PP = SQR (((RX - X(K,J)) ^ 2) + (RY - Y(K,J)) ^ 2)
560  P(K) = P(K) + PP
570  T = X(K,J) * RY - Y(K,J) * RX
580  A(K) = A(K) + T
590  T = RX * LY - RY * LX
600  A(K) = ( ABS (A(K) + T)) / 2
610  R(K) = A(K) / P(K)
620  B(K) = RX - LX
630  RETURN
640  REM
650  REM SUBROUTINE
660  REM TURN ON PRINTER
670  REM
680  PR# 1
690  REM
700  REM SET MARGINS
710  REM
720  PRINT  CHR$ (27);"J,90,$"
730  PRINT  CHR$ (27);"L,528,432,$"
740  RETURN
750  REM
760  REM INPUT DATA
770  REM
780  HOME : VTAB (5): PRINT "RIVER WATER SURFACE PROFILES"
790  PRINT : PRINT "FOR NATURAL CHANNELS WITH STEADY FLOW"
800  PRINT : PRINT "NO PROVISION IS MADE FOR OVERBANK FLOW"
810  PRINT : PRINT "DO YOU WISH TO USE THE PRINTER (Y/N)? "
820  INPUT PT$
830  IF PT$ = "Y" THEN PT = 1
840  HOME : VTAB (5)
850  PRINT "ENTER (1) FOR U.S. CUSTOMARY OR (2) FOR SI UNITS: ";
860  INPUT SI
870  IF SI < 1 OR SI > 2 THEN  PRINT : PRINT "ENTER 1 OR 2": GOTO 860
880  G = 32.174:MF = 1.486
890  IF SI > 1 THEN G = G * .3048:MF = 1
900  HOME : VTAB (5): PRINT "ENTER THE TOLERANCE FOR THE WATER SURFAC
     E CONVERGENCE"
910  PRINT "(0.001 IS RECOMMENDED): ";
920  INPUT TA
930  HOME : VTAB (5): PRINT "ENTER PRINTOUT LEVEL:"
940  PRINT "    (1) RESULTS ONLY"
950  PRINT "    (2) ADDITIONAL DETAILS"
960  PRINT "        (AVAILABLE WITH PRINTER ONLY)"
970  INPUT PL
980  IF PL < 1 OR PL > 2 THEN PL = 2
990  IF PT < > 1 THEN PL = 1
1000 HOME : VTAB (5): PRINT "ENTER THE NUMBER OF DISCHARGES TO BE EV
     ALUATED: ";
1010 INPUT QQ
1020 MF = MF * MF
1030 DIM Q(QQ),W1(QQ)
1040 FOR IQ = 1 TO QQ
```

Program 4 cont.

```
1050  PRINT : PRINT "ENTER Q(";IQ;") AND CORRESPONDING WS: ";
1060  INPUT Q(IQ),W1(IQ)
1070  NEXT IQ
1080  HOME : VTAB (5)
1090  PRINT "ENTER THE NUMBER OF RIVER CROSS SECTIONS: ";
1100  INPUT M
1110  HOME : VTAB (5)
1120  PRINT "ENTER THE MAX. NUMBER OF COORDINATE"
1130  PRINT "POINTS USED AT ANY CROSS SECTION: ";
1140  INPUT P
1150  DIM N(M),X(M,P),Y(M,P),LL(M),WS(M),A(M),P(M)
1160  DIM S(M),H1(M),H2(M),B(M),R(M),V(M)
1170  FOR J = 1 TO M
1180  HOME : VTAB (5)
1190  PRINT "ENTER DATA FOR RIVER CROSS SECTION (";J;") AS FOLLOWS:"
1200  PRINT
1210  PRINT "ENTER THE NUMBER OF COORDINATE POINTS"
1220  PRINT "REQUIRED FOR THIS SECTION: ";
1230  INPUT N(J)
1240  IF J = 1 THEN NN = N(J)
1250  IF N(J) < = P THEN 1290
1260  PRINT : PRINT "NUMBER OF COORDINATE POINTS EXCEEDS ";P
1270  PRINT "REDUCE NUMBER OR INCREASE ARRAY DIM": PRINT
1280  GOTO 1190
1290  IF J > 1 THEN 1360
1300  PRINT : PRINT "ENTER THE VALUE OF MANNING N: ";
1310  INPUT NQ
1320 NQ = 2 * G * NQ * NQ / MF
1330  PRINT
1340  GOTO 1390
1350  PRINT
1360  PRINT : PRINT "ENTER THE DISTANCE FROM THE DOWNSTREAM  SECTION:
      ";
1370  INPUT LL(J)
1380  PRINT
1390  PRINT "ENTER THE COORDINATE POINTS WORKING FROM LEFT TO RIGHT B
      ANK"
1400  PRINT
1410 N = N(J)
1420  FOR I = 1 TO N
1430  PRINT "ENTER X,Y VALUES FOR POINT ";I
1440  INPUT X(J,I),Y(J,I)
1450  IF X(J,I) > = X(J,I - 1) THEN 1500
1460  PRINT : PRINT "X VALUE LESS THAN PREVIOUS VALUE,"
1470  PRINT "REENTER X,Y"
1480  PRINT
1490  GOTO 1430
1500  PRINT
1510  NEXT I
1520  NEXT J
1530  REM
1540  REM END OF DATA INPUT
1550  REM START OF CALCULATIONS
1560  REM TURN ON PRINTER
1570  REM
1580  IF PT = 1 THEN  GOSUB 660
1590  REM
1600  REM LOOP FOR NUMBER OF DISCHARGES
1610  REM
1620  FOR IQ = 1 TO QQ
1630 WS(1) = W1(IQ)
1640  REM
```

Program 4 cont.

```
1650  REM CHECK THAT WS GREATER THAN BED ELEVATION
1660  REM
1670  I = 1
1680  IF Y(1,I) < WS(1) THEN 1750
1690  IF I = NN THEN 1720
1700  I = I + 1
1710  GOTO 1680
1720  PRINT "STOP!! THE WATER SURFACE IS BELOW THE BED"
1730  PRINT "ENTER RUN TO RESTART": END
1740  REM
1750  REM  CALCULATE CHARACTERISTICS AT SECTION 1
1760  REM
1770  PRINT : PRINT : PRINT "THE DISCHARGE IS ";Q(IQ);
1780  IF SI > 1 THEN 1820
1790  PRINT " CFS."
1800  PRINT : PRINT "ALL RESULTS ARE IN U.S. CUSTOMARY UNITS"
1810  GOTO 1840
1820  PRINT " CMS."
1830  PRINT : PRINT "ALL RESULTS ARE IN SI UNITS"
1840  PRINT
1850  REM
1860  REM OUTPUT FORMAT FOR PRINT USE
1870  REM FOR APPLE II+ ONLY
1880  REM MODIFY AS NEEDED
1890  REM
1900  A$ = "####.##"
1910  S$ = "####"
1920  W$ = "########.##"
1930  B$ = "#########."
1940  C$ = "####.#######"
1950  REM HEADING FOR RESULTS PRINTOUT
1960  PRINT "SECTION     WS ";
1970  IF PL = 1 THEN 1990
1980  PRINT "     A      R      B      V        S          H
1990  LL(1) = 0
2000  IF PL = 1 THEN  PRINT
2010  K = 1
2020  REM
2030  REM CALCULATE SECTION CHARACTERISTICS
2040  REM
2050  GOSUB 110
2060  V(K) = Q(IQ) / A(K)
2070  VH = V(K) * V(K) / (2 * G)
2080  H1(K) = WS(K) + VH
2090  S(K) = NQ * VH / (R(K) ^ (4 / 3))
2100  IF K = 1 THEN H2(1) = H1(K)
2110  IF K > 1 THEN 2250
2120  REM
2130  REM PRINTOUT OF RESULTS
2140  REM USING THE APPLE II+ PRINT USE
2150  REM
2160  &  PRINT USES$;K;
2170  &  PRINT USEW$;WS(K);
2180  IF PL = 1 THEN 2250
2190  &  PRINT USEB$;A(K);
2200  &  PRINT USEA$;R(K);
2210  &  PRINT USEA$;B(K);
2220  &  PRINT USEA$;V(K);
2230  &  PRINT USEC$;S(K);
2240  &  PRINT USEA$;H1(K);
2250  REM  GET TRIAL UPSTREAM WS
2260  PRINT
```

Program 4 cont.

```
2270  IF Z > 0 THEN 2540
2280  Z = K + 1
2290  WS(Z) = WS(K) + S(K) * LL(Z)
2300  TK = K
2310  K = Z
2320  GOSUB 110
2330  Z = K
2340  V(Z) = Q(IQ) / A(Z)
2350  VH = V(Z) * V(Z) / (2 * G)
2360  H1(Z) = WS(Z) + VH
2370  S(Z) = NQ * VH / (R(Z) ^ (4 / 3))
2380  K = TK
2390  SA = .5 * (S(K) + S(Z))
2400  H2(Z) = H2(K) + SA * LL(Z)
2410  HE = H2(Z) - H1(Z)
2420  K = Z
2430  REM
2440  REM TEST FOR CONVERGENCE
2450  REM
2460  IF ABS (HE) < = (TA) THEN 2160
2470  FR = 2 * VH * B(Z) / A(Z)
2480  REM
2490  REM CORRECTION TO WS ELEVATION
2500  FS = 1.5 * S(Z) * LL(Z) / R(Z)
2510  DY = HE / (1 - FR + FS)
2520  WS(Z) = WS(Z) + DY
2530  GOTO 2310
2540  IF K < M THEN 2280
2550  NEXT IQ
2560  PR# 0
2570  END
```

PROGRAM 5: SEDIMENT FALL VELOCITY[6]

This program goes beyond the material in Chapter 8, in that it calculates particle fall velocities on the basis of a new parameter, the *shape factor SF*, as well as the mean diameter and specific gravity.[7] Assuming that a particle can be represented by an ellipsoid, the shape factor *SF* is defined as the ratio of the shortest semi-axis to the square root of the product of the intermediate and longest semi-axes. Thus a sphere would have a shape factor of 1 (however an irregular sediment particle with *SF* = 1 will have a fall velocity slightly less than that of a sphere. In the other extreme, a particle with *SF* = 0.3 would be very flat. A value of approximately 0.7 is perhaps most typical for naturally worn sediments.

The computational procedure is based on a graph similar to Fig. 8–3 in which the curve for *SF* = 1 lies just above the sphere curve. The curves for lower values of *SF* are similar, but located still higher on the graph, to reflect their greater drag coefficients.[8] In the program *CD* and *RE* are the drag coefficient and particle Rey-

[6]Referenced in Chapter 8.
[7]A. L. Prasuhn and M. Knofczynski, "Improved Computation of Fall Velocity," *Proceedings* 21st Congress, International Association for Hydraulic Research, Melbourne, Australia, Aug. 1985.
[8]Inter-Agency Committee on Water Resources Report 12, "Some Fundamentals of Particle Size Analysis," 1957.

nolds number. The independent parameters are *SF* and *CS,* the latter equaling $F_R/\rho\nu^2$. This, it should be noted, is the auxiliary scale in Fig. 8–3. To develop the procedure, the curves were first rearranged with both C_D and Re plotted as functions of *CS* for shape factors of 0.3, 0.5, 0.7, 0.9, and 1.0. These two sets of curves are included in the data lines of Program 5.

The program provides a direct solution. The curves are entered with the appropriate values of *SF* and *CS* and both C_D and Re are calculated, interpolating as needed. A value of the fall velocity is determined from both parameters, however, they are usually almost identical. See footnote 7 for additional details.

Regardless of units chosen, the particle diameter must be entered in millimeters. The temperature is entered in either Fahrenheit or Celsius degrees and the kinematic viscosity calculated using Program 1. In addition, the program also requires the shape factor and particle specific gravity. The resulting fall velocities are in ft/s or cm/s units, accordingly.

The use of a printer is similar to the other programs written for the Apple II +. However, the program is easily modified for other machines or languages. This is a useful program to insert in any sediment transport program requiring fall velocities. Although the program may be used without great difficulty, the footnotes 7 and 8 should be consulted for a clear understanding of the procedure.

Program 5

```
10    DEF  FN LG(A) =  LOG (A) /  LOG (10)
20    GOTO 680
30    REM
40    REM SUBROUTINE
50    REM TURN ON PRINTER
60    REM SET MARGINS
70    REM
80    PR# 1
90    PRINT   CHR$ (27);"J,120,900,$"
100   PRINT   CHR$ (27);"L,528,432,$"
110   RETURN
120   REM
130   REM SUBROUTINE
140   REM CALCULATE KINEMATIC VISCOSITY
150   REM TEMPERATURE IS TF
160   REM
170 TC = TF
180   IF SI = 2 THEN 200
190   TC = (TF - 32) * (5 / 9)
200 XT = TC - 20
210   IF TC > 20 THEN 240
220 NU = 10 ^ (1301 / (998.333 + 8.1855 * XT + .00585 * XT * XT) - 3.
      30233)
230   GOTO 260
240 TT = 20 - TC
250 NU = .01002 * (10 ^ ((1.3272 * TT - .001053 * XT * XT) / (TC + 10
      5)))
260 NU = NU / 10000: REM IN SQ M/S
270   IF SI = 1 THEN NU = NU * 20.76391
280   RETURN
290   REM
300   REM SUBROUTINE
```

Program 5 cont.

```
310   REM INTERPOLATE SHAPE FACTOR
320   REM
330   IF M = 2 THEN 400
340   IF SF >  = .3 AND SF < .5 THEN ST = .3
350   IF SF >  = .5 AND SF < .7 THEN ST = .5
360   IF SF >  = .7 AND SF < .9 THEN ST = .7
370   IF SF >  = .9 AND SF < 1 THEN ST = .9
380   IF SF = 1 THEN ST = 1
390   RETURN
400   IF SF > .3 AND SF < .5 THEN ST = .5
410   IF SF > .5 AND SF < .7 THEN ST = .7
420   IF SF > .7 AND SF < .9 THEN ST = .9
430   IF SF > .9 AND SF < 1 THEN ST = 1
440   RETURN
450   REM
460   REM SUBROUTINE
470   REM INTERPOLATE TO GET CD
480   REM
490   IF ST = .3 THEN K = 1:A = 30000::B = 2.78:BB = .958
500   IF ST = .5 THEN K = 2:A = 300000:B = 1.72:BB = 1.216
510   IF ST = .7 THEN K = 3:A = 500000:B = 1.10:BB = 1.52
520   IF ST = .9 THEN K = 4:A = 600000:B = .67:BB = 1.949
530   IF ST = 1 THEN K = 5:A = 1000000:B = .51:BB = 2.235
540   IF CS > (A) THEN 600
550   FOR J = 1 TO 10
560   IF S(K,1,J) >  FN LG(CS) THEN 580
570   NEXT J
580   Y = 10 ' (S(K,I,J - 1) + ( FN LG(CS) - S(K,1,J - 1)) * (S(K,I,J) -
      S(K,I,J - 1)) / (S(K,1,J) - S(K,1,J - 1)))
590   RETURN
600   CA = CS
610   IF I = 2 THEN CS = 1
620   IF I = 3 THEN B = BB
630   Y = B *  SQR (CS)
640   CS = CA
650   RETURN
660   REM
670   REM MAIN PROGRAM TO CALCULATE FALL VELOCITY
680   REM
690   REM READ IN CURVE DATA
700   REM
710   DIM S(5,3,11)
720   REM
730   REM K=1,2,3,4,5 IS SF=.3,.5,.7,.9,1
740   REM  I=1,2,3 IS CS, CD AND RE
750   REM J=1,2....10 ARE LOG OF VALUES
760   FOR K = 1 TO 5
770   FOR I = 1 TO 3
780   FOR J = 1 TO 10
790   READ S(K,I,J)
800   NEXT J
810   NEXT I
820   NEXT K
830   REM
840   REM ENTER SEDIMENT AND WATER DATA
850   REM DELETE IN SUBROUTINES
860   REM
870   HOME : VTAB (5): INPUT "DO YOU WANT TO USE THE PRINTER (Y/N)? "
      PT$
880   IF PT$ = "Y" THEN PT = 1
890   HOME : VTAB (5)
900   INPUT "ENTER SHAPE FACTOR ";SF
```

Program 5 cont.

```
910   PRINT : INPUT "ENTER (1) U.S. CUSTOMARY OR (2) SI UNITS ";SI
920   PRINT : INPUT "ENTER SPECIFIC GRAVITY OF PARTICLE ";SG
930   PRINT : PRINT "ENTER TEMPERATURE IN DEG ";
940   IF SI = 2 THEN  PRINT "C ";:  INPUT TF: GOTO 960
950   PRINT "F ";:  INPUT TF
960 G = 32.17
970   IF SI = 2 THEN G = G * .3048
980   PRINT
990   INPUT "ENTER NUMBER OF SIZES TO BE ANALYZED ";NZ
1000   DIM D(NZ),W(NZ),WCD(NZ),WRE(NZ),DD(NZ)
1010   DIM CD(NZ),RE(NZ)
1020   FOR II = 1 TO NZ
1030   HOME : VTAB (5)
1040   PRINT "ENTER DIAMETER IN MM FOR SIZE ";II
1050   INPUT DD(II)
1060 D(II) = DD(II): REM SAVE ORIGINAL DIAMETERS
1070   IF SI = 2 THEN D(II) = D(II) / 1000: GOTO 1090
1080 D(II) = D(II) / 304.8
1090   NEXT II
1100   REM
1110   REM CALCULATE VISCOSITY
1120   REM
1130   GOSUB 140
1140   REM
1150   REM CALCULATION OF FALL VELOCITY
1160   REM
1170 PI = 3.141593
1180   FOR II = 1 TO NZ
1190 D = D(II)
1200 CS = ((PI / 6) * D * D * D * (SG - 1) * G) / (NU * NU)
1210 M = 1:JJ = 0
1220   REM
1230   REM INTERPOLATE SHAPE FACTOR
1240   REM
1250   GOSUB 310
1260   GOTO 1330
1270   IF SF = ST THEN 1390
1280 M = 2:JJ = 1
1290   GOSUB 310
1300   REM
1310   REM INTERPOLATE CD
1320   REM
1330 I = 2
1340   GOSUB 470
1350   IF JJ = 1 THEN 1390
1360 C1 = Y: REM  CD
1370 S1 = ST
1380   GOTO 1270
1390 C2 = Y: REM  CD
1400 S2 = ST
1410   IF C1 = C2 THEN 1440
1420 CD = 10 ^ ( FN LG(C1) - (( FN LG(C1) -  FN LG(C2)) * ( FN LG(SF)
      -  FN LG(S1)) / ( FN LG(S2) -  FN LG(S1))))
1430   GOTO 1450
1440 CD = C1
1450 WCD = ((4 * D * (SG - 1) * G) / (3 * CD))    .5
1460 WCD(II) = WCD
1470   REM
1480   REM USE RE TO GET W
1490   REM
1500 M = 1:JJ = 0
1510   GOSUB 310
```

Program 5 cont.

```
1520   GOTO 1560
1530   IF SF = ST THEN 1620
1540 M = 2:JJ = 1
1550   GOSUB 310
1560 I = 3
1570   GOSUB 470
1580   IF JJ = 1 THEN 1620
1590 R1 = Y: REM   RE
1600 S1 = ST
1610   GOTO 1530
1620 R2 = Y
1630 S2 = ST
1640   IF R1 = R2 THEN 1670
1650 RE = 10 ^ ( FN LG(R1) + (( FN LG(R2) -  FN LG(SF)
       -  FN LG(S1)) / ( FN LG(S2) -  FN LG(S1))))
1660   GOTO 1680
1670 RE = R1
1680 WRE = (RE * NU) / D
1690 WRE(II) = WRE
1700 D(II) = D
1710   IF SI = 1 THEN 1740
1720 WRE(II) = WRE(II) * 100:WCD(II) = WCD(II) * 100
1730 D(II) = D(II) * 1000
1740   REM
1750   REM NO ADJUSTMENT REQUIRED FOR U.S. CUSTOMARY UNITS
1760   REM
1770 CD(II) = CD:RE(II) = RE
1780   NEXT II
1790   REM
1800   REM PRINT OUT RESULTS
1810   REM PRINTOUT FORMAT
1820   REM FOR APPLE II+ WITH PRINT USE ONLY
1830   REM MODIFY AS NEEDED FOR OTHER MICROS
1840   REM
1850 A$ = "###.##"
1860 B$ = "####.#"
1870 C$ = "#####.##"
1880 D$ = "  #.##^^^^"
1890 E$ = "####.###"
1900   REM
1910   REM TURN ON PRINTER
1920   REM
1930   IF PT = 1 THEN   GOSUB 80
1940   PRINT " DIA.     SF      SP     TEMP      CD        RE       W (CD)   W (R
       E)"
1950   IF SI = 1 THEN 1980
1960   PRINT " MM               GR      C                            CM/S      CM/
       S"
1970   GOTO 1990
1980   PRINT " MM               GR      F                            FT/S      FT/
       S"
1990   PRINT "------------------------------------------------------------
       --"
2000   FOR II = 1 TO NZ
2010   &   PRINT USEA$;DD(II);
2020   &   PRINT USEA$;SF;
2030   &   PRINT USEA$;SG;
2040   &   PRINT USEB$;TF;
2050   &   PRINT USEC$;CD(II);
2060   &   PRINT USED$;RE(II);
2070   IF SI = 1 THEN 2110
2080   &   PRINT USEC$;WCD(II);
```

Program 5 cont.

```
2090    &   PRINT USEC$;WRE(II)
2100    GOTO 2130
2110    &   PRINT USEE$;WCD(II);
2120    &   PRINT USEE$;WRE(II)
2130    NEXT II
2140    PRINT "-------------------------------------------------------------
        --"
2150    IF SI = 1 THEN 2180
2160    PRINT : PRINT : PRINT "KINEMATIC VISCOSITY = ";NU;" SQ M/S"
2170    END
2180    PRINT : PRINT : PRINT "KINEMATIC VISCOSITY = ";NU;" SQ FT/S"
2190    END
2200    REM
2210    REM CURVE DATA FOLLOWS
2220    REM
2230    DATA .556,1.204,1.813,2.602,2.903,3.114,3.301,3.602,3.954,4.477
2240    DATA 1.903,1.477,1.146,.778,.668,.602,.556,.505,.477,.444
2250    DATA -.464,.065,.536,1.114,1.321,1.459,1.575,1.752,1.942,2.22
2260    DATA .532,.929,1.398,1.778,2.255,2.813,3.342,4.255,5,5.477
2270    DATA 1.903,1.602,1.301,1.079,.845,.602,.447,.279,.218,.21
2280    DATA -.482,-.133,.251,.551,.908,1.308,1.651,2.191,2.594,2.837
2290    DATA .505,1.114,1.58,1.954,2.663,3.114,3.699,4.114,4.602,5.699
2300    DATA 1.903,1.415,1.114,.903,.58,.398,.204,.114,.079,.041
2310    DATA -.496,.052,.436,.728,1.244,1.561,1.95,2.203,2.464,3.031
2320    DATA .477,1,1.623,2.699,3.146,3.929,4.301,4.875,5.301,5.778
2330    DATA 1.903,1.477,1.041,.462,.279,.041,-.046,-.125,-.155,-.174
2340    DATA -.51,-.021,.494,1.322,1.637,2.14,2.376,2.703,2.931,3.179
2350    DATA .447,1.114,1.763,2.301,2.845,3.532,4.114,4.532,5.114,6
2360    DATA 1.903,1.322,.903,.602,.342,.079,-.097,-.187,-.252,-.292
2370    DATA -.525,.099,.623,1.053,1.455,1.929,2.301,2.556,2.887,3.349
2380    REM END OF DATA
```

PROGRAM 6: ANALYSIS OF CULVERTS

This program is a modification of a program originally written by Jeff Lewendowski to determine the minimum circular culvert diameter using the procedures of Section 11–3. Two diameters are determined based on inlet and outlet control and the larger selected. The program is written in Basic and the output is left unformatted. Either U.S. customary or SI units may be used. The input, controlled by prompts, requires the discharge, head water depth, culvert length, slope, and Manning n. In addition, estimates of the entrance loss coefficient and discharge coefficient are required.

The tailwater may be entered if known. If not known, the user enters zero. In the latter event, or if the tailwater is less than the diameter, the tailwater is assumed equal to the diameter. This should always be on the conservative side.

The calculation of diameter assuming inlet control uses the secant method. The calculation for outlet control uses a slower incremental iteration scheme. The outlet control calculations are initially based on the assumption that the tailwater is equal to the diameter. If the resulting diameter is less than the specified tailwater a larger diameter is determined using the actual tail water.

Program 6

```
10    REM CULVERT PROGRAM
20    REM DETERMINES CULVERT DIAMETER
```

Program 6 cont.

```
30    REM ORIGINALLY WRITTEN IN FORTRAN BY JEFF LEWANDOWSKI
40    REM WHILE A GRADUATE STUDENT
50    REM AT SOUTH DAKOTA STATE UNIVERSITY
60    REM PROCEDURES BASED ON SECTION 11-3
70    REM A CIRCULAR CULVERT IS ASSUMED
80    REM FREE SURFACE FLOW IS TESTED FOR
90    REM BUT NOT ANALYZED
100   REM
110   GOTO 170
120   REM
130   REM SUBROUTINE FOR INLET CONTROL EQUATION
140   REM
150 F = C1 * D * D * SQR (G2 * (HW - D / 2)) - Q
160   RETURN
170   REM
180   REM ENTERING OF DATA
190   REM
200   HOME : VTAB (5)
210   PRINT "CULVERT PROGRAM"
220   PRINT : PRINT "TO DETERMINE CULVERT DIAMETER"
230   PRINT : PRINT "IS THE PROBLEM SPECIFIED IN (1) U.S. CUSTOMARY OR
      (2) SI UNITS ";
240   INPUT SI
250   IF SI < > 1 AND SI < > 2 THEN  PRINT : PRINT "ENTER 1 OR 2 ONL
      Y": PRINT : GOTO 230
260   HOME : VTAB (5)
270   PRINT "ENTER THE DISCHARGE IN ";
280   IF SI = 1 THEN  PRINT "CFS": GOTO 300
290   PRINT "CUBIC METERS/S"
300   INPUT Q
310   PRINT : PRINT "ENTER THE CULVERT LENGTH IN ";
320   IF SI = 1 THEN  PRINT "FEET": GOTO 340
330   PRINT "METERS"
340   INPUT L
350   PRINT : INPUT "ENTER THE SLOPE OF THE CULVERT ";S
360   PRINT : INPUT "ENTER THE MANNING N FOR THE CULVERT ";N
370   PRINT : PRINT "ENTER THE HEAD WATER DEPTH (ABOVE THE INLET INVER
      T) IN ";
380   IF SI = 1 THEN  PRINT "FEET": GOTO 400
390   PRINT "METERS"
400   INPUT HW
410   PRINT : PRINT "ENTER THE TAIL WATER DEPTH (ABOVE THE OUTLET INVE
      RT) IN ";
420   IF SI = 1 THEN  PRINT "FEET": GOTO 440
430   PRINT "METERS"
440   PRINT "NOTE: IF THE TAIL WATER IS UNKNOWN ENTER 0"
450   INPUT TW
460   HOME : VTAB (5)
470   PRINT "ESTIMATE AND ENTER THE INLET LOSS COEFFICIENT CL"
480   PRINT : PRINT "GENERALLY 0.05 < CL < 0.50"
490   PRINT : PRINT "A LOW VALUE IS ASSOCIATED WITH A SMOOTH EFFICIENT
      ENTRANCE"
500   INPUT CL
510   HOME : VTAB (5)
520   PRINT "ESTIMATE AND ENTER THE DISCHARGE COEFFICIENT FOR THE ENTR
      ANCE CD"
530   PRINT : PRINT "GENERALLY 0.62 < CD < 1.0"
540   PRINT : PRINT "A HIGH VALUE IS ASSOCIATED WITH A SMOOTH EFFICIEN
      T ENTRANCE"
550   INPUT CD
560   HOME : VTAB (5)
570   REM
580   REM ASSUME INLET CONTROL
```

Program 6 cont.

```
590  REM
600  G = 32.174:MF = 1.486
610  IF SI = 2 THEN G = G * .3048:MF = 1
620  G2 = 2 * G
630  PI = 3.14159265
640  II = 0
650  CO = PI / 4
660  C1 = CD * CO
670  REM
680  REM DIAMETER FOUND BY SECANT METHOD
690  REM
700  D1 = 1:D2 = 2: REM STARTER DIAMETERS
710  D = D1: GOSUB 150
720  F1 = F
730  D = D2: GOSUB 150
740  F2 = F
750  SF = (F2 - F1) / (D2 - D1)
760  D3 = D2 - F2 / SF
770  II = II + 1
780  REM
790  REM TEST FOR CONVERGENCE
800  REM
810  IF  ABS ((D3 - D2) / D2) < .00001 THEN 840
820  D1 = D2:D2 = D3
830  GOTO 710
840  DI = D3
850  I1 = II
860  REM
870  REM ASSUME OUTLET CONTROL
880  REM
890  REM IF TW IS UNKNOWN OR LESS THAN D PROGRAM ASSUMES TW = D
900  REM THIS WOULD BE CONSERVATIVE FOR TW < D
910  REM
920  II = 0:D = 0.00001:DD = 1:D1 = 0.00001
930  HS = HW + S * L
940  C2 = 1 + CL
950  C3 = G2 * N * N * L / (MF * MF)
960  C4 = Q * Q / (CO * CO * G2)
970  C5 = 4 / 3
980  REM
990  REM PROGRAM FIRST ASSUMES THAT TW = D
1000  REM
1010  R = D / 4
1020  HC = (C2 + (C3 / R ^ C5)) * (C4 / D ^ 4) + D
1030  II = II + 1
1040  REM
1050  REM TEST FOR CONVERGENCE
1060  REM
1070  IF  ABS ((HS - HC) / HS) < .00001 THEN 1120
1080  IF HC > HS THEN D1 = D:D = D + DD: GOTO 1010
1090  DD = DD / 10
1100  D = D1
1110  GOTO 1010
1120  REM
1130  REM TEST FOR TW
1140  REM
1150  IF TW <  = D THEN 1320
1160  REM
1170  REM IF TW > D THEN SOLVE FOR LARGER D
1180  REM
1190  D = .00001:D1 = .00001:DD = 10
1200  HL = HS - TW
```

Program 6 cont.

```
1210 R = D / 4
1220 HC = (C2 + (C3 / R ^ C5)) * (C4 / D ^ 4)
1230 II = II + 1
1240  REM
1250  REM TEST FOR CONVERGENCE
1260  REM
1270  IF  ABS ((HL - HC) / HL) < .00001 THEN 1320
1280  IF HC > HL THEN D1 = D:D = D + DD: GOTO 1210
1290 DD = DD / 10
1300 D = D1
1310  GOTO 1210
1320 DO = D
1330 I2 = II
1340  REM
1350  REM TEST FOR INLET OR OUTLET CONTROL
1360  REM
1370  IF DO > DI THEN A$ = "OUTLET CONTROL":DA = DO: GOTO 1390
1380 A$ = "INLET CONTROL":DA = DI
1390  PRINT "RESULTS OF THE CULVERT ANALYSIS:"
1400  PRINT : PRINT "ON THE BASIS OF INLET CONTROL:"
1410  PRINT "DIAMETER EQUALS ";DI;
1420  IF SI = 1 THEN  PRINT " FT": GOTO 1440
1430  PRINT " M"
1440  PRINT "AND "I1" ITERATIONS WERE REQUIRED"
1450  PRINT : PRINT "ON THE BASIS OF OUTLET CONTROL:"
1460  PRINT "DIAMETER EQUALS ";DO;
1470  IF SI = 1 THEN  PRINT " FT": GOTO 1490
1480  PRINT " M"
1490  PRINT "AND "I2" ITERATIONS WERE REQUIRED"
1500  PRINT : PRINT "THE CULVERT HAS "A$
1510  PRINT "AND THE REQUIRED DIAMETER IS "DA;
1520  IF SI = 1 THEN  PRINT " FT": GOTO 1540
1530  PRINT " M"
1540  IF HW > = DA * 1.2 THEN  END
1550  PRINT : PRINT "HW IS NOT GREATER THAN 1.2D"
1560  PRINT "CULVERT WILL PROBABLY NOT FLOW FULL"
1570  END
```

Answers To Selected Problems

(Many of the problems were worked on a computer and the answers will differ slightly from those of a hand calculator.)

CHAPTER 2

2–1 2.03 slug/ft^3
2–3 1009 kg/m^3
2–5 1.76 × 10^{-4} ft^2/s
2–7 0.0062 slug/ft^3 (increase)
2–9 98,000 lb
2–11 134,700 ft-lb
2–13 2400 lb
2–15 3400 lb
2–17 216.6 kN
2–19 $q = g^{0.5}H^{1.5}\phi(P/H)$
2–21 $P = \rho d^2 V^3 \phi(w/d,\ L/d,\ V/\sqrt{dg},\ Vd/v)$

CHAPTER 3

3–1 1.95 ft
3–3 36.62 in., 36.59 in.
3–7 Horton eq. 0.17 in./day, 0
3–9 $E = 0.134V(e_w - e_a)$

3–11 No evaporation
3–13 Net loss = 86.6 cfs
3–15 Net loss = 117.5 cfs
3–17 0.16 in./hr
3–19 3.6 mm/hr
3–21 0.83 ft/s
3–23 35,400 ac-ft, 2.15 in.
3–25 4.54 mm
3–27 1238 ac-ft
3–29 26,050 ac-ft, 0.604 in.
3–31 4675 cfs
3–33 5092 cfs
3–35 18.15 m^3/s
3–39 13,000 cfs
3–41 232 m^3
3–43 Peak 24,650 cfs
3–45 Peak 29,410 cfs
3–47 Peak 1895 m^3/s
3–49 Peak 2500 cfs
3–51 Peak 2110 cfs
3–53 Peak 268 m^3/s

3–55 Peak 265 m³/s
3–57 Peak 210 m³/s
3–59 Peak 1290 cfs
3–61 Peak 5635 cfs
3–63 Peak 4915 cfs
3–65 Peak 35,000 cfs
3–67 Peak 661 m³/s
3–69 Peak 843 m³/s
3–71 12.1 m³/s
3–73 1.98 cfs, 1820 ac-ft

CHAPTER 4

4–1 35%
4–3 25%
4–5 0.14, 0.21
4–7 440 ft/day
4–9 410 l/day/m²
4–11 131,600 gpd/ft
4–13 1.13×10^{-4} ft/s
4–15 0.057 ft³
4–17 1840 l/day/m²
4–21 1.55 l/s
4–23 0.97 l/s
4–25 29.4 ft
4–27 4.3%
4–29 79,000 l/day/m, 12.12 m
4–31 3.6%
4–39 29.4 ft
4–41 12.12 m
4–55 10.82 m³/day/m, 11.07 m³/day/m
4–57 0.200, 0.209

CHAPTER 5

5–1 0.0021%, 33.5%
5–3 0.02, 0.98, 0.4545, 0.5455
5–5 0.010, 0.990, 0.260, 0.740
5–7 0.0539, 0.426, 0.0625
5–9 0.00287, 0.0284, 0.866
5–11 24,600 cfs, 27,300 cfs
5–13 3027 m³/s, 3163 m³/s
5–15 339,600 cfs, 600,000 cfs, 3.30 yr
5–17 316,500 cfs, 518,300 cfs, 3.10 yr

5–19 11.1 yr, 41.3 yr, 0.914
5–22 7040 yr, 14.95 m³/s
5–24 28 yr, 24,800 cfs
5–28 1.7 yr, 3670 cfs
5–29 365 m³/s, 51 yr, 0.632
5–31 238 yr, 736,000 cfs,

CHAPTER 6

6–1 35.23 ft
6–3 518 psf
6–5 4.582 m
6–7 0.455 cfs, 0.452 cfs
6–9 1799 ft
6–11 0.733 m
6–15 0.461 cfs
6–17 1.56 cfs
6–19 0.175 m³/s, 293 kN/m²
6–25 1.34 m³/s, 0.80 m³/s, 0.41 m³/s
6–29 17.07 cfs out of no. 1
6–31 *AB* 9.893 cfs, *BE* 3.107 cfs
6–33 *AB* 11.04 cfs, *BE* −1.48 cfs
6–35 16.92 hr
6–37 17.97 hr
6–39 13.92 hr
6–41 34.83 hr
6–43 41.44 hr, 92.44 hr
6–45 23.07 min
6–47 8.89 s, 9.51 s
6–49 33.7 s
6–51 17.23 ft/s
6–53 8.02 ft/s, 11.35 ft/s
6–55 4.591 kN/m²
6–57 $Q = 0.0836 \Delta h^{0.5}$
6–59 0.225 m³/s, 0.242 m³/s
6–61 1600 psi T, 4800 psi T
6–63 18,000 kN/m²
6–65 152.4 psi C
6–67 4.02 in., −3.22 in.
6–69 764,000 lb (F_V)
6–71 197.2 kN
6–73 90,900 lb
6–75 2727 psi, 2754 psi, 2692 psi

6–77 136.2 psi, 108 psi, 53.8 psi

6–79 20.45 MN/m^2, 6.90 MN/m^2, 1.10 MN/m^2

6–81 1.12 MN/m^2, 377 kN/m^2, 60.4 kN/m^2

CHAPTER 7

7–3 0.23 psf

7–5 2676 m^3/s

7–7 1.793 N/m^2, 4.00 m^3/s

7–9 3.49 ft

7–11 36.0 m^3/s

7–13 34.6 m^3/s

7–15 1.654 m

7–17 0.030, 0.33 psf

7–21 5.17 ft

7–23 5.82 ft

7–25 6.10 ft

7–27 5.91 ft

7–29 1.63 ft

7–31 1.732 m

7–33 0.900 m

7–35 1.818 m

7–37 2.931 m

7–39 $y_1 = 4.222$ m, $y_2 = 2.842$ m

7–41 0.48 ft

7–43 3.38 ft

7–45 4.209 m

7–47 7.07 ft

7–51 4246 cfs, 4.17 ft, 2007 hp

7–53 3.518 m, 5.384 m, 21.8 m

7–55 133.8 m^3/s, 1.109 m

7–57 1.28 ft, 3.42 ft

7–59 2.16 m, 5.56 m/s, 9.09 m/s

7–61 Mild, Supercritical, Increase

7–69 S-2 curve

7–71 M-3 curve with hydraulic jump

7–75 Hydraulic jump upstream

7–79 S-3 curve

7–85 2.747 m

7–87 16,100 cfs, 18.88 ft

7–89 0.20 ft increase

7–95 7.53 ft

7–97 2.21 cfs

7–99 0.0061

7–101 by, $b + 2y$

7–103 Section 3: $WS = 106.76$ ft

7–105 Section 2: $WS = 104.501$ m

7–107 0.00275 m^3/s, 0.0870 m^3/s

7–109 0.150 cfs, 0.334 cfs, 11.90 cfs

7–111 0, 1.83%, 7.21%

7–113 0.0075 cfs, 0.234 cfs, 7.00 cfs

7–115 1.76%, 1.73%

7–117 2.76 cfs

7–119 15.83 cfs, 35.73 cfs

7–121 0.204 m^3/s

7–123 23.43 cfs, 68.59 cfs

CHAPTER 8

8–1 0.32 mm, 0.42 mm, 1.25 mm

8–3 0.23 mm, 0.30 mm, 0.71 mm

8–7 0.0034 s, 0.0054 s, 0.014 ft/s

8–9 0.15 mm, 0.1 mm

8–11 Re = 62, $w = 8.13$ cm/s

8–13 0.16 N/m^2, 0.46 N/m^2, 8.53 N/m^2

8–15 CS 1.09 m^3/s, VCS 3.81 m^3/s

8–17 8700 tons/day

8–19 By sizes 3190 tons/day

8–21 905 tons/day

8–23 7080 tons/day

8–26 323 tons/day

8–30 5500 tons/day

8–32 776 tons/day

8–35 10,700 tons/day

8–38 400 N/s

8–42 2810 N/s

8–46 12,100 tons/day

8–48 5760 tons/day

8–50 568 N/s

8–53 2130 N/s

8–55 VFG 6.05 m^3/s

8–57 1.26 m^3/s

8–59 10,000 tons/day

8–61 11,300 tons/day

8–73 $y = 4.56$ ft, $b = 53.77$ ft

8–75 $y = 1.005$ m, $b = 5.756$ m

8–77 $y = 4.15$ ft, $b = 4.25$ ft

8–79 $y = 1.592$ m, $b = 6.411$ m

8–81 2.75 ft
8–83 5.83 ft, 16.78 ft, 22.61 ft
8–87 0.17 psf, 0.39 in.
8–89 241 yr
8–91 273 yr
8–93 288 yr

CHAPTER 9

9–1 $L_r^3 V_r^2$
9–7 8, 0.0024
9–9 0.043
9–11 15.24 m/s
9–13 402 hp
9–15 0.0118, 0.0190
9–17 48,800 ton
9–19 970 MN
9–21 0.0512 m³/s, 1.5 m, 2.35 MW
9–23 84.5
9–25 6.45 ft/s
9–27 8
9–29 0.712 m³/s
9–31 1.587, 1.260, 4.00, 5.04
9–33 126,900 lb, 6920 hp
9–35 0.00475
9–37 0.224
9–39 0.022
9–43 0.1414, 0.0177
9–45 0.0028
9–47 0.16 ft, 0.98 ft
9–49 27.0

CHAPTER 10

10–1 89.3%
10–3 501 cfs
10–5 83.7%
10–7 77.4%
10–9 450 rpm, 4.69 ft, 1974 hp
10–11 300 rpm, 1.984 m, 2.254 MW
10–17 478.3 rpm, 8.933 MW, 91.3%
10–19 125.9 deg, 135.1 deg, 7.574 kW
10–21 2739 kW, 128.2 rpm, Francis turbine

10–23 375 rpm
10–25 130.4 rpm
10–27 300 rpm, 4166 hp
10–29 789 gpm, 30.23 ft
10–31 759 rpm, 52.7 hp
10–33 1739 rpm
10–35 66.7, 66.7, Francis turbine
10–43 N_s (SI) = 3.813 N_s (U.S. customary)
10–45 0.664 m
10–47 −11.27 ft
10–49 −10.58 ft
10–51 17.77 ft
10–53 −0.20 ft
10–61 320.6 rpm, 93.8%, 99.6 m³/s
10–65 0.81 ft
10–69 0.88 m³/s, 244 kW, 65%
10–71 0.46 m³/s, 170 kW, 70%
10–73 1100 gpm, 201 ft, 78 hp
10–77 2860 gpm, 57 ft, 54 hp

CHAPTER 11

11–1 2.05 in., 3.1 in.
11–3 881 cfs
11–5 86.7 m³/s
11–7 653 cfs
11–9 9.2 m³/s
11–11 360 cfs, 990 cfs
11–13 267 cfs, 725 cfs, 719 cfs, 2042 cfs
11–17 31.5 cfs
11–19 0.603 m³/s
11–21 2.77 ft
11–23 2.451 m
11–25 2.32 ft
11–27 2.35 ft
11–29 3.78 ft
11–33 7.03 ft
11–37 0.23 ft
11–39 0.21 ft
11–41 0.270 m
11–43 0.174 m

INDEX